T0213578

# Lecture Notes in Computer Science　　　10584

Commenced Publication in 1973
Founding and Former Series Editors:
Gerhard Goos, Juris Hartmanis, and Jan van Leeuwen

More information about this series at http://www.springer.com/series/7409

Niall Adams · Allan Tucker
David Weston (Eds.)

# Advances in Intelligent Data Analysis XVI

16th International Symposium, IDA 2017
London, UK, October 26–28, 2017
Proceedings

 Springer

*Editors*
Niall Adams
Imperial College London
London
UK

David Weston
Birkbeck, University of London
London
UK

Allan Tucker
Brunel University London
Uxbridge
UK

ISSN 0302-9743        ISSN 1611-3349   (electronic)
Lecture Notes in Computer Science
ISBN 978-3-319-68764-3        ISBN 978-3-319-68765-0   (eBook)
DOI 10.1007/978-3-319-68765-0

Library of Congress Control Number: 2017955669

LNCS Sublibrary: SL3 – Information Systems and Applications, incl. Internet/Web, and HCI

Printed on acid-free paper

This Springer imprint is published by Springer Nature
The registered company is Springer International Publishing AG
The registered company address is: Gewerbestrasse 11, 6330 Cham, Switzerland

# Preface

We are proud to present the proceedings of the 16th International Symposium on Intelligent Data Analysis, which took place during October 26–28 in London, UK. The series started in 1995 and was held biennially until 2009. In 2010, the symposium re-focused to support papers that go beyond established technology and offer genuinely novel and game-changing ideas, while not always being as fully realized as papers submitted to other conferences.

IDA 2017 continued this approach and sought first-look papers that might elsewhere be considered preliminary, but contain potentially high-impact research. It has been 20 years since the first independent IDA symposium. To celebrate this event, authors of the best papers presented at IDA 2017 were invited to submit a full paper for possible publication for a special issue in the Wiley Series *Statistical Analysis and Data Mining: The ASA Data Science Journal.*

The IDA Symposium is open to all kinds of modelling and analysis methods, irrespective of discipline. It is an interdisciplinary meeting that seeks abstractions that cut across domains. IDA solicits papers on all aspects of intelligent data analysis, including papers on intelligent support for modelling and analyzing data from complex, dynamical systems.

Intelligent support for data analysis goes beyond the usual algorithmic offerings in the literature. Papers about established technology were only accepted if the technology was embedded in intelligent data analysis systems, or was applied in novel ways to analyzing and/or modelling complex systems. The conventional reviewing process, which favors incremental advances on established work, can discourage the kinds of papers that were selected for IDA 2017. The reviewing process addressed this issue explicitly: Referees evaluated papers against the stated goals of the symposium, and any paper for which at least one Program Committee (PC) advisor wrote an informed, thoughtful, positive review was accepted, irrespective of other reviews. Indeed, this had a notable impact on what papers were included in the program.

We were pleased to see a very strong program. We received 70 submissions by 200 authors from 27 different countries, out of which 18 were accepted as regular papers, 10 as regular poster papers. All proceedings track submissions were reviewed by three PC members and one PC advisor. We were also pleased to include for oral presentation only four Horizon Track submissions. The Horizon Track aims to provide a forum for potentially ground-breaking research that is, however, not yet mature enough for archival dissemination.

We were honored to have the following distinguished invited speakers at IDA 2017:

- Professor Paul Cohen, University of Pittsburgh, USA, on the topic "Mining for Causal Results"
- Dr. Aldo Faisal, Imperial College London, UK, on the topic "Reverse Engineering Human Decision-Making from High-Resolution Analysis of Behavior"

– Professor Niels Peek, University of Manchester, UK, on the topic "Opportunities and Challenges of Learning Health Systems."

The conference was hosted by the Department of Computer Science and Information Systems at Birkbeck College, University of London. It was held at Woburn House in Bloomsbury.

We wish to express our gratitude to all authors of submitted papers for their intellectual contributions; to the PC members and advisors and additional reviewers for their effort in reviewing, discussing, and commenting on the submitted papers and to the members of the IDA Steering Committee for their ongoing guidance and support. We thank Alan Mosca for running the conference website. We gratefully acknowledge Stephen Swift and Tara Orlanes-Angelopoulou for the local organization of the symposium. We thank the following who ensured the smooth running of the symposium: Abul Hasan, Thomas Nealon, and Manni Singh from Birkbeck; Awad Alyousef, Mahir Arzoky, Bashir Dodo, and Leila Yousefi from Brunel; and Chris Beyer from Magdeburg University. We are grateful to our sponsors: The Department of Computer Science and Information Systems at Birkbeck and Springer. We are especially indebted to KNIME, who funded the IDA Frontier Prize for the most visionary contribution presenting a novel and surprising approach to data analysis in the understanding of complex systems.

August 2017

Niall Adams
Allan Tucker
David Weston

# Organization

## General Chair

David Weston — Birkbeck, University of London, UK

## Program Committee Chairs

Niall Adams — Imperial College, UK
Allan Tucker — Brunel University London, UK

## Local Chair

Stephen Swift — Brunel University London, UK

## Sponsorship and Publicity Chairs

David Weston — Birkbeck, University of London, UK
Matthijs van Leeuween — Leiden University, The Netherlands

## Advisory Chairs

Panagiotis Papapetrou — Stockholm University, Sweden
Joost Kok — Leiden University, The Netherlands
Jaakko Hollmen — Aalto University School of Science, Finland

## Webmaster

Alan Mosca — Birkbeck, University of London, UK

## Program Chair Advisors

Michael Berthold — University of Konstanz, Germany
Hendrik Blockeel — K.U. Leuven, Belgium
Elizabeth Bradley — University of Colorado, USA
Tijl De Bie — Ghent University, Data Science Lab, Belgium
Elisa Fromont — University of Saint-Etienne, France
Jaakko Hollmén — Aalto University School of Science, Finland
Frank Klawonn — Ostfalia University of Applied Sciences, Germany
Joost Kok — LIACS, Leiden University, The Netherlands
Nada Lavrač — Jozef Stefan Institute, Slovenia
Xiaohui Liu — Brunel University London, UK
Panagiotis Papapetrou — Stockholm University, Sweden

Arno Siebes                  Universiteit Utrecht, The Netherlands
Hannu Toivonen              University of Helsinki, Finland
Matthijs van Leeuwen        Leiden University, The Netherlands

## Program Committee

Niall Adams                 Imperial College London, UK
Ana Aguiar                  FEUP, Portugal
Fabrizio Angiulli           DEIS, University of Calabria, Italy
Martin Atzmueller           Tilburg University, The Netherlands
José Luis Balcázar          Universitat Politècnica de Catalunya, Spain
Gustavo Batista             University of São Paulo, Brazil
Maria Bielikova             Slovak University of Technology in Bratislava, Slovakia
Dean Bodenham               Imperial College London, UK
Christian Borgelt           Otto-von-Guericke-University of Magdeburg, Germany
Henrik Boström              Stockholm University, Sweden
Paula Brito                 LIAAD/INESC-Porto L.A., Universidade do Porto,
                            Portugal
Edward Cohen                Imperial College London, UK
Paulo Cortez                University of Minho, Portugal
Bruno Cremilleux            Université de Caen, France
Andre de Carvalho           University of Sao Paulo, Brazil
José Del Campo-Ávila        Universidad de Málaga, Spain
Brett Drury                 National University of Ireland, Galway, Ireland
Wouter Duivesteijn          TU Eindhoven, The Netherlands
Saso Dzeroski               Jozef Stefan Institute, Slovenia
Nuno Escudeiro              Instituto Superior de Engenharia do Porto, Portugal
Ad Feelders                 Universiteit Utrecht, The Netherlands
Ingrid Fischer              University of Konstanz, Germany
Johannes Fürnkranz          TU Darmstadt, Germany
João Gama                   University Porto, Portugal
Tias Guns                   Vrije Universiteit Brussel, Belgium
Lawrence Hall               University of South Florida, USA
Nick Heard                  Imperial College London, UK
Frank Höppner               Ostfalia University of Applied Sciences, Germany
Francois Jacquenet          Laboratoire Hubert Curien, France
Norbert Jankowski           Nicolaus Copernicus University, Poland
Ulf Johansson               Jönköping University, Sweden
Alipio M. Jorge             University of Porto, Portugal
Irena Koprinska             The University of Sydney, Australia
Wojtek Kowalczyk            Leiden University, The Netherlands
Rudolf Kruse                University of Magdeburg, Germany
Niklas Lavesson             Blekinge Institute of Technology, Sweden
Jose A. Lozano              The University of the Basque Country, Spain
George Magoulas             Birkbeck College, University of London, UK
Mohamed Nadif               University of Paris Descartes, France

| | |
|---|---|
| Andreas Nuernberger | Otto-von-Guericke University of Magdeburg, Germany |
| Vera Oliveira | University of Porto, Portugal |
| Kaustubh Raosaheb Patil | Massachusetts Institute of Technology, USA |
| Mykola Pechenizkiy | Eindhoven University of Technology, The Netherlands |
| Ruggero G. Pensa | University of Turin, Italy |
| Marc Plantevit | LIRIS, Université Claude Bernard Lyon, France |
| Lubos Popelinsky | Masaryk University, Czech Republic |
| Alexandra Poulovassilis | Birkbeck College, University of London, UK |
| Miguel A. Prada | Universidad de Leon, Spain |
| Ronaldo Prati | Universidade Federal do ABC, Brazil |
| Alessandro Provetti | Birkbeck College, University of London, UK |
| Fabrizio Riguzzi | University of Ferrara, Italy |
| George Roussos | Birkbeck College, University of London, UK |
| Stefan Rueping | Fraunhofer IAIS, Germany |
| Antonio Salmeron | University of Almería, Spain |
| Vítor Santos Costa | University of Porto, Portugal |
| Carlos Soares | University of Porto, Portugal |
| Christine Solnon | INSA de Lyon, France |
| Stephen Swift | Brunel University London, UK |
| Frank Takes | LIACS, Leiden University, The Netherlands |
| Allan Tucker | Brunel University London, UK |
| Melissa Turcotte | LANL, USA |
| Antti Ukkonen | University of Helsinki, Finland |
| Antony Unwin | University of Augsburg, Germany |
| Peter van der Putten | LIACS, Leiden University and Pegasystems, The Netherlands |
| Jan N. van Rijn | Leiden University, The Netherlands |
| Maarten Van Someren | University of Amsterdam, The Netherlands |
| Veronica Vinciotti | Brunel University London, UK |
| David Weston | Birkbeck College, University of London, UK |
| Albrecht Zimmermann | Université Caen, France |
| Indre Zliobaite | University of Helsinki, Finland |

## Additional Reviewers

Alberti, Marco
Argento, Luciano
Bellodi, Elena
Cardoso, Douglas
Cunha, Tiago
Dockhorn, Alexander
Doell, Christoph
Garcia-Bernardo, Javier
Gossen, Tatiana
Held, Pascal

Huang, Alexander
Kaššák, Ondrej
Mantovani, Rafael
Oikarinen, Emilia
Paurat, Daniel
Schulze, Sandro
Schwerdt, Johannes
Sousa, Ricardo
Ullah, Ihsan
Ševcech, Jakub

# Contents

# Improving Chairlift Security with Deep Learning

Kevin Bascol[1,2]([⊠]), Rémi Emonet[1], Elisa Fromont[1], and Raluca Debusschere[2]

[1] Univ Lyon, UJM, Lab Hubert Curien, CNRS UMR 5516,
42000 Saint-Etienne, France
kevin.bascol@univ-st-etienne.fr
[2] Bluecime Inc., 445 rue Lavoisier, 38330 Montbonnot Saint Martin, France

**Abstract.** This paper shows how state-of-the-art deep learning methods can be combined to successfully tackle a new classification task related to chairlift security using visual information. In particular, we show that with an effective architecture and some domain adaptation components, we can learn an end-to-end model that could be deployed in ski resorts to improve the security of chairlift passengers. Our experiments show that our method gives better results than already deployed hand-tuned systems when using all the available data and very promising results on new unseen chairlifts.

**Keywords:** Deep learning · Convolutional neural networks · Image classification · Domain adaptation

## 1 Introduction

Ski resorts are all equipped with different and numerous chair and ski-lifts. To ensure the security at each boarding terminal, one person (at least) is in charge of continuously monitoring the lift traffic and ensuring safe boarding for all its passengers. Possible hazardous situations include: improper seating, restraining bar not pulled down, loss of a gliding equipment, unaccompanied child, etc. On average, all these situations lead each year to 20 serious injuries in France alone, out of which more than a third occur during or just after boarding the lift. Thus the chairlift administrators are in search of viable solutions to prevent the possible accidents. All the chairs could be equipped with sensors that could monitor the number of persons on a given chair, whether they are well seated, whether the position of the restraining bar (or security railing) is correct a few seconds after leaving the boarding station (which is very important to prevent falls), whether a child is not alone on the chair, whether the skiers have not lost a ski or a snowboard etc. Such sensors should trigger an alarm that would stop or slow down the chairlift if a particular anomaly is detected. However, in the extreme weather conditions that ski-resorts can face, small mechanical sensors are subject to wear and often prone to failure, thus other more robust solutions are required.

The work presented in this article is made in collaboration with a start-up called Bluecime, which has installed video cameras filming the boarding platforms in several ski-resorts. Each of these cameras is linked to a computer that

© Springer International Publishing AG 2017
N. Adams et al. (Eds.): IDA 2017, LNCS 10584, pp. 1–13, 2017.
DOI: 10.1007/978-3-319-68765-0_1

processes the video stream in real time. The company has developed a software (called SIVAO) that takes as input the video of a running chairlift, focuses on one moving chair and automatically (using signal processing techniques) detects if the security railing is up or down a few seconds after boarding. Since its creation, the company has successfully shown that video cameras represent a reliable solution which needs almost no maintenance. Besides, they have proved that the real-time image processing of the captured images leads to better results for the cited tasks than mechanical sensors, even in extreme weather conditions (e.g. the artifacts created by the presence of snow flakes are compensated by various pre-processing methods applied to the input images). If the results obtained by the company are already very good, they are currently limited to one particular task (triggering an alarm if the restraining bar position is not as expected). Moreover, their system needs to be calibrated on a per-lift basis, which is time consuming and makes it more difficult to scale the system to a very large number of lifts.

We believe that intelligent data analysis techniques and, in particular, deep learning, can be used to solve this problem in a more precise and adaptable way. We explain in this article how state-of-the-art deep learning architectures designed for general image classification tasks as well as dedicated domain adaptation methods can be used to improve and generalize the results obtained by the company.

## 2  Background on Neural Networks and Related Work

Deep learning is nowadays almost a synonym for learning with deep (more than 2 layers) convolutional neural networks (CNN, see [4] for some insights on neural networks). This learning technique has become tremendously popular since 2012 when a deep architecture, called ALEXNET, proposed by A. Krizhevsky et al. [7] was able to win the ImageNet Large Scale Visual Recognition Challenge (ILSVRC) with an outstanding improvement of the classification results over the existing systems: the error rate was 15% compared to 25% for the next ranked team. Since then, countless different architectures have shown excellent performance in many domains such as computer vision, natural language processing or speech recognition [4].

The amount of labeled data available, the systematic use of convolutions, the better optimization techniques and the advances in manufacturing graphical processing units (GPU) have contributed to this success. However, **deep** networks were supposed to suffer from (at least) three curses. (1) The deeper the network, the more difficult it is to update the weights of the first layers (the ones closer to the input): deep networks are subject to *vanishing and exploding gradient* that leads to convergence problems. (2) The bigger the network, the more weights need to be learned. In statistical machine learning, it is well known that more complex models require more training examples to avoid overfitting phenomena and guarantee relatively good test accuracy. (3) Neural network training aims at minimizing a loss which is a measure of the difference between the computed output of the network and the target. The computed outputs depend on

(i) the inputs, (ii) all the weights of the network and (iii) the non linear activation functions that are applied at each layer of the network. The function to minimize is thus high-dimensional and non-convex. As the minimization process is usually achieved using stochastic gradient descent and back-propagation, it can easily get trapped in local optimum.

In this paper, we are interested in deep learning approaches developed for image classification. In this context, the use of convolutions (i.e., a reduced number and a better configuration of the weights) and the huge size of the current image datasets [2,8] partially solved the second and third curses mentioned above. In the following, we will show other architectures and learning "tricks" that practically alleviate all the remaining problems with deep learning for image classification.

*Learning Tricks.* The first improvement concerns the *initialization of the weights*. It was shown [1] that image classification tasks could all benefit from a pre-training step on a large dataset such as ImageNet [2]. The ImageNet dataset contains images grouped in 1000 different classes. Practically, the output of a network trained on ImageNet consists in a 1000-D vector which gives the probability for an input image to be of each of the 1000 classes. To use such a network on a new task (with much fewer and different target classes), the idea is to "cut" the last layer(s) of the pre-trained network. Only the convolutional part is kept and its output is considered as a generic image descriptor. A new network can be built from this pre-trained generic feature extractor by training new additional "classification" layers (usually fully connected). The whole network can then be further *fine-tuned*, i.e., trained end-to-end on the new problem of interest. This procedure is called "transfer learning".

The second improvement is the *batch normalization* [6]. This simple yet very effective idea consists in normalizing the activation outputs of each layer of the network according to the current input batch. Its main benefit is to reduce the internal covariate shift, which corresponds to the amplification of small changes in the input distribution after each layer, and so creates high perturbations in the inputs of the deepest layers.

*CNN Architectures.* Simonyan and Zisserman [9] created VGGNET 16 and VGGNET 19 composed of respectively 13 and 16 convolutional layers followed by 3 fully-connected layers. As for the previously mentioned ALEXNET, the spatial dimensions were reduced with max-pooling layers, the inputs of the convolutions being typically padded so as to keep the spatial size. However, each time the spatial dimensions were divided by two, the number of filters in the next layers was doubled to compensate for the information loss. With this method, Simonyan et al. managed to obtain an error rate of 7% at ILSVRC 2014.

Following these results, several other techniques have been designed to be able to train deeper networks. For instance, He et al. [5] introduced the concept of *residual mapping* which allowed them to create a network with 18 to 152 convolutional layers called a Residual Network (ResNet). Their architecture is divided into blocks composed of 2 or 3 convolutional layers (depending on the total depth of the network). At the end of a block, its input is added to the output of the layers. This sum of an identity mapping with a "residual mapping" has proven effective at overcoming the vanishing gradient phenomenon during the back-propagation phase and allows to train very deep networks. An example of such blocks is shown in Fig. 2. Using residual blocks, the network is learned faster and with better performance. At ILSVRC 2015, ResNet 152 showed the best performance with less than 4% of error rate. They also used *batch normalization* [6] on the first (or first two) layer(s) of each residual block. We will base our system on this architecture.

*Domain Adaptation.* Some applications require to be able to learn a model on some data (e.g. images of cats) called the *source domain* and deploy it on similar but different data (images of tigers) called the *target domain.* This well-known machine learning problem is called *Domain Adaptation.* Many interesting approaches have been proposed in deep learning to tackle this problem such as the one from Ganin et al. [3]. Their idea (as shown in Fig. 1) is to train two networks that share the same first (convolutional) layers called the "feature extractor". The first network is dedicated to the classification task on one particular domain. The second network aims at predicting the domain of an input example. Note that to train this second network, the examples of the target domain do not need to be labeled (and are not in [3]). The only information needed to train the second network is whether an example belongs to the source or to the target domain. The two networks are trained in an adversarial way according to the shared layers (using a mechanism called gradient reversal on the second network optimization). As a consequence, the shared features of the networks are discriminative for the classification task as well as domain invariant.

## 3    Proposed Deep ResNet with Domain Adaptation

Based on the state-of-the art study presented in Sect. 2, we propose an image classification architecture using domain adaptation and a convolutional residual network pre-trained on ImageNet. This architecture is shown in Fig. 1 and is divided into three parts: (1) a feature extractor which learns a new image representation; (2) a classifier which predicts the class of an image, and (3) a domain discriminator which ensures that the feature extractor is domain invariant, to improve classification accuracy on unseen data.

### 3.1    Network Architecture

Our network inputs are RGB images of size 224 × 224 that can be viewed as three-dimensional tensors (two spatial dimensions and one dimension for the RGB channels). After some trial and error, we decided to use the ResNet architecture presented in the previous section with 50 layers and pre-trained on the ImageNet dataset (see the "learning tricks"). Networks with more layers gave a slightly better accuracy but were longer to train and the biggest ones

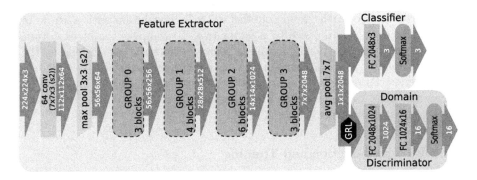

**Fig. 1.** The proposed ResNet architecture with domain adaptation, for a 3-class classification problem with 16 domains (chairlifts).

(with more than 101 layers) did not allow us to process test images in real time on a CPU. The 50 layers architecture (shown in Fig. 1) gave us a good trade-off between computational efficiency and accuracy. This network is composed of 49 convolutional layers and only one fully connected layer at the output of the network. This last layer is preceded by an average pooling layer which reduces the size of the feature maps to a 1D-vector. The last layer of the original ResNet 50 architecture was changed to fit our 3-classes classification problem (instead of the 1000 classes of the ImageNet classification problem). We used our chairlift dataset to train this last layer and fine-tune the entire network.

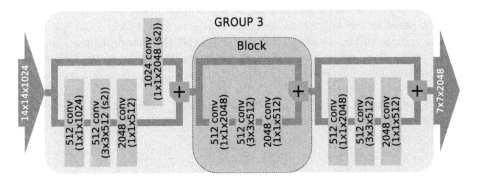

**Fig. 2.** Details of the last group of layers in our architecture (see Fig. 1), which is composed of three residual blocks and different convolution filters.

In the feature extractor, each group is composed of a set of blocks. Each block is composed of three layers: a $1 \times 1$ convolution that acts as a learnable dimensionality reduction step, a $3 \times 3$ convolution that extracts some features and a $1 \times 1$ convolution that restores the dimensionality. As explained in Sect. 2, we use the residual connections between each block to ease the training phase.

The first block of each group contains a $3 \times 3$ convolution with a stride of 2 which reduces the spatial size by a factor two. To compute the block output with the sum operator, the input needs to have the same dimensions as the output. Therefore, a $1 \times 1$ convolutional layer with a stride of 2 is added to reduce the spatial size and also to match the number of feature maps which changes according to the group. In Fig. 2, we show in more details the last group of layers in the feature extractor.

### 3.2   Objective Function and Training

Different losses are classically used in deep learning. The best known for image classification is the cross-entropy [4]. After experimenting with different losses, we decided to use the more robust multi-class hinge loss (Eq. 1) commonly used in SVM:

$$loss(W, X, t) = \frac{1}{b} \sum_{i=1}^{b} \frac{1}{|C|} \sum_{c \in C \setminus \{t\}} max(0, o_c^i - o_t^i + m) \qquad (1)$$

where $W$ is the set of all the parameters in the network, $X$ is the subset of the input dataset (mini-batch) given to the network during a forward pass, $b$ is the mini-batch size, $C$ is the set of classes, and $t$ is the ground-truth label corresponding to each example of the mini-batch. The margin $m$ can be seen as the minimal "distance" required between the computed probability of the prediction of the target class ($o_t^i$) and the other classes ($o_c^i$), since the loss value is equal to 0 when $o_t^i \geq o_c^i + m$. This loss presents several advantages: (i) it is defined on $p(x) = 0$ which makes it more robust than the cross-entropy; (ii) it does not penalize the weights on well classified examples which could speed up the convergence of the network; (iii) the slope and the margin of the function can be easily changed to give a custom weight to the different type of errors (false positive, false negative). This last advantage will be explored in our future work.

To update the weights of the network, a stochastic gradient descent algorithm is applied on the loss function. The weight update rule is given by:

$$W(\tau + 1) = W(\tau) - \eta \cdot \frac{\partial\, loss(W(\tau), X, t)}{\partial\, W(\tau)} \qquad (2)$$

where $\tau$ is the current iteration and $\eta$ is *the learning rate* which needs to be tuned.

## 4   Experiments

SIVAO, the current system deployed by Bluecime (introduced in Sect. 1), can be decomposed into two parts: a *vehicle detector* and a *situation classifier*. The purpose of the vehicle detector is to determine the vehicle position, scale, and orientation so as to track it each time it passes in front of the camera.

The situation classifier detects different vehicle parts with respect to the position given by the vehicle detector. Its purpose is to detect whether the vehicle is empty or not and also to infer the position of the restraining bar, so that it can classify a frame into *Safe* or *Unsafe*. A detection zone is pre-configured and, at the end of the zone, the system computes a final decision based on a configured number of frame-level decisions made during the tracking of the vehicle. Depending on this decision (predicted label), an alarm is triggered or not.

Our proposed network could replace the situation classifier to enhance the classification performance while keeping the efficient vehicle detector designed by Bluecime. In the following, we evaluate the performance of our deep learning approach applied on images provided by the vision-based vehicle detector.

**Fig. 3.** Training images from three different chairlifts (labeled from left to right *Empty*, *Safe*, and *Unsafe*)

### 4.1 Chairlift Dataset

Bluecime has already equipped 16 chairlifts (here named A, B, ...) from 3 different ski-resorts with the system presented in Sect. 1. To build the dataset, each time a vehicle (chair) passes in front of the camera, we use SIVAO to record 2 images that are approximately centered on the chair: one far from and one close to the camera. Examples of such images are given in Fig. 3. A total of 50 000 such images have been recorded and manually labeled in 3 categories: *Empty*, *Safe* and *Unsafe*. In the first case (*Empty*), the vehicle does not carry any passengers. The system should not trigger an alarm in this situation. In the second case (*Safe*), the vehicle carries passengers who closed the restraining bar completely. In the last case (*Unsafe*), the vehicle carries passengers and the restraining bar is slightly or completely open.

The dataset is unbalanced with respect to the classes: around 28 000 images are labeled *Empty*, 18 000 are labeled *Safe*, and 4 000 are labeled *Unsafe*. The dataset is also unbalanced with respect to the different chairlifts (which will be called *domains* in the following): the least represented chairlift has 1 800 examples, whereas the most represented one has 6 200 examples.

As they were designed by different manufacturers at different times, all the 16 chairs in the dataset are different, though some are similar. There are even some unique cases, for instance a chairlift having a glass bubble (chairlift D) as a second protection, or another whose vehicles do not have a complete frame (see Fig. 3, left).

## 4.2    Evaluation Procedure

SIVAO is a commercialized product that aims at improving the work of chairlift operators. We use its current performance as an indication of the minimal result requirements for an acceptable quality of service that we aim to surpass with our automatic system. To compare SIVAO and our method, the results of our multiclass approach are cast into a binary classification problem. To do that, we merge the classes *Empty* and *Safe* as Negative, and the class *Unsafe* becomes Positive.

To study the behavior of our approach, we considered six different experimental settings:

1. **OOC** ("Only One Chairlift"). 16 independent experiments are averaged: each chairlift is considered independently (as it is done by SIVAO), thus the training (resp. the test) set contains 85% (resp. 15%) of the images of a single chairlift. Obviously, with only one chairlift per experiment, the domain adaptation component is not used on this setting.
2. **ALL.** One experiment is performed using only the features extractor and the classifier, with 85% of all the available images (from all the chairlifts) for training and 15% for testing. The same experiment is done 16 times with 16 randomly chosen training/test sets and the results are averaged in Table 1.
3. **ALL DA** ("Domain Adaptation"). Same as ALL but using the domain adaptation component described in Sect. 2.
4. **LOCO** ("Leave One Chairlift Out"). 16 experiments are averaged: in each, we use only the feature extractor and the classifier, with the images of all the chairlifts but one (thus 15) mixed in the training set and all the images of the remaining chairlift as test set.
5. **LOCO DA.** Same as LOCO but with domain adaptation using *no examples* from the target chairlift.
6. **LOCO DA.** Same as LOCO but with classical domain adaptation where some *unlabeled* examples of the target chairlift are used by the domain adaptation component.

In setting 1 (OOC), due to time constraints and to be closer to a realistic setting, we do not tune the hyper-parameters for each domain but we use one single hyper parameter setting (given in the next section) for all the chairlifts. In this setting, we expect the network to quickly overfit our data and also to be penalized by the lack of examples especially for the least represented chairlifts.

In settings 2 and 3 (ALL and ALL DA), we train our network using all the training data available. We only build one model for all the chairlifts which makes this setting easier to deploy in practice. However, the hyper-parameters of the system are also global which may harm the final performance. These settings could be used with the current cameras installed by the company but do not evaluate the real ability of our system to work on new chairlifts.

Settings 4 to 6 (LOCO, LOCO DA-, and LOCO DA) really show the potential of our approach. Ideally, our method should show good enough performance on these settings to allow Bluecime to deploy their system equipped with this model on any new chairlift with no manual labeling. In these settings, we expect a performance drop compared to the OOC or ALL settings because of the variability of the different domains.

To evaluate their system, the company relies on numerous measures. Among these, 4 statistical measures assess the overall performance of the system: recall, precision, F-measure, and accuracy. *Unsafe* examples are considered positive (the alarm has to be triggered), and safe examples negative (no alarm needed), thus the classes *Empty* and *Safe* are both considered negative. The *recall* gives the proportion of examples correctly detected positive among all the examples expected positive. In our case, that is the ability of the system to trigger an alarm in unsafe situations. The *precision* gives the proportion of examples correctly detected positive among all the examples detected positive. Thus it indicates the ability of the system to avoid useless alarms. The *F-measure* (harmonic mean between precision and recall) and the accuracy (percentage of correct predictions) give a more global view on the performance of the system. In the following, we report these 4 measures.

### 4.3   Training Details

In our experiments, we use mini-batches of 84 images (guided by GPU memory limitations), enforcing a balance over classes and domains: we randomly select a chairlift, then randomly select a class, then randomly select an example. We tune our hyper-parameters using a grid search. In all the reported experiments, the learning rate is set to $10^{-5}$ (with an ADAM optimizer), the hinge loss margin to 0.33 for the domain adaptation component (with the reversal layer) and to 0.01 for the domain discriminator. The gradient reversal layer used for the domain adaptation induces a hyper-parameter $\gamma$ set to 10 as defined by Ganin et al. [3].

Our dataset is composed of images of size $237 \times 237$. When an image is loaded, we randomly select a $224 \times 224$ crop in the image to fit the size of the original ResNet architecture. To maximally exploit the small number of examples available in the positive class, we decided not to use a validation set and to stop the learning process after a constant (high enough) number of iterations set to 12 000. Training our model for 12 000 iterations takes about 4 to 5 h. The classification of one image is done in approximately 15 ms which allows our method to be used in real time.

## 4.4   Experimental Results

In Table 1, we present the results on all the datasets for all the experimental settings. The first line of the table gives the performance of the current hand-tuned
method developed by Bluecime. In the OOC setting (the closest to what the
company is currently doing), the proposed network brings a 7 point improvement in precision, recall and F-measure and +1pt in accuracy. This validates
the relevance for the company of the proposed machine learning-based method.
With the ALL setting, a 5pts gain in F-measure and 1pt gain in accuracy are
observed. While the gain is smaller than with OOC, the *one-model-fits-all* nature
of ALL makes it a very attractive solution from an industrial standpoint. Adding
domain adaptation (ALL DA) has a very limited impact which can be explained
by the fact that all the domains are already in the training set.

**Table 1.** Performance results of our deep learning system for the three main group
of settings (OOC, ALL, LOCO) compared to the hand-tuned existing system of the
Bluecime company.

| Experiment | F-measure | Precision | Recall | Accuracy |
|---|---|---|---|---|
| Bluecime | 84.72 | 87.88 | 81.78 | 97.40 |
| OOC | **91.70** | **94.64** | **88.94** | **98.68** |
| ALL | 89.47 | 93.49 | 85.78 | 98.33 |
| ALL DA | **89.71** | **93.88** | **85.89** | **98.36** |
| LOCO | 76.23 | 76.71 | 75.75 | 96.07 |
| LOCO DA- | 72.36 | 70.84 | 73.93 | 95.30 |
| LOCO DA | **84.24** | **82.76** | **85,77** | **97.33** |

   As expected, during the LOCO experiment, performance losses occur for
all the measures, with −8pts of F-measure and −1pts of accuracy. In this setting, performance is too low for industrial deployment. However, adding domain
adaptation (LOCO DA) brings back the performance to a level comparable with
the Bluecime system, with −0.5pts of F-measure and a similar accuracy. These
results emphasize that domain adaptation is relevant and allows to create models
that have competitive results even on a chairlift with non-labeled examples.

   In the case when we do not have any examples of a specific chairlift (e.g.,
launching the system immediately after its installation) adding domain adaptation (LOCO DA-) causes a 4pts loss of F-measure (compared to plain LOCO).
This behavior may be unexpected but it has an intuitive explanation: domain
adaptation encourages the learning of features that are domain-invariant but
also specializes these features for the domains it has seen, causing a detrimental
effect on an unseen chairlifts. These results show that in a LOCO setting, domain
adaptation should be used only if we can retrain the model using new unlabeled
images (setting LOCO DA), and not if the goal is to produce an off-the-shelf
model.

**Table 2.** Comparison of OOC and LOCO experiments on 4 specific chairlifts

| Chairlift | Size | Experiment | F-measure | Precision | Recall | Accuracy |
|---|---|---|---|---|---|---|
| A | 2842 | OOC | 95.68 | 97.08 | 94.33 | 97.06 |
|  |  | LOCO DA | 92.98 | 93.56 | 92.40 | 95.21 |
| B | 6306 | OOC | 88.89 | 91.67 | 86.27 | 98.83 |
|  |  | LOCO DA | 72.19 | 66.50 | 78.95 | 96.70 |
| C | 1822 | OOC | 76.92 | 78.95 | 75.00 | 96.79 |
|  |  | LOCO DA | 77.78 | 83.05 | 73.13 | 96.93 |
| D | 2736 | OOC | 85.71 | 100.00 | 75.00 | 99.76 |
|  |  | LOCO DA | 82.76 | 85.71 | 80.00 | 99.63 |
| Overall average (Table 1) |  | OOC | 91.70 | 94.64 | 88.94 | 98.68 |
|  |  | LOCO DA | 84.24 | 82.76 | 85,77 | 97.33 |

In the Table 2, we focus on 4 different chairlifts in order to illustrate how performance varies in function of the specificity of each chair model.

Chairlift A – In both settings, results are better that the average with +4pts (OOC) and +9pts (LOCO) of F-measure. This gain in performance is possible, for OOC, thanks to a good balance in the classes of the images of this chairlift, and for LOCO, thanks to the chairlift configuration similar to several other ones.

Chairlift B – OOC shows a slightly lower performance with −3pts of F-measure but +0.5pts of accuracy. This is explained by the over-representation of the *Safe* class (4 600 images versus 340 for the *Unsafe* class). With the LOCO setting, we observe a considerable performance drop with −12pts of F-measure. This poor performance is mostly due to the sun which casts different shadows depending of the time of the day. Besides, a high number of images present flares decreasing the image quality.

Chairlift C – OOC has a F-measure of 15pts below the overall result. It is even outperformed by LOCO (+1pts of F-measure and +0.1 of accuracy). Chairlift C is the domain with the fewest number of examples (around 1 800). This number is probably not sufficient to train our deep architecture. In the LOCO experiment, the poor performance is mostly due to the shadow cast by the chairlift tower in the afternoon.

Chairlift D – This is the only chairlift with a glass bubble, and also the only one which has the restraining bar (and also the bubble) closed on *Empty* images. This implies that, even if its data is unbalanced (2 000 *Empty*, 670 *Safe*, and 30 *Unsafe*), on the OOC setting, the network has perfect precision and retrieves 3 of the 4 positives examples, as only 15% of the 30 *Unsafe* are used for testing. The F-measure is 6pts lower than the average result, yet it is highly impacted by this lack of positive testing examples. In LOCO, we could have expected an important performance drop considering the chairlift unique features (the bubble), but when the restraining bar is open, the bubble is open too, and so the configuration in that case is not dramatically different to other chairlifts. This explains why the loss of F-measure is limited to 2pts.

These results underline the different factors limiting the performance of the proposed system. When training and testing is done on the same domain, we need a consequent amount of data to reach good performances. When testing on a new domain, if this domain is very different from the others, we can expect worse results with DA since the domain adaptation will not be able to bridge the domain gap.

For Bluecime, collecting data is easy but annotating them is not because of human resource limitations. Besides, several new chairlifts are equipped each season. Thus the LOCO setting with DA seems the most promising one in a first deployment phase, since it avoids the need for numerous annotations for the new chairlift. This setting also provides a basis for an active learning system that would limit the number of needed annotations. In a second phase, with enough annotated data, the OOC setting could be used for each resort.

## 5   Conclusion

We have presented an end-to-end deep learning system that can improve the security on chairlifts using visual information. Leveraging domain adaptation techniques, the model is able to achieve competitive performance even with (new) chairlifts for which no labeled data is available. Overall, this ability to generalize, its competitive accuracy and its operation in real-time make this system well suited for industrial deployment.

**Acknowledgment.** The authors acknowledge the support from the *SoLStiCe* project ANR-13-BS02-0002-01.

## References

1. Chatfield, K., Simonyan, K., Vedaldi, A., Zisserman, A.: Return of the devil in the details: delving deep into convolutional nets. In: BMVC (2014)
2. Deng, J., Dong, W., Socher, R., Li, L.J., Li, K., Fei-Fei, L.: ImageNet: a large-scale hierarchical image database. In: CVPR (2009)
3. Ganin, Y., Ustinova, E., Ajakan, H., Germain, P., Larochelle, H., Laviolette, F., Marchand, M., Lempitsky, V.: Domain-adversarial training of neural networks. JMLR (2016)
4. Goodfellow, I., Bengio, Y., Courville, A.: Deep Learning. MIT Press (2016). http://www.deeplearningbook.org
5. He, K., Zhang, X., Ren, S., Sun, J.: Deep residual learning for image recognition. In: Computer Vision and Pattern Recognition (CVPR) (2016)
6. Ioffe, S., Szegedy, C.: Batch normalization: accelerating deep network training by reducing internal covariate shift. In: ICML, pp. 448–456 (2015)
7. Krizhevsky, A., Sutskever, I., Hinton, G.E.: Imagenet classification with deep convolutional neural networks. In: NIPS (2012)

8. Lin, T.-Y., Maire, M., Belongie, S., Hays, J., Perona, P., Ramanan, D., Dollár, P., Zitnick, C.L.: Microsoft COCO: common objects in context. In: Fleet, D., Pajdla, T., Schiele, B., Tuytelaars, T. (eds.) ECCV 2014. LNCS, vol. 8693, pp. 740–755. Springer, Cham (2014). doi:10.1007/978-3-319-10602-1_48
9. Simonyan, K., Zisserman, A.: Very deep convolutional networks for large-scale image recognition. arXiv preprint arXiv:1409.1556 (2014)

# Discovering Motifs with Variants in Music Databases

Riyadh Benammar[1,2]($^{(\boxtimes)}$), Christine Largeron[1,3], Véronique Eglin[1,2],
and Myléne Pardoen[4]

[1] Université De Lyon, Lyon, France
[2] CNRS INSA-Lyon, LIRIS, UMR5205, 69621 Lyon, France
{riyadh.benammar,veronique.eglin}@insa-lyon.fr
[3] UJM-Saint-Etienne, CNRS, Institut d'Optique Graduate School,
Laboratoire Hubert Curien UMR 5516, 42023 Saint-Etienne, France
christine.largeron@univ-st-etienne.fr
[4] Institut des Sciences de l'Homme (FRE 3768), 14 Avenue Berthelot,
69363 Lyon cedex 07, France
mylene.pardoen@wanadoo.fr

**Abstract.** Music score analysis is an ongoing issue for musicologists. Discovering frequent musical motifs with variants is needed in order to make critical study of music scores and investigate compositions styles. We introduce a mining algorithm, called CSMA for **C**onstrained **S**tring **M**ining **A**lgorithm, to meet this need considering symbol-based representation of music scores. This algorithm, through motif length and maximal gap constraints, is able to find identical motifs present in a single string or a set of strings. It is embedded into a complete data mining process aiming at finding variants of musical motif. Experiments, carried out on several datasets, showed that CSMA is efficient as string mining algorithm applied on one string or a set of strings.

**Keywords:** Music scores analysis · Music motif mining · String mining

## 1 Introduction

Everyone is able to tell sometimes: 'This song seems to be from Supertramp or this music from Chopin'. By listening to a musical excerpt one can recognize the singer or the music's style, even if the singer is unknown. For an instrumental piece we are able to recognize the composer. This is probably due to the motifs appearing in the music score. Musical motifs are pieces of music that can define a signature for a composer, a music score or a music style. They correspond to identical repeating music chunks or variations applied on a part of music that is modelled by a string. As music notes are characterized by three kinds of information (melodic, rhythmic and harmonic), musical motifs can also be melodic and/or rhythmic and/or harmonic. To the best of our knowledge, only few works tried to extract musical motifs where the mining process is applied on one or many sequences of musical symbols. However, musical motif extraction

© Springer International Publishing AG 2017
N. Adams et al. (Eds.): IDA 2017, LNCS 10584, pp. 14–26, 2017.
DOI: 10.1007/978-3-319-68765-0_2

with variants, except transposed motifs, is a case-study that was not addressed before. In our work we are interested in melodic and/or rhythmic motifs mining with variants.

Our goal is to extract musical motifs from music scores transcriptions. In data mining, this task corresponds to motif mining from a single sequence or a set of sequences called strings. In this paper, we propose an algorithm, called Constrained String Mining Algorithm (CSMA), able to solve this task. In addition, in order to find musical motif variants, we present a preprocessing of the music data. This preprocessing provides different representations of a music score offering to CSMA the possibility to identify three variants of an initial motif: transposed, inverted and mirror forms. Experiments carried out on synthetic and real datasets have confirmed the interest of the proposed approach.

The rest of the paper is organized as follows. Section 2 is dedicated to the state of the art, while the proposed algorithm CSMA is presented in Sect. 3. The experiments are described in Sect. 4 and Sect. 5 concludes the paper.

## 2   Related Work

Agrawal and Srikant introduced sequence mining methods, based on their well-known algorithm Apriori [1], initially designed for itemset mining in a transactional database where each transaction consists in a customer-id, the transaction time, and the items bought [2]. Formally, the set of items is defined by $I = \{i_1, ..., i_n\}$ and a sequence s over I is an ordered list $< s_1...s_l >$, where $s_i \subset I$ $(1 \leq i \leq l, l \in \mathbb{N})$ is an itemset. $l = |s|$ is called the size of the sequence [3]. A pattern is a subsequence with multiple occurrences in the database. It is considered as frequent if it appears at least *minFrequency* time where *minFrequency* is a threshold fixed by the user.

Several algorithms were proposed to solve efficiently this task like FreeSpan and PrefixSpan, based on pattern-growth methods [4,5] or SPADE which implements a vertical format-based mining approach [6]. Among the most recent sequence mining algorithms that outperform the other approaches, we can mention CM-SPADE, an extended version of SPADE that can incorporate constraints. This algorithm is based on co-occurrence MAP structure allowing a good pruning strategy during the candidate generation. Some other works introduced constraint-based sequential pattern mining approaches, like cSpade [7] and Prefix-Growth [8], where a set of constraints related to pattern length, gaps inside pattern and other parameters are set by the user.

In our framework, we are more interested in motif mining on strings than by pattern mining in sequences. Indeed, motif mining is applied on a string dataset, which is equivalent to a sequence dataset with itemsets of length one. This corresponds to our case where we consider that a music score can be represented by one or several sequences, one sequence per instrument, and where at each timestamp there is only one note in the sequence; the harmonic information is not considered.

In such sequences, three types of motifs can be extracted: *contiguous motifs*, *non-contiguous motifs* and *approximate contiguous motifs*. Contiguous motifs are motifs where the elements are lined up behind each other whereas in non-contiguous motifs, jumps between elements positions are allowed. In approximate contiguous motifs, introduced by Floratou et al., the presence of a certain controlled noise is permitted but this notion is out of the scope of this work which focuses only on contiguous and non-contiguous motifs [9].

Finally, we can mention episode mining which exploits sequences of events where each event has an associated time of occurrence. To detect events occurring frequently close to each other, episode mining considers a sliding overlapping window of fixed size. An event is considered as frequent if it occurs in a minimal number of windows [10]. For more details, we refer the interested reader to [11].

In our framework, a musical motif corresponds to a music chunk appearing at minimal number of positions through the music score. This motif can be melodic, rhythmic or both, depending on the nature of the music event features. Among the first works in the music domain, Hsu *et al.* introduced a method to identify frequent motifs in a music score, by considering only the melodic information in a single sequence [12]. Liu *et al.* [13] proposed an improvement of this method aiming at finding all non-trivial motifs (*i.e.* motifs that do not have sub motifs with the same frequency). These algorithms are not suited for our task since we are interested in identifying motifs with gaps whereas they consider only contiguous motifs. Moreover they only process one sequence while we consider music scores with one or several instruments corresponding to a set of sequences, one per instrument. However, we retain from the algorithm proposed by Liu *et al.* the structure for coding the motifs, that we adapt to handle gaps.

Besides, other works study the representation of the music scores. When music data is in audio format, pitch values can be firstly extracted. Then, melodic motifs can be identified using for instance, episode mining approach like in [13]. To exploit both melodic and rhythmic information, Béatrice Fuchs suggests to transform the input music data into a data stream and then, mining frequent itemset [14]. Finally, the work the most related to ours has been done by Jiménez *et al.* who designed an algorithm able to find transposed musical motifs by exact matching [15]. In this approach, the song is transformed into a sequence of notes and the motifs are extracted by a sequence mining algorithm, called SSMiner. In the same way than in this last work, we are interested in finding motifs as well as their variations. For a given motif we can have three possible variations: transposed, inverted and mirror forms. All these forms are interesting for the musicologist since they reflect the style of a composer. Consequently, they can define a signature for the composer, the score or the music style that can be used as features for other mining tasks such as supervised or non supervised clustering.

We propose in the next section a new algorithm, **CSMA**, able to extract motifs from one or several strings with constraints related to the minimal frequency, gaps inside motifs, and motif length. In **CSMA**, the motif positions in the music score are saved. Those positions are helpful for the musicologist to analyse the score. They are also exploited to build new representations of the music score which are used to extract variants of the musical motifs.

## 3   Contribution

In this section, we introduce a new algorithm, called **CSMA** (**C**onstrained **S**tring **M**ining **A**lgorithm), for discovering all frequent motifs in a string. CSMA performs motifs search according to constraints related to frequency, gaps between motifs, minimal and maximal length of motifs. It uses the same structure as the algorithm of Liu *et al.* for coding motifs but with some modifications to take gaps into account [13].

Hence, a motif $m_i$ is defined by three elements $m_i = (X, freq(X), P_i = [(p_{i1}, len_{i1}), \ldots, (p_{in}, len_{in})])$ such that $X$ corresponds to the motif value (ordered list of items), $freq(X)$ corresponds to its frequency and $P_i$ its positions and lengths. In the set of positions, called $P_i$, the $j^{th}$ position of the $i^{th}$ motif is denoted $p_{ij}$ and its length at this position is denoted $len_{ij}$.

*Example 1.* Given the sequence $S = <ABABCDCABDCE>$:

- The motif $m$ corresponding to $A$ is defined by $m$=(A,freq(A)=3, P = [(1,1), (3,1), (8,1)]);

The pseudo-code of **CSMA** is given in Algorithm 1. This algorithm takes in input a sequence $S$, a minimum frequency threshold $minFrequency$, a maximum allowed gap length inside motifs $maxGap$, a minimum motif length $minLength$ and a maximum motif length $maxLength$.

Table 1 shows an example of the process on the sequence of *Example 1* with $minFrequency = 2$, $maxGap = 1$, $minLength = 1$ and $maxLength = 4$.

The first step of CSMA (Line 4, Algorithm 1) consists in computing the set $\mathcal{F}_1$ containing the frequent motifs of length one. It is performed using the function **COMPUTE** which starts by enumerating all possible items from $S$ and computing the frequency of each item. The items with frequency greater than or equals to $minFrequency$ are added to $\mathcal{F}_1$.

In the example of Table 1, the set of items is $I = \{A, B, C, D, E\}$ and $\mathcal{F}_1 = \{m_1 = (A, 3, P_1=[(1,1),(3,1),(8,1)]), m_2 = (B, 3, P_2=[(2,1),(4,1),(9,1)]), m_3 = (C, 3, P_3=[(5,1),(7,1),(11,1)]), m_4 = (D, 2, P_4=[(6,1),(10,1)])\}$; the motif containing $E$ does not belong to $\mathcal{F}_1$ because it does not appear at least $minFrequency$ times. In order to get the set $\mathcal{F}_K$ containing the motifs of length equals to $(K = 2)$, a joining operation **JOIN** is considered (line 7) between each element $m_i$ of $\mathcal{F}_{K-1}$ and each item $m_j$ belonging to $\mathcal{F}_1$. The joining operation is $\mathcal{O}(|P_i| \times |P_j|)$. So, in order to prune the search space, we compute the position on which the motif $m_i$ is considered as frequent (line 8). This position, called $frequentPosition$, corresponds to the sum of the index of $m_i \in \mathcal{F}_{K-1}$ at the $minFrequency^{th}$ position and the length of $m_i$ for the same position.

*Example 2.* In Example 1, as $minFrequency = 2$, the $minFrequency^{th}$ position will be the second position. Consequently, for the motif $m_1 = (A,3,[(1,1), (3,1), (8,1)])$ knowing that its second position corresponds to $p_{12} = 3$ and its length at this position is $len_{12} = 1$, $frequentPosition$ of $m_1$ equals $p_{12}+len_{12} = 3+1 = 4$.

---

**Algorithm 1:** Constrained String Mining Algorithm (CSMA)

---

**Input** : Sequence $S$, $minFrequency$, $maxGap$, $minLength$ and $maxLength$
**Output:** $\mathcal{F}$: The set of frequent motifs respecting constraints

```
1  begin
2  |    K = 1;
3  |    F₁ = ∅;
4  |    COMPUTE(F₁); /* Compute frequent motifs of length 1*/
5  |    while F_K ≠ ∅ do
6  |    |    K = K + 1;
7  |    |    for m_i = (X, freq(X), P_i) ∈ F_{K-1} do
8  |    |    |    frequentPosition = p_{iminFrequency} + len_{iminFrequency};
9  |    |    |    C = GEN_CANDIDATES(frequentPosition, F₁);
10 |    |    |    for m_j = (Y, freq(Y), P_j) ∈ C do
11 |    |    |    |    m_l = JOIN(m_i, m_j, maxGap, maxLength);
   |    |    |    |    /*m_l = (Z, freq(Z), P_k) is a new motif built by joining m_i and
   |    |    |    |    m_j*/
12 |    |    |    |    if freq(Z) ≥ minFrequency then
13 |    |    |    |    |    F_K = F_K ∪ {m_l};
14 |    |    |    |    end
15 |    |    |    end
16 |    |    end
17 |    end
18 |    F = ⋃_{k≤K} F_k
19 |    FILTER(F, minLength); /* This function removes motifs that violates
   |    minLength frequency constraints and keeps only frequent ones */
20 |    return F;
21 end
```

---

Then, candidate motifs are generated using the **GEN_CANDIDATES** function. Given the position $frequentPosition$ and the set of all frequent motifs of length one already extracted, this function computes a set $\mathcal{C} \subseteq \mathcal{F}_1$ of candidate motifs that could be joined to $m_i$ starting from $frequentPosition$. Our pruning strategy is based on the fact that a motif $m_j \in \mathcal{F}_1$ cannot be a candidate for $m_i \in \mathcal{F}_{K-1}$ if it does not appear after $frequentPosition$ since the joining result of $m_i$ and $m_j$ cannot be frequent.

For instance, in example of Table 1, for the motif $m_4(D)$, as the motifs $A$, $B$ and $D$ do not appear after its $frequentPosition$, which equals to 11, $DA$, $DB$ and $DD$ could not be frequent and, the only candidate for this motif is $C$.

Once the selection of candidate motifs is done, a set $\mathcal{C} \subseteq \mathcal{F}_1$ is computed. Then, the joining operation is performed for the selected motif $m_i$ with each element $m_j \in \mathcal{C}$. The motif joining (concatenation) is defined as follows:

Let be two motifs $m_1 \in \mathcal{F}_{K-1}$ and $m_2 \in \mathcal{F}_1$ defined as $m_1 = (X, freq(X), P_1 = [\bigcup_{i \leq freq(X)} (p_{1i}, len_{1i})])$ and $m_2 = (Y, freq(Y), P_2 = [\bigcup_{j \leq freq(Y)} (p_{2j}, len_{2j})])$, $m_1$ join $m_2$ gives $m_3 \in \mathcal{F}_K$ defined as $m_3 = (Z, freq(Z), P_3)$ such that $Z$ is the concatenation of $(X, Y)$ and $P_3$ is a set of positions $p_{3k}$ and lengths $len_{3k}$.

**Table 1.** Processing of $S = <ABABCDCABDCE>$ with $minFrequency = 2$, $maxGap = 1$ and $maxLength = 4$

| $K$ | $k$ | $\mathcal{F}_{K-1}$ | Frequent Position | $\mathcal{C}$ | New motif | Accepted | Violations |
|---|---|---|---|---|---|---|---|
| - | - | $\emptyset$ | - | - | - | | |
| 1 | - | $A, 3, [(1,1), (3,1), (8,1)]$ | - | - | - | ✓ | |
| 1 | - | $B, 3, [(2,1), (4,1), (9,1)]$ | - | - | - | ✓ | |
| 1 | - | $C, 3, [(5,1), (7,1), (11,1)]$ | - | - | - | ✓ | |
| 1 | - | $D, 2, [(6,1), (10,1)]$ | - | - | - | ✓ | |
| 2 | 1 | $A, 3, [(1,1), (3,1), (8,1)]$ | 4 | $A$ | $A * A, 1, [(1,3)]$ | X | $minFrequency$ |
| 2 | 1 | $A, 3, [(1,1), (3,1), (8,1)]$ | 4 | $B$ | $A * B, 3, [(1,2), (3,2), (8,2)]$ | ✓ | |
| 2 | 1 | $A, 3, [(1,1), (3,1), (8,1)]$ | 4 | $C$ | $A * C, 1, [(3,3)]$ | X | $minFrequency$ |
| 2 | 1 | $A, 3, [(1,1), (3,1), (8,1)]$ | 4 | $D$ | $A * D, 1, [(8,3)]$ | X | $maxGap$, $minFrequency$ |
| 2 | 2 | $B, 3, [(2,1), (4,1), (9,1)]$ | 5 | $A$ | $B * A, 1, [(2,2)]$ | X | $maxGap$, $minFrequency$ |
| 2 | 2 | $B, 3, [(2,1), (4,1), (9,1)]$ | 5 | $B$ | $B * B, 1, [(2,3)]$ | X | $maxGap$, $minFrequency$ |
| 2 | 2 | $B, 3, [(2,1), (4,1), (9,1)]$ | 5 | $C$ | $B * C, 2, [(4,2), (9,3)]$ | ✓ | |
| 2 | 2 | $B, 3, [(2,1), (4,1), (9,1)]$ | 5 | $D$ | $B * D, 2, [(4,3), (9,2)]$ | ✓ | |
| 2 | 3 | $C, 3, [(5,1), (7,1), (11,1)]$ | 8 | $A$ | $C * A, 1, [(7,2)]$ | X | $minFrequency$ |
| 2 | 3 | $C, 3, [(5,1), (7,1), (11,1)]$ | 8 | $B$ | $C * B, 1, [(7,3)]$ | X | $minFrequency$ |
| 2 | 3 | $C, 3, [(5,1), (7,1), (11,1)]$ | 8 | $C$ | $C * C, 1, [(5,3)]$ | X | $maxGap$, $minFrequency$ |
| 2 | 3 | $C, 3, [(5,1), (7,1), (11,1)]$ | 8 | $D$ | $C * D, 1, [(5,2)]$ | X | $maxGap$, $minFrequency$ |
| 2 | 4 | $D, 2, [(6,1), (10,1)]$ | 11 | $C$ | $D * C, 2, [(6,2), (10,2)]$ | ✓ | |
| 3 | 1 | $A*B, 3, [(1,2), (3,2), (8,2)]$ | 5 | $A$ | $A * B * A, 1, [(1,3)]$ | X | $maxGap$, $minFrequency$ |
| 3 | 1 | $A*B, 3, [(1,2), (3,2), (8,2)]$ | 5 | $B$ | $A * B * B, 1, [(1,4)]$ | X | $maxGap$, $minFrequency$ |
| 3 | 1 | $A*B, 3, [(1,2), (3,2), (8,2)]$ | 5 | $C$ | $A * B * C, 2, [(3,3), (8,4)]$ | ✓ | |
| 3 | 1 | $A*B, 3, [(1,2), (3,2), (8,2)]$ | 5 | $D$ | $A * B * D, 2, [(3,4), (8,3)]$ | ✓ | |
| 3 | 2 | $B * C, 2, [(4,2), (9,3)]$ | 12 | - | - | | |
| 3 | 2 | $B * D, 2, [(4,3), (9,2)]$ | 11 | $C$ | $B * D * C, 2, [(4,4), (9,3)]$ | ✓ | |
| 3 | 3 | $D * C, 2, [(6,2), (10,2)]$ | 12 | - | - | | |
| 4 | 1 | $A * B * C, 2, [(3,3), (8,4)]$ | 12 | - | - | | |
| 4 | 2 | $A * B * D, 2, [(3,4), (8,3)]$ | 11 | $C$ | $A*B*D*C, 2, [(3,5), (8,4)]$ | X | $maxLength$, $minFrequency$ |
| 4 | 3 | $B * D * C, 2, [(4,4), (9,3)]$ | 12 | - | - | | |

A position $p_{3k} \in P_3$ equals to $p_{1i}$ if and only if $\exists j \leq freq(Y)$ such that the three conditions are verified:

$$\begin{cases} 0 \leq p_{2j} - (p_{1i} + len_{1i}) \leq maxGap & (1) \\ i = \arg\min_{l \leq freq(X)} (p_{2j} - (p_{1l} + len_{1l})) & (2) \\ p_{2j} + len_{2j} - p_{1i} \leq maxLength & (3) \end{cases}$$

The first condition guarantees the constraint associated to $maxGap$. It allows to compute all possible gap values for a given $p_{2j}$ considering all $p_{1i}$. Only positive gap values that are lower or equals to $maxGap$ are retained. Then, with the second condition, the index $i$ of $p_{1i}$ which has a minimal value for the gap is recovered. Finally, in the third condition, the length of the motif is computed in order to check if it respects the $maxLength$ constraint.

More precisely, the positions $p_{1i}$ from $m_1$ and $p_{2j}$ from $m_2$ verifying the previous conditions allow to define the position $p_{3k}$ corresponding to $p_{1i}$ for $m_3$, and the length $len_{3k}$ is equal to $p_{2j} + len_{2j} - p_{1i}$. The frequency of $m_3$ is equal to the number of positions in $P_3$.

It can be noticed that the frequency of each new motif is lower or equal to its sub-motifs. This means that the joining operation verifies the anti-monotony property which allows to prune the search space.

Once $\mathcal{F}_2$ is obtained, the other sets $\mathcal{F}_K$ of length $K > 2$, are computed and the while loop stops when no new motif is generated. In conclusion, in this example we obtain the following result: $\mathcal{F}_1 = \{A, B, C, D\}$, $\mathcal{F}_2 = \{A * B, B * C, B * D, D * C\}$, $\mathcal{F}_3 = \{A * B * C, A * B * D, B * D * C\}$ and $\mathcal{F}_4 = \emptyset$, where $*$ denotes a gap.

In the next step, (line 19 in Algorithm1), all frequent motifs of order $k \leq K$ are put in $\mathcal{F}$. Then, motifs that do not respect the $minLength$ constraint are removed from $\mathcal{F}$. The **FILTER** function scans each motif $m = (X, freq(X), P = [\bigcup_{i \leq freq(X)} (p_i, len_i))]) \in \mathcal{F}$ and if it finds a position for which $len_i$ is lower than $minLength$ it removes it from the set $P$. In the end, the value $freq(X)$ is updated and if it is lower than $minFrequency$ the motif is removed from $\mathcal{F}$. In the example, as $minLength$ equals to 1, $\mathcal{F} = \mathcal{F}_1 \cup \mathcal{F}_2 \cup \mathcal{F}_3$. In Table 1, the column $violation$ shows violated constraints that fails the joining operation.

As Algorithm 1 makes a breath first search to build motifs, it needs an exponential running time which is estimated to $O((max(|P|) \times |F_1|)^{maxLength})$; with $max(|P|)$ the maximal size of positions sets.

## 3.1   Extension of CSMA

The previous algorithm searches motifs in a single string. However, for a given piece of music, we can be interested in identifying motifs for different instruments namely in several sequences, one per instrument. One can noticed, that CSMA can be extended, in a easy way, to extract motifs within a string database. The adaptation is done in the joining operation by adding to the first conditions a new one according to which "$p_{1i}$ and $p_{2j}$ should belong to the same string". In the sequel, to make difference between the two versions, we call the first one CSMA1 and the adapted one CSMA2.

## 4   Evaluation of CSMA

Our experiments aims to evaluate the results provided by CSMA on synthetic and real datasets.

## 4.1   Experiments on Synthetic Dataset

In order to generate data that can be interpreted as musical data, we consider that a note is an item, a measure is a sequence and a score is obtained by concatenation of the sequences belonging to the database. The strategy, that we used to control the motifs belonging to the sequence, consists in generating a dataset without motif, then, generating motifs and introduces them into the dataset. SPMF generator has been used to build sequences with large vocabulary (maximum distinct items) and small itemsets (item count per itemset) with the following parameters [16]. Sequence count, maximum distinct items ($\mathcal{I}$), item count by itemset and itemset count per sequence ($\mathcal{S}$) have been fixed respectively to 500, 1000, 1, and 10.

Thus, with these settings, in the generated dataset, called *D500I1KT10*, the probability of having *freq* occurrences of a motif of size $l$ is very low, estimated to $(\mathcal{A}_l^{|\mathcal{I}|}/\mathcal{A}_{|\mathcal{S}|}^{|\mathcal{I}|})^{freq}$.

Then, a set of *contiguous motifs* has been generated as well as a set of gaps which have been inserted into the motifs to form non-contiguous motifs. Both types of motifs have been inserted independently in D500I1KT10 leading to two databases, each containing ten datasets: the datasets in Database1 contain sequences with contiguous motifs whereas those in Database2 contain sequences with non contiguous motifs.

These datasets allow to evaluate CSMA2, the version of our algorithm able to handle a set of sequences. For this reason, CSMA2 has been compared to CM-SPADE, proposed by Philippe Fournier-Viger [17]. As expected, CSMA2 and CM-SPADE have found the same motifs with exactly the same frequencies. The first conclusion is that CSMA is able to extract contiguous and non-contiguous motifs from databases composed of different sequences but one of its advantage is to provide the positions of the detected motifs.

The second part of the evaluation concerns CSMA1, the version of our algorithm which searches motifs from a single sequence. For this experiment, each dataset has been transformed into one string, by concatenating all the sequences together. Thus, we obtain two other databases built from Database 1 and Database 2, called respectively string1, with contiguous motifs, and string2, with non contiguous motifs. The results are evaluated according to the number of identified distinct motifs and the frequency (number of occurrences corresponding to these motifs). The average values $\mu$ and standard deviations $\sigma$ computed over the 10 datasets for each database are reported as final results. CSMA1 is compared with the algorithm of Liu. For CSMA, minFrequency, minLength, maxLength have been set respectively to 2, 2, 20 with maxGap equals to 0 for string1 and to 3 for string2. The results are presented in Table 2.

Concerning the datasets containing a sequence without gap (string1), the results show that CSMA1 and Liu algorithm find approximatively the same numbers of distinct motifs with almost the same frequencies. The difference comes from the fact that Liu's algorithm identifies only non-trivial motifs when our algorithm detects all the motifs verifying the constraints. For the datasets with gaps (string2), we observe the same phenomena but the difference is more

**Table 2.** String mining results

| Algorithms | String1 ($\mu(\sigma)$) | | String2 ($\mu(\sigma)$) | |
|---|---|---|---|---|
| | Number of distinct motifs | Frequencies | Number of distinct motifs | Frequencies |
| CSMA1 | 1100.6 (41.45) | 7893.2(1210.14) | 6964.8 (2669.22) | 25514.8 (10035,26) |
| Liu [13] | 1003.4(38.94) | 7088.4 (869.72) | 1073.1 (21) | 6891.4 (914) |

important. This is explained by the fact that long motifs are broken into smaller ones, due to the gaps, but they remain frequent. However, since CSMA1 allows gaps of length 3, the number of motifs detected by CSMA is higher than with Liu algorithm. So, CSMA is able to extract contiguous and non-contiguous motifs from a single sequence and it can take into account the gaps, which is not the case of the Liu's algorithm.

In conclusion, with its both versions, CSMA can find contiguous and non-contiguous motifs, with or without gaps, from a single string or a set of strings.

### 4.2    Real Music Data

**Preprocessing of the Music Data.** Music symbol-based features are usually represented by pitch values. These values contain two kinds of informations: melodic and rhythmic. Melodic information corresponds to an alphabetic letter, followed by the octave value that can be encoded in MIDI format. For instance, a C note on the $5^{th}$ octave, written C5 corresponds to 72 in MIDI format. The rhythmic information corresponds to the note duration. The aim of our work is to identify musical motifs which can be melodic, rhythmic or both (pitch sequence) and our experiments have shown that CSMA is able to do it. Moreover, we search also three musical variants of a motif: its transposed, inverted or mirror form as illustrated on Fig. 1. A transposed motif is a music part that contains the same tonal variation than an other part in the music score, for instance in Fig. 1, $m_2$ is a transposed form of $m_1$. An inverted motif is a part that is characterized by inverted tonal variation of an other part, like for example, $m_4$ and $m_5$. Finally, a mirror motif corresponds to a symmetric positioning of notes such as $m_5$ and $m_6$. These forms are useful to characterize a composer or a music score. They can be used as a signature for other data mining tasks such as supervised and non supervised clustering or composer identification.

We introduce this section with the presentation of a method based on a preprocessing of the initial melodic sequence $S$, able to detect their melodic variants. To this end, we consider the sequence of intervals between consecutive notes from $S$. By this way, three new sequences are built. The first one, denoted $V$ corresponds to the melodic variation between two notes. For instance in Fig. 1, the variation between the note in the first position (60) and the note in the second one (61) equals 1 (61-60). The second sequence, denoted $-V$ is obtained by taking the opposite of each value in the sequence $V$, so each positive value in $V$ becomes negative and vice-versa. Finally, the last sequence, called inverse of $-V$ and denoted $\overline{-V}$ is obtained by taking the sequence $-V$ in reverse order beginning by its last element. Once the primary and secondary sequences have

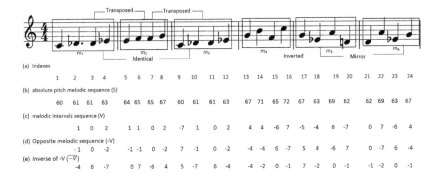

**Fig. 1.** musical motifs and their variants

been defined, CSMA can be applied to them to extract musical motifs variants, as explained below.

To detect transposed and inverted motifs, $-V$ is put after $V$ (i.e. the last element of $V$ is followed by the first element of $-V$) this makes a sequence $< V, -V >$ of size $2l_v$ on which CSMA is applied. Then, if CSMA generates a new motif having two positions $(i, len_i), (j, len_j)$ such that *if* $i \leq l_v \wedge j \leq l_v \wedge len_i = len_j \wedge$ *there is no identical motif occurrence in that positions* then we have a *transposed form* at positions $i$ and $j$. For example, as we can see in Fig. 1(c), the motif with the value $<1\ 0\ 2>$ has three positions which correspond to 1, 5 and 9. As the positions 1 and 9 correspond to the identical motifs $m_1$ and $m_3$ previously found, it remains the motif at position 5 which corresponds to $m_2$. So we conclude that $m_2$ is a transposed motif of $m_1$ and $m_3$.

If CSMA extracts a motif having two positions $i$ and $j$, if $i \leq l_v, j > l_v$ and $S_{i-1} = S_{j-l_v}$ then the subsequence $S_1 =< S_{j-l_v}, ..., S_{j-l_v+len_j} >$ is an *inverted form* of the subsequence $S_2 =< S_{i-1}, ..., S_{i-1+len_i} >$. For example, when we concatenate sequence (c) and (d) from Fig. 1, the motif with the value $<4\ -6\ 7>$ is present at position 13, which is lower than 23 ($l_v$), and at position 40, which is greater than 23. Moreover, $S_{i-1} = S_{13-1} = 67$ and $S_{j-l_v} = S_{40-23} = S_{17} = 67$. So the motif $m_5$, which starts from position 17, is an inverted form of the motif $m_4$, which starts from position 13.

To detect mirror form, $\overline{-V}$ is put after $V$ such that the last element of $V$ is followed by the first element of $\overline{-V}$. This makes again a new sequence $< V, \overline{-V} >$ of size $2l_v$. If CSMA finds a motif $m = (X, freq(X), P)$ with $(i, len_i), (j, len_j) \in P$ such that $(i \leq l_v) \wedge (j > l_v) \wedge (V_{i+len_i/2} = 0) \wedge (len_i = len_j)$ then the subsequence from position $i - 1$ to $i + len_i - 1$ in the melodic sequence $S$ is a *mirror motif*.

For example, in Fig. 1(c)(e), the motif with value $<-4\ 6\ -7>$ appears at two positions in $< V, \overline{-V} >$, one before $l_v$ and another one after. As the middle value variation is equal to 0, then the motif $<-4\ 6\ -7\ 0\ 7\ -6\ 4>$ starting from position 17 to 24, that includes $m_5$ and $m_6$, is a mirror motif.

**Table 3.** Number of distinct musical motifs (and variants) in real music scores

| Music score | Part | Melodic sequence | | Melodic variations | | | Rhythmic sequence | | Pitch sequence |
|---|---|---|---|---|---|---|---|---|---|
| | | Size | Simple motifs | Transposed motifs | Inverted motifs | Mirror motifs | Size | Simple motifs | Simple motifs |
| Score 1 | P1 | 287 | 13 | 5 | 3 | 3 | 307 | 58 | 39 |
| | P2 | 295 | 17 | 17 | 1 | 0 | 313 | 56 | 61 |
| | P3 | 239 | 19 | 2 | 2 | 4 | 270 | 66 | 5 |
| | P4 | 217 | 41 | 5 | 1 | 14 | 240 | 56 | 143 |
| Score 2 | P1 | 66 | 56 | 55 | 0 | 2 | 141 | 34 | 68 |
| | P2 | 166 | 7 | 11 | 0 | 1 | 244 | 77 | 28 |
| | P3 | 518 | 55 | 24 | 6 | 9 | 586 | 138 | 154 |
| | P4 | 129 | 16 | 21 | 0 | 4 | 182 | 55 | 47 |
| | P5 | 458 | 73 | 21 | 0 | 1 | 475 | 98 | 90 |

**Evaluation on Music Scores.** CSMA has been tested on two music scores 'The art of fugue Bach BWV 1080' (score 1) and 'Johann Pachelbel hexachordum apollinis' (score 2) in midi format. Firstly, each music score has been transformed into a set of symbolic sequences: a sequence per instrument. Thus we obtained four sequences, P1 to P4, for score 1 and five for score 2 (P1 to P5). Then, different types of sequences have been extracted: absolute pitch sequence (MIDI value-based melodic sequence), duration sequence (rhythmic sequence) and pitch sequence (melodic and rhythmic sequence). Melodic sequences allow to extract simple motifs with or without variations (transposed, inverted and mirror forms) whereas only simple motifs can be detected in the other sequences.

The parameters *minFrequency*, *maxLength*, *maxGap* and *minLength* were respectively fixed to 2, maximum Java integer value (no constraint for *maxLength*), 0 and 4 for simple motifs and 3 for variants.

The number of motifs extracted for each sequence is given in Table 3.

We can notice that both music scores contain all types of motifs even if parts in score 1 are of the same size whereas score 2 contains mix long and short parts. However, melodic variants appear across the different parts in score 1. We can conclude that there is a general theme hidden through the parts in score 1. The distribution of the motifs in score 2 is different. Score 2 uses more transposed motifs, notably in part 1 (P1) even if this part is the shortest. Part 3 contains the different forms. That is not the case of part 5 even if they have approximatively the same size. This difference in the motif distribution confirms our hypothesis concerning the discriminant power of these forms used as descriptive features of music scores for composers work identification.

# 5    Conclusion and Future Work

In this paper, we introduced an original motif mining algorithm, called CSMA, able to find contiguous and non-contiguous motifs. This algorithm incorporates constraints related to frequency, gap size and motif length. With its two versions, CSMA can find motifs from one or multiple strings. Even if it has been designed to extract musical motifs, CSMA can be used in other contexts.

One of its advantage is that it saves the motif positions in the string and offers the possibility to find motifs with gaps. Those positions are, then, useful to extract musical motifs variants such as transposed, inverted and mirror forms.

It should be pointed out that these positions are also useful for the expert in his analysis of the music scores. A software is under designing. It allows to display the extracted motifs on the music sheet and it will be used for an evaluation by the expert. Moreover, in future works, we plan to optimize CSMA in order to improve the running time and to use the motifs and their variants as features representing composers in the context of clustering tasks.

**Acknowledgement.** The funding for this project was provided by a grant from *la région Rhone Alpes*.

# References

1. Agrawal, R., Srikant, R., et al.: Fast algorithms for mining association rules. In: Proceedings of 20th International Conference on Very Large Data Bases, VLDB, vol. 1215, pp. 487–499 (1994)
2. Agrawal, R., Srikant, R.: Mining sequential patterns. In: Proceedings of the Eleventh International Conference on Data Engineering, pp. 3–14. IEEE (1995)
3. Mooney, C.H., Roddick, J.F.: Sequential pattern mining-approaches and algorithms. ACM Comput. Surv. (CSUR) **45**(2), 19 (2013)
4. Han, J., Pei, J., Mortazavi-Asl, B., Chen, Q., Dayal, U., Hsu, M.C.: Freespan: frequent pattern-projected sequential pattern mining. In: Proceedings of the sixth ACM SIGKDD, pp. 355–359. ACM (2000)
5. Han, J., Pei, J., Mortazavi-Asl, B., Pinto, H., Chen, Q., Dayal, U., Hsu, M.: Prefixspan: mining sequential patterns efficiently by prefix-projected pattern growth. In: Proceedings of the 17th ICDE, pp. 215–224 (2001)
6. Zaki, M.J.: Spade: an efficient algorithm for mining frequent sequences. Mach. Learn. **42**(1), 31–60 (2001)
7. Zaki, M.J.: Sequence mining in categorical domains: incorporating constraints. In: Proceedings of the ninth ICIKM, pp. 422–429. ACM (2000)
8. Pei, J., Han, J., Wang, W.: Constraint-based sequential pattern mining: the pattern-growth methods. J. Intell. Inf. Syst. **28**(2), 133–160 (2007)
9. Floratou, A., Tata, S., Patel, J.M.: Efficient and accurate discovery of patterns in sequence data sets. IEEE Trans. Knowl. Data Eng. **23**(8), 1154–1168 (2011)
10. Mannila, H., Toivonen, H., Verkamo, A.I.: Discovery of frequent episodes in event sequences. DMKD **1**(3), 259–289 (1997)
11. Fournier-Viger, P., Lin, J.C.W., Kiran, R.U., Koh, Y.S.: A survey of sequential pattern mining. Data Sci. Pattern Recogn. **1**(1), 54–77 (2017)
12. Hsu, J.L., Chen, A.L., Liu, C.C.: Efficient repeating pattern finding in music databases. In: Proceedings of the seventh ICIKM, pp. 281–288. ACM (1998)
13. Liu, C.C., Hsu, J.L., Chen, A.L.: Efficient theme and non-trivial repeating pattern discovering in music databases. In: Proceedings, 15th International Conference on Data Engineering, pp. 14–21. IEEE (1999)
14. Fuchs, B.: Co-construction interactive de connaissances, application à l'analyse mélodique. In: IC 2011, 22èmes Journées francophones d'Ingénierie des Connaissances, pp. 705–722 (2012)

15. Jiménez, A., Molina-Solana, M., Berzal, F., Fajardo, W.: Mining transposed motifs in music. J. Intell. Inf. Syst. **36**(1), 99–115 (2011)
16. Fournier-Viger, P., Lin, J.C.-W., Gomariz, A., Gueniche, T., Soltani, A., Deng, Z., Lam, H.T.: The SPMF open-source data mining library version 2. In: Berendt, B., Bringmann, B., Fromont, É., Garriga, G., Miettinen, P., Tatti, N., Tresp, V. (eds.) ECML PKDD 2016. LNCS, vol. 9853, pp. 36–40. Springer, Cham (2016). doi:10.1007/978-3-319-46131-1_8
17. Fournier-Viger, P., Gomariz, A., Campos, M., Thomas, R.: Fast vertical mining of sequential patterns using co-occurrence information. In: Tseng, V.S., Ho, T.B., Zhou, Z.-H., Chen, A.L.P., Kao, H.-Y. (eds.) PAKDD 2014. LNCS, vol. 8443, pp. 40–52. Springer, Cham (2014). doi:10.1007/978-3-319-06608-0_4

# Biclustering Multivariate Time Series

Ricardo Cachucho$^{(\boxtimes)}$, Siegfried Nijssen, and Arno Knobbe

LIACS, Leiden University, Leiden, The Netherlands
r.cachucho@liacs.leidenuniv.nl

**Abstract.** Sensor networks are able to generate large amounts of unsupervised multivariate time series data. Understanding this data is a non-
trivial task: not only patterns in the time series for individual variables
can be of interest, it can also be important to understand the relations between patterns in different variables. In this paper, we present
a novel data mining task that aims for a better understanding of the
prominent patterns present in multivariate time series: *multivariate time
series biclustering*. This task involves the discovery of subsets of variables
that show consistent behavior in a number of shared time segments. We
present a biclustering method, BiclusTS, to solve this task. Extensive
experimental results show that, in contrast to several traditional biclustering methods, with our method the discovered biclusters respect the
temporal nature of the data. In the spirit of reproducible research, code,
datasets and an experimentation tool are made publicly available to help
the dissemination of the method.

## 1 Introduction

Multivariate time series (MTSs) are becoming increasingly available, mostly
through the rapid development of measurement systems. More and more,
machinery, infrastructures and even humans use sensors, collecting data syn-
chronously over time. The continuous measuring of these systems also means
that most datasets produced by them are unsupervised. These datasets contain
a range of phenomena, from recurring phenomena recognizable across all the
sensors, to some phenomena that are only recognizable in some of the signals or
even random events that do not reoccur at all. The nature of MTSs data offers a
big opportunity for pattern recognition. As a consequence, there is a considerable
need for unsupervised methods that can provide insight in this data.

Among unsupervised tasks that have been studied in the time series field
are the *segmentation* and *motif discovery* tasks [11]. These tasks aim to either
identify recurring patterns in the time series, or to partition the time series into
segments. Most of this earlier work is however biased towards univariate time
series. Unlike the univariate setting, mining MTSs poses the following challenges.
For a start, understanding the relationship between different variables is highly
important. Additionally, the relationships between different variables may alternate at different moments in time. As a result, there is a need for new methods
that identify subsets of variables with consistent behavior over subsets of time.

© Springer International Publishing AG 2017
N. Adams et al. (Eds.): IDA 2017, LNCS 10584, pp. 27–39, 2017.
DOI: 10.1007/978-3-319-68765-0_3

Which methods can identify such subsets? We consider *biclustering* to be a good approach to solve this pattern recognition problem.

*Biclustering* is a well-known task in the literature [2–10]. Essentially, it aims at discovering subsets of rows with corresponding subsets of columns, such that the submatrix selected by these rows and columns satisfies a certain cohesiveness requirement. Biclustering can be seen as a generalization of clustering where not only the rows are clustered, but the columns as well. Please note that one bicluster is the result of the biclustering task. Furthermore, biclusters can overlap each other and not all rows or columns need to be included in the biclusters. However, as we will point out throughout the paper, traditional biclustering methods are not well-suited for analysing multivariate time series, as they ignore the temporal aspect of the data in the analysis. They can include an arbitrary subset of time points in a bicluster; there is no guarantee that contiguous time points are included in the bicluster. As a result, very small segments of the time series, or even individual time points, may be selected in the bicluster.

In this paper, we propose a new biclustering method that finds recurring patterns over time for MTSs. The distinguishing feature of our biclustering method is that it finds biclusters spanning sufficiently long segments of time. As a result, the biclusters that are found can be thought of as recurring motifs in multivariate time series data. Key choices in our algorithm include (1) a search strategy for finding biclusters in time, and (2) the definition of a coherence measure to score the quality of biclusters in time. Key advantages of our method compared to other methods include the simple elegance of the algorithm (eg., no discretization is needed) and the interpretability of the results.

In the following sections we define the biclustering problem for MTSs, present our biclustering method and experimental results and at last a conclusion. Our experiments show promising results, both in terms of the quality of the results and scalability of the method. For reproducibility purposes, we make our code [14] and datasets [15–18] freely available. Additionally, we also developed and published an experimentation tool [12] with intuitive GUI, where all the experiments described can be easily tested and applied to various datasets.

## 2 Preliminaries

In this section we define the main concepts of our problem: multivariate time series (Sect. 2.1), followed by the problem statement of biclustering multivariate time series (Sect. 2.2).

### 2.1 Multivariate Time Series

We assume that we are given a multivariate time series of length $n$ over $m$ variables. This time series is represented in an $n \times m$ matrix $\mathbf{T}$. In this matrix, $\mathbf{T}_{ij}$ represents the measurement at time point $i$ for variable $j$.

We will often need to identify parts of this data matrix. We introduce some notation to make this easier. Let $I \subseteq \{1, \ldots, n\}$ be a subset of time points and let $J \subseteq \{1, \ldots, m\}$ be a subset of variables, then $\mathbf{T}_{IJ}$ defines the following submatrix:

$$
\mathbf{T}_{IJ} = \begin{pmatrix}
\mathbf{T}_{I_1 J_1} & \mathbf{T}_{I_1 J_2} & \cdots & \mathbf{T}_{I_1 J_k} \\
\mathbf{T}_{I_2 J_1} & \mathbf{T}_{I_2 J_2} & \cdots & \mathbf{T}_{I_2 J_k} \\
\vdots & \vdots & \ddots & \vdots \\
\mathbf{T}_{I_\ell J_1} & \mathbf{T}_{I_\ell J_2} & \cdots & \mathbf{T}_{I_\ell J_k}
\end{pmatrix},
$$

where $I_u$ is the $u$th element in the set $I$, $k = |J|$ and $\ell = |I|$. Furthermore, we will use $\mathbf{T}_{i\bullet}$ as a shorthand for $T_{IJ}$ with $I = \{i\}$ and $J = \{1, \ldots, m\}$ and $\mathbf{T}_{\bullet j}$ as a shorthand for $T_{IJ}$ with $I = \{1, \ldots, n\}$ and $J = \{j\}$.

## 2.2   Problem Statement

As discussed in the introduction, we are interested in identifying biclusters in time series data. More formally, we define the problem at hand as follows.

**Definition 1.** *One bicluster in time series data consists of:*

- *selected segments $\mathscr{I} = \{I_1, \ldots, I_q\}$, where each segment consists of contiguous measurements $I_x = \{a_x, a_x + 1, \ldots, b_x\}$, for some $a_x, b_x \in \{1, \ldots, n\}$ and $I_x \cap I_y = \emptyset$ if $x \neq y$;*
- *selected variables $J \subseteq \{1, \ldots, m\}$,*

*such that:*

- *the selected segments and variables satisfy the following requirement:*

$$
H(\mathscr{I}, J) < \delta,
$$

*where*

$$
H(\mathscr{I}, J) = \frac{1}{|\mathscr{I}||J|} \sum_{j \in J} \sum_{I \in \mathscr{I}} d(\mathbf{T}_{Ij}, \mathbf{T}_{\cup \mathscr{I}_j}); \tag{1}
$$

*here $\cup \mathscr{I} = \cup_{I \in \mathscr{I}} I$ and $d(\mathbf{t}_1, \mathbf{t}_2)$ measures how similar two time series $\mathbf{t}_1$ and $\mathbf{t}_2$ are; i.e., it is required that each selected segment is similar to the union of the other selected segments, for all chosen variables;*
- *$J$ is maximal: no variable can be added such that the similarity constraint remains satisfied;*
- *$\mathscr{I}$ is maximal: no segment can be added such that the similarity constraint remains satisfied;*
- *each segment $I \in \mathscr{I}$ has a sufficient length, i.e., $|I| \geq \ell$.*

Notice that $H(I, J)$ is set to capture the coherence for each column independently. This allows biclusters to be coherent, even if two variables do not share their distribution. This way, two variables can measure the same phenomenon in different ways, depending on the system characteristics and measuring system setup.

Many alternatives are possible for the definition of the similarity, $d(\mathbf{t}_1, \mathbf{t}_2)$, between two subsequences. In this paper, we propose to measure the similarity of two subsequences by determining the divergence between the *distributions* of measurements in these subsequences.

**Definition 2.** *Given two probability densities $p(x)$ and $p'(x)$ over the same domain, a* divergence score $d(p, p')$ *is a score with the following properties:*

– $d(p, p') \geq 0$;
– $d(p, p') = 0$ *iff $p(x) = p'(x)$ for all $x$.*

**Definition 3.** *Given two time series subsequences $\mathbf{t}_1$ and $\mathbf{t}_2$, where $\mathbf{t}_1$ contains samples from probability density $p(x)$ and $\mathbf{t}_2$ contains samples from probability density $p'(x)$, a* divergence score *for these subsequences is an* estimate *of the divergence of these distributions, i.e., $d(\mathbf{t}_1, \mathbf{t}_2)$ is an estimate of $d(p, p')$ as calculated from $\mathbf{t}_1$ and $\mathbf{t}_2$.*

## 3   Biclustering Multivariate Time Series

The problem formalized in the previous section is hard to solve exactly. In order to solve it, we propose a heuristic method. Our algorithm iteratively removes segments or columns until a bicluster is obtained with the desired quality. Unlike the traditional biclustering algorithm of Cheng and Church [2], our method is designed for multivariate time series, stressing solutions for the temporal aspect of MTSs data in a bicluster.

In our setting, given the length of most time series, removing individual rows is not a feasible approach; converging on a bicluster would take too long. Furthermore, it is unlikely that a heuristic that removes rows one by one is likely to lead to segments of good quality. For these reasons, we propose a new approach that works as follows. Firstly, identify segment boundaries in the time series of each variable. Secondly, we combine the segment boundaries of the different attributes to obtain segment boundaries across all variables. At last, we perform a node deletion biclustering algorithm where either columns and all rows between segment boundaries are removed at the same time.

### 3.1   Time Series Segmentation

The process we propose for segmentation is detailed in Algorithm 1. This algorithm computes the density differences between consecutive subsequences of time series using a sliding window approach. The user-defined parameters for this algorithm are as follows. The size of each subsequence (window) is defined by the parameter $w$. The window is moved over the time series in steps of size *jump*. This parameter is intuitively bounded by $1 \leq jump \leq w$. Note that by setting $jump \geq \ell$, we can ensure that segments are never of length shorter than $\ell$.

Having decided on how to go through the data, one needs to calculate the divergence scores between consecutive windows of data, $d(\mathbf{t}_1, \mathbf{t}_2)$. A solution for

---

**Algorithm 1.** Segmentation.

---

**Input:** multivariate time series $\mathbf{t}$, threshold for local maximum selection $\theta$, window
  size $w$, jump between consecutive windows $jump, 1 \le jump \le w$.
  $\mathscr{C} \leftarrow \{1\}$
  **for** each column of $\mathbf{T}$ **do**
    **for** each $i \in \{0, 1, \ldots, \lfloor (n - w)/jump - 1 \rfloor\}$ **do**
      $s \leftarrow i \cdot jump$
      $d[i] \leftarrow d(\mathbf{t}[s + 1, \ldots, s + w], \mathbf{t}[s + jump + 1, \ldots, s + jump + w])$
    **end for**
    **for** each $i \in \{2, \ldots, \lfloor (n - w)/jump - 2 \rfloor\}$ **do**
      **if** $d[i - 1] \le d[i] \wedge d[i + 1] \le d[i] \wedge d[i] \ge \theta$
      $\mathscr{C} \leftarrow \mathscr{C} \cup \{i \cdot jump + w\}$
    **end for**
  **end for**
  **return**  segment start indices $\mathscr{C}$

---

this divergence score is proposed in Sect. 3.3. After calculating the divergence
score for all consecutive windows, we extract all local maxima to find the segment
boundaries. The risk of using local maxima is that too many boundaries might
be found, creating too many segments. For this purpose, a threshold $\theta$ for the
selection of local maximum divergences is introduced. Only local maxima above
$\theta$ will be selected as segment boundaries. A reasonable solution for this threshold
is to normalize each variable and have the boundaries rescaled between 0 and 1
such that $0 \le \theta \le 1$.

  Algorithm 1 returns a set $\mathscr{C}$ consisting of the segment boundaries. This set
is the union of all the segments found for each of the $m$ variables. Note that
we segment the variables independently of each other; this is important as in
biclustering, we assume that the behavior of different variables over time may
be different.

## 3.2  Biclustering

In this section, we present the *BiclusTS* algorithm to solve the problem of finding
multiple biclusters given a multivariate time series. We will discuss the main
challenge of recognizing interesting subsets of rows and columns $(\mathbf{T}_{\mathscr{I} J})$, while
respecting the temporal nature of time series. We then move into the details of
how to find a bicluster and how to explore the data in order to find multiple
biclusters.

**BiclusTS: Single Segment/Variable Deletion.** The BiclusTS algorithm is
described in Algorithm 2. The algorithm assumes an initial set of segments in
the data, as computed by Algorithm 1. Then, a greedy process removes segments
and variables (columns) that present the largest divergence to the bicluster.
During this repeated process, the difference $H(\mathscr{I}, J)$ reduces monotonically until

---

**Algorithm 2.** BiclusTS: Find One Bicluster.

---

**Input:** initial set of segments $\mathscr{I}$, acceptability bound $\delta$.
  let $J$ be the set of all variables
  calculate $H(\mathscr{I}, J)$
  **while** $H(\mathscr{I}, J) > \delta$ **do**
    **for** all segments $I \in \mathscr{I}$ and variables $j \in J$ **do**
      calculate $d(\mathbf{T}_{Ij}, \mathbf{T}_{\cup \mathscr{I} j})$
    **end for**
    **for** each segment $I \in \mathscr{I}$ **do**
      calculate $\frac{1}{|J|} \sum_{j \in J} d(\mathbf{T}_{Ij}, \mathbf{T}_{\cup \mathscr{I} j})$
    **end for**
    **for** each variable $j \in J$ **do**
      calculate $\frac{1}{|\mathscr{I}|} \sum_{I \in \mathscr{I}} d(\mathbf{T}_{Ij}, \mathbf{T}_{\cup \mathscr{I} j})$
    **end for**
    find maximum margin divergence; remove the corresponding segment or variable
    recalculate $H(\mathscr{I}, J)$
  **end while**
  **return** $\mathbf{T}_{\mathscr{I} J}$, a bicluster that is a submatrix of $\mathbf{T}$

---

---

**Algorithm 3.** Find $k$ biclusters.

---

**Input:** initial set of segments $\mathscr{I}$, acceptability bound $\delta$, the desired number of biclusters $k$.
  $\mathscr{B} \leftarrow \varnothing$, an empty set of biclusters
  **while** $|\mathscr{B}| \leq k$ and $\mathscr{I} \neq \varnothing$ **do**
    $\mathscr{B} \leftarrow \mathscr{B} \cup BiclusTS(\mathscr{I}, \delta)$
    Remove all segments in $\mathscr{B}$ from $\mathscr{I}$
  **end while**
  **return** $\mathscr{B}$, a set of biclusters found in matrix $T$

---

it drops below the acceptability bound $\delta$. The remaining subset of segments $\mathscr{I}$ and columns $J$ is returned as a bicluster.

Notice that $H(I, J)$ is a probabilistic measure, not interested in the correlation between variables. Instead, it is set to account coherence of each column independently (see Definition 1). This allows biclusters to be coherent, even if the behavior between variables is not correlated, as we consider desirable. This is because time series can measure the same phenomenon in different ways, depending on the system characteristics and measuring system setup. As an example, high levels of body activity can be measured both by high levels of heart rate and sinusoidal patterns of acceleration.

The requirements of our method to find a bicluster is to provide a segmented time series, composed of the time series $\mathbf{T}$ itself and the boundaries of each segment (as produced by Algorithm 1). This will ensure a faster biclustering procedure and results consistent with the temporal aspects of the MTSs. Another requirement of Algorithm 2 is a parameter $\delta$ that ensures a certain similarity for all segments within each column of the bicluster.

**Finding a Given Number of Biclusters.** Having described the process of finding one bicluster, the challenge of finding a number of biclusters remains. For this task, we propose Algorithm 3, which finds $k$ non-overlapping biclusters by iteratively looking for biclusters in those segments that have not been selected yet. As input, we must have an initial segmented multivariate time series $T$ and acceptability bound $\delta$ already introduced in Sect. 3.2. $k$ is the number of potential biclusters to be found. We use the word "potential" because there are situations where $k$ is not reached, due to unavailable segments to bicluster, or when the acceptability bound $\delta$ is too low to produce any results.

## 3.3   Density-Difference Estimation (LSDD)

An important choice that remains to be specified is which divergence score to use. Time series can assume many different shapes, depending on the phenomena and measurement system, making the comparison between subsequences a non-trivial problem. Obvious solutions for comparing time series would be two-step approaches. For instance, one could first estimate the probability density distributions (PDFs) of both subsequences, and then compare these using an $f$-divergence measure. As pointed out by [1], the drawback of such approaches is that good estimations will smoothen the PDFs and thus result in underestimations of density-differences.

Instead of taking such a step-wise approach, we here propose to use a more direct approach, based on calculating a least-squares density-difference (LSDD) [1]. LSDD measures the similarity between two time series by directly estimating density-differences $(f(x))$ between time series subsequences. This method does not require a separate estimation of the time series distributions. LSDD directly estimates the density-difference between two samples, $f(x)$, by fitting a density-difference model $g_\theta(x)$ that minimizes:

$$\operatorname*{argmin}_{\theta} \int \Big(g_\theta(x) - f(x)\Big)^2 dt + \lambda\theta^T\theta. \tag{2}$$

Note that the second term in this formula is a regularization term. The model $g_\theta(t)$ that is used to estimate the difference is a mixture model of Gaussians:

$$g(x) = \sum_{\ell=1}^{|c|} \theta_\ell \exp\left(-\frac{||x - c_\ell||}{2\sigma^2}\right),$$

where $\mathbf{c}$ is a random sample of measurements in both time series $\mathbf{t}_1$ and $\mathbf{t}_2$.

To fit the model, the equivalence of Eq. 2 with the following formula is exploited:

$$\operatorname*{argmin}_{\theta} \int g_\theta^2(x)\, dx - 2 \int g_\theta(x) f(x)\, dx + \lambda\theta^T\theta. \tag{3}$$

Given that $f(x)$ is unknown, an empirical estimate is used for the second term:

$$\sum_{\ell=1}^{|c|} \frac{\theta_\ell}{|\mathbf{t}_1|} \sum_{i=1}^{|\mathbf{t}_1|} \exp\left(-\frac{||t_{1i} - c_\ell||}{2\sigma^2}\right) - \frac{\theta_\ell}{|\mathbf{t}_2|} \sum_{i=1}^{|\mathbf{t}_2|} \exp\left(-\frac{||t_{2i} - c_\ell||}{2\sigma^2}\right).$$

The resulting minimization problem can be solved analytically, as shown by the authors of LSDD [1]. Note that this model fitting procedure has two parameters: the Gaussian kernel width $\sigma$ and the regularization parameter $\lambda$. The authors of LSDD propose to optimize these parameters using cross-validation. When LSDD is used on a large scale, this cross-validation becomes too demanding.

We argue that the parameters of LSDD are very stable (Sect. 4.1), i.e., the optimal choice for the parameters values does not change very often for the same time series. Thus, we propose that the parameters are estimated with cross-validation a limited number of times at the beginning of the segmentation task (Algorithm 1). Then, we fix these parameters for the rest of the LSDD calculations. Fixing the parameters of LSDD is also beneficial to speed up the biclustering process, due to the extensive amount of LSDD calculations in Algorithm 2 to compute the mean density difference scores ($H(\mathscr{I}, J)$).

## 4   Experiments

We evaluate our method on four datasets, details of which are given in the table below. The datasets were selected for their length and their multivariate nature (with datasets having up to 119 variables). Except for *Accelerometry*, the datasets also have variables that can be grouped in different categories, such that each group will show considerably different behaviour. For example, the *InfraWatch* data [11] is collected from three types of sensor (each sensitive to different phenomena and time scales): strain gauges, vibration sensors, and temperature sensors.

All experiments were performed with an implementation in R. To demonstrate this implementation, an experimentation tool [12] is available online [13]. Additionally, a tutorial, the code [14] and the datasets [15–18] are also made available.

| Dataset | # Variables | # Time points | Sampling rate | Duration |
|---|---|---|---|---|
| Accelerometry [15] | 3 | 176700 | 85 Hz | 34.6 min |
| Snowboarding [16] | 21 | 21180 | 1 Hz | 5.88 h |
| Running [17] | 6 | 951200 | 100 Hz | 2.64 h |
| InfraWatch [11, 18] | 119 | 17996 | 1/3600 Hz | 749.8 days |

### 4.1   Segmentation

In Sect. 3.1, we proposed a solution to segment MTSs in order to bicluster them. Here, we present the experimental results that support our decisions on how to solve this segmentation task. As suggested in Sect. 3.1, all the datasets were normalized to ease the decision about parameter $\theta$, which was fixed at 0.75.

**LSDD Estimation.** The method we used to compare consecutive subsequences is a single-shot estimation of the difference between probability densities (LSDD), defined in Sect. 3.3. This method is based on fitting a Gaussian model where two parameters, $\lambda$ and $\sigma$, need to be estimated. As was mentioned in Sect. 3.3, we estimate reasonable values for these parameters prior to the LSDD estimation for the entire time series, and then work with these fixed values. Clearly, we are trading off computational speed over accuracy of setting $\lambda$ and $\sigma$ always with cross-validation.

Here, the experiments consist of running Algorithm 1 for all the datasets, with non-overlapping data windows of 100 samples ($w = jump = 100$), and considering two setups. In one setup, for each iteration we estimate LSDD with cross-validation. All the parameter estimations and computation times are saved. In the other setup, we cross-validate LSDD the first 100 iterations and then fix the parameters, by choosing the median $\sigma$ and $\lambda$.

| Dataset | Penalty | Deviation |
|---|---|---|
| Accelerometry | 172 | 1.2% |
| Snowboarding | 62 | 1.5% |
| Running | 91 | 0.7% |
| InfraWatch | 93 | 3.1% |

First, we consider the penalty in time produced by the cross-validation at each subsequence, compared to fixing the parameters at the start. For each dataset, the penalty of computing LSDD with cross-validation was calculated. For each setup of running Algorithm 1, all the LSDD estimation times are summed up over all LSDD estimations per variable, and over all variables. Thus, the penalty is the time ratio of computing LSDD with cross-validation to LSDD with fixed parameters (so how many times faster the second is). Clearly, doing cross-validation at each subsequence is prohibitively expensive, while our choice to fix parameters at the start of each LSDD estimation is much more realistic.

While clearly being much faster, there is the risk of producing sub-optimal values for $\lambda$ and $\sigma$. We test this end of the trade-off by examining the stability of parameter values when LSDD is estimated using cross-validation. If these values remain mostly the same throughout the LSDD estimation, then we can safely pre-compute the values and fix them. For this experiment, we test parameter settings from a fixed range, identical to that proposed originally [1]: $\lambda \in \{0.001, 0.003, 0.01, 0.031, 0.1, 0.316, 1, 3.162, 10\}$ and $\sigma \in \{0.25, 0.5, 0.75, 1, 1.2, 1.5, 2, 3, 5\}$. For the four datasets and both parameters, the percentage of values found that deviate from the median were obtained, and averaged over all variables (see Deviation in table above). The parameters are extremely stable and can be safely estimated and fixed prior to LSDD estimation, with considerable efficiency gains.

## 4.2   Biclustering

As a reference point for reproduction of experiments, the results and parameter settings of our proposed method is presented in the table below. For all the experimental settings, all parameters defaults can be found in the online tool. As a comparison between methods, we observe how many segments were created and the segment's average size. Notice that we want a relatively small number of segments with rather large sizes.

| Datasets | Results | | | Parameter settings | | |
|---|---|---|---|---|---|---|
| | Duration | # Segments | Segment size | Window | Jump | Delta |
| Accelerometry | 7.55 s | 14 | 642.5 (0.36%) | 500 | 100 | 0.01 |
| Snowboarding | 81.5 s | 24 | 81.5 (0.38%) | 90 | 30 | 0.01 |
| Running | 16.7 s | 103 | 1671.8 (0.18%) | 500 | 100 | 0.01 |
| InfraWatch | 36.2 h | 21 | 36.2 (0.20%) | 120 | 24 | 0.01 |

At this point, it is important to understand the differences between the results produced by our approach and those produced by traditional biclustering algorithms, such as the Cheng & Church algorithm. In this experiment, we apply to our four datasets all the available algorithms in the `biclust` package [9] in R. This means that experiments were run to compare BiclusTS with the following biclustering algorithms: Cheng & Church [2], the Xmotifs biclustering algorithm [8], the Plaid model [7], Bimax [4], and Questmotif [8,10].

In the table below, we present the number of segments created by each algorithm. Please note that some algorithms are not even able to deal with large datasets (represented by -). The numbers of segments are in most case more than those resulting from BiclusTS, showing that fragmentation is a systematic problem of the traditional biclustering algorithms.

| Datasets | # Segments | | | | | BiclustTS |
|---|---|---|---|---|---|---|
| | Cheng & Church | Xmotifs | Plaid | Bimax | Questmotif | |
| Accelerometry | 4034 | 1863 | 1641 | 141 | 3338 | 14 |
| Snowboarding | 22 | 547 | 1 | - | 13 | 24 |
| Running | 3852 | 3852 | 9 | - | 34791 | 103 |
| InfraWatch | 17 | 64 | - | 1 | 98 | 21 |

Complementary to the number of segments, one should look at the average size of the segments biclustered. The average size of the segments created by the traditional biclustering algorithms is in most cases very small. Interestingly, in two cases (indicated by the *), the biclustering algorithms produced segments stretching the entire length of the time series, thus failing to identify any meaningful segmentation into different activities. Notice that Plaid has some exceptions. Still, take as an example the *Snowboarding* dataset. For this dataset, Plaid created a bicluster containing all the data.

| Datasets | Average segments size | | | | | BiclustTS |
|---|---|---|---|---|---|---|
| | Cheng & Church | Xmotifs | Plaid | Bimax | Quest | |
| Accelerometry | 13.6 | 16.8 | 44.5 | 1.1 | 20.4 | 642.5 |
| Snowboarding | 2.8 | 15.1 | 21180* | - | 1420.4 | 81.5 |
| Running | 1.5 | 2.8 | 105688 | - | 2.7 | 1671.8 |
| InfraWatch | 5.7 | 2.3 | - | 17996* | 83.1 | 36.2 |

**Demonstration.** Our method was designed to find biclusters that avoid very short segments of consecutive time points. However, having good subsets of time periods represented in the bicluster is not enough. BiclusTS also aims to

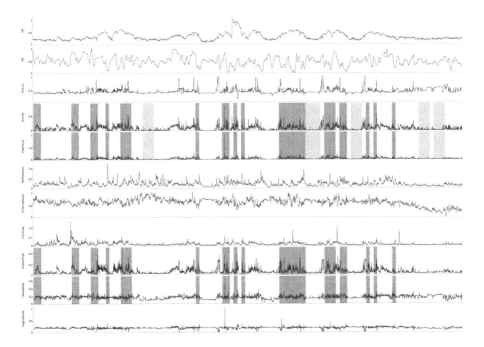

**Fig. 1.** Examples of biclusters produced with BiclusTS. The first bicluster (orange) concerns four sensors related to activity levels as captured by the accelerometer, specifically high peak accelerations observed in various directions. The second bicluster (yellow) uses a subset of these sensors, in order to characterize periods where the subject was taking the ski lift. (Color figure online)

capture interesting phenomena involving complex patterns. These patterns can be observed visually.

In order to show what can be expected from BiclusTS, we applied it to the *Snowboarding* dataset. With 21 variables, this dataset measures a person using a BioHarness chest sensor while riding a snowboard. The sensor measures vital signs such as heart rate, breath rate and body temperature, as well as acceleration. The dataset covers various sorts of activities common for a day of snowboarding in the high mountains.

Figure 1 shows two examples of biclusters related to alternating time periods of rest and downhill snowboard activities. As for the selected variables, one bicluster shows high levels of physical activity and different postures, during the periods of snowboarding. The other bicluster identifies resting periods in between snowboarding. These resting periods are recognized only using the activity levels measured by the sensor system (*Activity* and *PeakAcceleration*). As expected, our biclustering method makes use of different subsets of variables to describe different phenomena.

## 5   Conclusions

This paper introduced the task of biclustering multivariate time series. In this context, respecting the temporal order is absolutely critical. Given that the traditional biclustering setting assumes that all samples are independent, the shuffling process of finding subsets of rows and columns will lead to useless biclusters in the time series context. We showed the importance of an algorithmic solution to bicluster MTSs. First, we proposed the creation of segments of sufficient length; second, we argued for the use of an LSDD divergence score [1] to ensure that for each selected variable, all selected segments have a similar distribution.

We presented an algorithm for solving this biclustering task. It consists of two stages: first, a stage in which the time series are segmented in segments of sufficient length; second, a stage in which a selection of segments and variables is made. In both the first and second stage we used LSDD with fixed parameters. Experiments showed that LSDD produces stable results, and that the probabilistic descriptions of each segment can be used to accept, or reject a bicluster during the node deletion process.

## References

1. Sugiyama, M., Kanamori, T., Suzuki, T., Plessis, M., Liu, S., Takeuchi, I.: Density-difference estimation. In: Proceedings of NIPS, pp. 683–691 (2012)
2. Cheng, Y., Church, G.: Biclustering of expression data. In: Proceedings of Eighth International Conference on Intelligent Systems for Molecular Biology, pp. 93–103 (2000)
3. Madeira, S., Oliveira, A.: Biclustering algorithms for biological data analysis: a survey. J. IEEE/ACM Trans. Comput. Biol. Bioinform. **1**, 24–45 (2004)

4. Prelić, A., Bleuler, S., Zimmermann, P., Wille, A., Bühlmann, P., Gruissem, W., Hennig, L., Thiele, L., Zitzler, E.: A systematic comparison and evaluation of biclustering methods for gene expression data. J. Bioinform. **22**, 1122–1129 (2006)
5. Kluger, Y., Basri, R., Chang, J.T., Gerstein, M.: Spectral biclustering of microarray cancer data: co-clustering genes and conditions. J. Genome Res. **13**, 703–716 (2003)
6. Barkow, S., Bleuler, S., Prelić, A., Zimmermann, P., Zitzler, E.: BicAT: a biclustering analysis toolbox. J. Bioinform. **22**, 1282–1283 (2006)
7. Turner, H., Bailey, T., Krzanowski, W.: Improved biclustering of microarray data demonstrated through systematic performance tests. J. Comput. Stat. Data Anal. **48**, 235–254 (2003)
8. Murali, T., Kasif, S.: Extracting conserved gene expression motifs from gene expression data. In: Proceedings of Pacific Symposium on Biocomputing, pp. 77–88 (2003)
9. Kaiser, S., Leisch, F.: A toolbox for bicluster analysis in R. In: Proceedings of UseR, p. 101 (2008)
10. Kaiser, S., Leisch, F.: A generalized motif bicluster algorithm. In: Proceedings of UseR, p. 101 (2009)
11. Vespier, U., Nijssen, S., Knobbe, A.: Mining characteristic multi-scale motifs in sensor-based time series. In: Proceedings of CIKM, pp. 2393–2398 (2013)
12. Cachucho, R., Liu, K., Nijssen, S., Knobbe, A.: Bipeline: a web-based visualization tool for biclustering of multivariate time series. In: Proceedings of ECML PKDD, pp. 12–16 (2016)
13. http://fr.liacs.nl:7000
14. https://github.com/kainliu/ShinyDashboard
15. www.openml.org/data/download/1854941/accelerometry.csv
16. www.openml.org/data/download/1854944/snowboard.csv
17. www.openml.org/data/download/1854943/running.csv
18. www.openml.org/data/download/1854942/infrawatch.csv

# Visualization of Topic-Sentiment Dynamics in Crowdfunding Projects

Rafael A.F. do Carmo[1,2(✉)], Soong Moon Kang[3], and Ricardo Silva[2,4]

[1] Instituto Universidade Virtual, Universidade Federal do Ceará, Fortaleza, Brazil
carmorafael@virtual.ufc.br
[2] Department of Statistical Science, University College London,
London WC1E 6BT, UK
ricardo.silva@ucl.ac.uk
[3] School of Management, University College London, London WC1E 6BT, UK
smkang@ucl.ac.uk
[4] The Alan Turing Institute, London NW1 2DB, UK

**Abstract.** We develop a model that connects the ideas of topic modeling and time series via the construction of topic-sentiment random variables. By doing so, the proposed model provides an easy-to-understand topic-sentiment relationship while also improving the accuracy of regression models on quantitative variables associated with texts. We perform empirical studies on crowdfunding, which has gained mainstream attention due to its enormous penetration in modern society via a variety of online crowdfunding platforms. We study Kickstarter, one of the major players in this market and propose a model and an inference procedure for the amount of money donated to projects and their likelihood of success by capturing and quantifying the importance (sentiment) that possible donors give to the subjects (topics) of the projects. Experiments on a set of 45 K projects show that the addition of the temporal elements adds valuable information to the regression model and allows for a better explanation of the overall temporal behavior of the whole market in Kickstarter.

**Keywords:** Topic models · Time series · Regression

## 1 Introduction

Online platforms such as Kickstarter and Indiegogo have amplified the range and impact of crowdfunding projects around the world. The removal of geographic barriers between independent entrepreneurs and a multitude of possible donors (the crowd) enables the funding of a larger range of possible projects compared to traditional markets, a novel kind of exchange that is still not fully understood. Such a market has gained much interest from the general public and the scientific community [1], which aims to understand the dynamics of these projects and to create tools that help creators to maximize the odds of success of their enterprises [2].

© Springer International Publishing AG 2017
N. Adams et al. (Eds.): IDA 2017, LNCS 10584, pp. 40–51, 2017.
DOI: 10.1007/978-3-319-68765-0_4

In this paper, we propose an algorithmic approach to the problem of modeling the amount of money donated to projects and assessing the general state of the market to these projects. Unlike existing methods, the proposed approach makes use of time-dependent latent features derived from the textual description of the projects and past donations as explanatory variables of project success. These features capture the current importance donors give to the different topics addressed by existing projects. The experiments on this paper show empirically the importance of inferring latent information in the regression model we use, improving its performance and making a clear contribution to the explanation of the observed data. The proposed approach connects topic models which model the descriptions of projects to state-space time-series models which describes the dynamics of donations to projects.

## 2    Background

In this work, lowercase letters represent unitary elements $x$, column vectors are represented by bold letters $\boldsymbol{x}$, uppercase letters are matrices $\mathbf{X}$, $^\top$ denotes transposition, $\odot$ element-wise products, and $\otimes$ outer products, $\mathbf{I}_K$ is a $k$-th dimensional identity matrix, $\mathbb{1}$ is the indicator function, $[x, y]$ means the concatenation of elements $x$ and $y$, and $\mathrm{E}[f(x, y)]_{q(y)}$ refers to the expected value of the function $f(x, y)$ regarding the $q$ distribution of $y$.

### 2.1    Topic Models

Topic models (TM) are a class of mixture models for discrete data, where each mixture component describes a distribution over a possible set of discrete outcomes. One of the most common topic models is latent Dirichlet allocation (LDA) [3], where each mixture component is itself random, following a Dirichlet prior. Topic models are generative statistical tools that allow sets of high dimensional observations to be explained by lower dimensional latent groups. The idea behind this generative model in the context of text data is that topics define distributions over vocabulary, and texts are generated via a choice of topics proportions and words picked in the different topics. The generative process may be written as

1. For each topic $k$, sample $\boldsymbol{\beta}_k \sim \mathrm{Dirichlet}(\boldsymbol{\tau})$
2. For a text document $p$, draw topic proportion $\boldsymbol{\theta}_p \sim \mathrm{Dirichlet}(\boldsymbol{\eta})$
3. For each slot $i$ in document $p$
   (a) Draw topic allocation $z_{i,p} \sim \mathrm{Multinomial}(1, \boldsymbol{\theta_p})$
   (b) Draw word $w_{i,p} \sim \mathrm{Multinomial}(1, \boldsymbol{\beta}_{z_{i,p}})$

where $\tau$ and $\eta$ are model parameters on the Dirichlet priors of per-topic word distribution and per-document topic distributions, respectively.

## 2.2   Latent State-Space Models

Latent State-Space Models (LSSM) [4] are the workhorse of an enormous variety of models in different fields such as signal processing and econometrics. They provide a framework which assumes the observed sequence was generated from an underlying sequence of continuous latent states that follow a Markov process. For a sequence of states $\alpha_{1:T} = \{\alpha_1, \ldots, \alpha_T\}$ and observations $y_{1:T} = \{y_1, \ldots, y_T\}$, the generative process may be written as:

1. Draw initial state $\alpha_1 \sim p(\alpha_1)$
2. Draw observations $y_1 \sim p(y_1|\alpha_1)$
3. For each time-point $t$:
    (a) Draw $\alpha_t \sim p(\alpha_t|\alpha_{t-1})$
    (b) Draw $y_t \sim p(y_t|\alpha_t)$

The most usual parametrization for this system is fully Gaussian, which is facilitates the computation of quantities of interest such as the posterior distribution of the latent variables. In this work, we use Gaussianity in the Markov state-space evolution and use fully factorized chains. That is, the model is given at starting time 1 by $p(\alpha_1) = \text{Normal}(0, \mathbf{I})$. For each sequential elements, we define the evolution of the chain $p(\alpha_t|\alpha_{t-1}) = \text{Normal}(\mathbf{A}\alpha_{t-1}, \mathbf{I})$, where $\mathbf{A}$ is the parameters known as state (or system) matrix that drives the latent process. Usually, in LSSM we observe fixed-size (either univariate or multivariate) $y$ elements. However, for the problem under consideration, there will be a collection of elements (projects) which vary in time. Additionally, due to the high number of zeros in the dataset, we may parametrize the observations via a "hurdle" model for zero-inflated data.

## 2.3   Hurdle Models

Our definition of a hurdle model [5] is based on a two-stage model that defines a distribution on non-negative variables. In our case, each variable $Y$ is continuous for $Y > 0$ but with a positive probability for the event $Y = 0$. The mixture component that generates the choice between $Y = 0$ and $Y > 0$ is given by a model for Bernoulli outcomes based on the sign of a latent Gaussian variable. If the sign of the latent Gaussian is positive, this is followed by generating a numeric positive value following a log-Normal distribution:

$$y^\star \sim N(m^\star, 1), y = 0 \text{ if } y^\star \leq 0 \text{ else } \exp(z) \tag{1}$$

where $z \sim N(n, \delta)$ and $m^\star$ is a mean element which is going to be defined a posteriori. This model is going to be used to model the amount of money pledged for a given project $p$ at time $t$.

# 3   Model Definition

We assemble all the previous parts into a model that takes into consideration time-dependencies and latent factors related to the topics of the projects. Topics are inferred using topic models, and extra latent factors are introduced to account for the degree of attention a topic is receiving at any given time. We call these latent time-dependent factors "topic heats". The motivation for introducing these factors is illustrated in the context of movie projects as follows: there may be periods in which people are primarily interested in projects that involve about cinema and environmental questions, but in other periods of time the mix could be cinema and politics. These "interests" are not directly recorded in the data, but we indirectly capture them by modeling on-going dependencies between the amount of money people donate to projects and the topics inferred from the (e.g. Kickstarter) webpages of the projects.

In the following, let $p$ index any particular project and let $t$ index time. Given a pre-defined number $K$ of topics $\{\beta_1, \ldots, \beta_K\}$, let $\boldsymbol{\theta}_p$ be the corresponding $K$ topic proportions of $p$, regardless of time, and $\boldsymbol{\alpha}_{k,t}$ be the topic heat for topic $k$ at time $t$. Let $z_{i,p}$ and $w_{i,p}$ be the topic allocation and word for position $i$ in project $p$ as in a standard topic model. Finally, let $\boldsymbol{c}_{p,t}$ and $y_{p,t}$ be, respectively, fixed covariates (such as the amount pledged by the project) and donations received (in e.g. dollars) for project $p$ at time $t$. Projects start and end at different time-points, with the fixed covariates and the times of birth/death of a project assumed to be given instead of random.

1. Draw project's textual descriptions as in Sect. 2.1
2. For each time-point $t$:
   - Draw topic heat $\boldsymbol{\alpha}_t$ according to the Markov process in Sect. 2.2
   - For each project $p$ active at time $t$:
     (a) $m^\star_{p,t} = \boldsymbol{\lambda}^\mathsf{T}_{y^\star}(\boldsymbol{\theta}_p \odot \boldsymbol{\alpha}_t) + \rho^\star_y + \boldsymbol{\lambda}^{\star\mathsf{T}}_c \boldsymbol{c}_{p,t}$
     (b) $n_{p,t} = \boldsymbol{\lambda}^\mathsf{T}_y(\boldsymbol{\theta}_p \odot \boldsymbol{\alpha}_t) + \rho_y + \boldsymbol{\lambda}^\mathsf{T}_c \boldsymbol{c}_{p,t}$
     (c) Draw $y_{p,t}$ according to the hurdle model in Sect. 2.3 with parameters $(m^\star_{p,t}, n_{p,t}, \delta_y)$

where all new symbols are model parameters. By project active at time $t$, we mean any project $p$ which is open to receiving donations at time-point $t$. As said before, projects can last up to 60 days on Kickstarter and for different time-points there will be a different number of projects running. Inference in our model means capturing this information of variable dimensionality at time $t$, reducing it to the fixed-size latent elements, and transferring such information across time.

To finish the definition of the model, let $F$ be the full set of projects, $N_p$ the length of the text description of project $p$, $A_t$ the set of active projects at time $t$, and $1 : T$ the whole history of observations. We then define the **complete log-likelihood**

$$\ell(\boldsymbol{\eta}, \mathbf{A}, \boldsymbol{\rho}, \boldsymbol{\delta}) = \sum_{p \in F} \left[ \log p(\boldsymbol{\theta}_p; \boldsymbol{\eta}) + \sum_{n=1}^{N_p} \log p(z_{p,n} | \boldsymbol{\theta}_p) + \log p(w_{p,n} | z_{p,n}) \right] + \log p(\boldsymbol{\alpha}_1) +$$

$$\sum_{t=2}^{T} \log p(\boldsymbol{\alpha}_t | \boldsymbol{\alpha}_{t-1}; \mathbf{A}, \mathbf{I}_K) + \sum_{t=1}^{T} \sum_{p \in A_t} \log p(y_{p,t}, y_{p,t}^\star | \boldsymbol{\theta}_p, \boldsymbol{\alpha}_t; \boldsymbol{\lambda}_{y^\star}, \rho_{y^\star}, \boldsymbol{\lambda}_{c^\star}, \boldsymbol{\lambda}_y, \rho_y, \boldsymbol{\lambda}_c, \delta_y).$$

This assumes topics $\{\beta_1, \dots, \beta_K\}$ have been pre-defined by first fitting the standard variational latent Dirichlet allocation algorithm of [3] which can either be done with the text of all projects or a separate set of projects, which was the solution used in this paper (due to the availability of such separate set).

### 3.1 Inference and Estimation

Given the definition of the complete model and the characteristics of it, we turn our focus to defining the procedures for inference of the latent variables and estimation of the unknown parameters of the model.

In order to do so, we make use of Variational inference, which is a general deterministic approximation to intractable integrals or expectations which appear in complex models [6]. In a maximum likelihood (ML) or maximum a posteriori (MAP) setting, we are interested in estimating parameters based on the marginal likelihood of the observed variables $\boldsymbol{y}$ in a graphical model also containing the latent variables $\boldsymbol{x}$. Such a marginal is approximated as follows,

$$\log p(\boldsymbol{y}) = \log \int p(\boldsymbol{y}, \boldsymbol{x}) d\boldsymbol{x} \geq \int q(\boldsymbol{x}) \log p(\boldsymbol{y}, \boldsymbol{x}) d\boldsymbol{x} - \int q(\boldsymbol{x}) \log q(\boldsymbol{x}) d\boldsymbol{x},$$

where this lower bound holds for any $q(\boldsymbol{x})$. This approximation is usually called the Evidence Lower Bound (ELBO) and provides an optimal approximation (in terms of KL-Divergence) to the desired distribution $p(\boldsymbol{y} \mid \boldsymbol{x})$ and the target log-marginal likelihood $\log p(\boldsymbol{y})$. Equality is achieved at $q(\boldsymbol{x}) = p(\boldsymbol{x} \mid \boldsymbol{y})$, which is intractable to compute.

In this modeling we are dealing with, the key quantity of interest is the posterior distribution of the latent variables, including topic heats $\alpha_t$. Unfortunately this posterior is intractable to compute due to the non-linearity of the observation distribution in the time-series part of the model and to the Dirichlet structure of the TM. On the top of that, the parameters of the model are unknown and must be estimated from data. To obtain these quantities we develop a Variational Bayes Expectation-Maximization (VBEM) algorithm [7] in which a *structured* approximation to the posterior distribution is considered:

$$\log p(\theta, y^\star, \alpha, z | y, w) \approx q(\alpha_{1:T}) \prod_{p \in F} q(\boldsymbol{\theta}_p) q(\boldsymbol{z}_p) \prod_{t=1}^{T} \prod_{p \in A_t} q(y_{p,t}^\star)$$

By doing so, we maintain the temporal dependency among the topic heats, preventing the loss of crucial temporal dependency of these latent variables. This structure and the Gaussianity of the explicit dependency of $y$ and $y^\star$ on $\alpha$ allows us to perform exact (given the structure defined) forward-backward passes to infer the variational parameters of $q(\alpha)$ in a similar way to the Variational Kalman Smoother (VKM) algorithm [8].

We provide a summarized explanation of the VBEM algorithm starting by describing the more complicated E-Step and following the M-Step, which is straightforward to derive and makes use of expectations of the latent variables as replacements for their actual values. We provide the equations that are specific to the model under consideration and redirect the reader to [8] so that one reproducing this paper may plug the provided equations to the canonical algorithm. To maintain a reasonable computational cost on the learning procedure, we divide the procedure into two steps. In the first one, we perform a canonical LDA fitting [3] to the descriptions of the projects of the dataset and, given the posterior distributions of the topics, we proceed to fit the time-series part of the model.

## 3.2  Topic Heat Variational Distribution

To infer the posterior distribution of the topic heats, we make use of forward-backward messages to calculate the marginal variational distributions $q(\alpha_t)$ and pairwise ones $q(\alpha_t, \alpha_{t-1})$, adapting the VKM algorithm. We briefly explain the message parsing schema, focusing that the major differences of it to the algorithm presented in [8] are that instead of taking expectations with respect to the parameters of the model, we take expectations on the values of $y^\star$ and $\theta$ variables and the emission component of the model contains two parts. Also, $\log p(y_{p,t}, y^\star_{p,t} | \theta_p, \alpha_t) = \log p(y^\star_{p,t} | \theta_p, \alpha_t)$ when $y^\star_{p,t} < 0$ and $y_{p,t} = 0$, namely $y_{p,t}$ is not random in this case. We make this clear so that we can perform the derivations without having to make this fact explicit.

**Messages:** For the forward and backward messages, we must define the part of these messages related to the join over the latent state at time $t$ and the set of observed $y$ and the approximate $y^\star$ and $\theta$. Taking as example the forward message $f(\alpha_t)$ which must be defined as:

$$f(\alpha_t) = \int \mathrm{E}[p(\alpha_{t-1}, y_{1:t-1}, y^\star_{1:t-1}) p(\alpha_t | \alpha_{t-1}) \prod_{p \in A_t} p(y_{p,t}, y^\star_{p,t} | \theta_p, \alpha_t) d\alpha_{t-1}]_{q(-\alpha)}$$

$$= \int \mathrm{E}[\mathrm{Normal}(\alpha_{t-1}; \mu_{t-1}, \Sigma_{t-1}) \mathrm{Normal}(\alpha_t; \mathbf{A}\alpha_{t-1}, \mathbf{I}_K)$$

$$\prod_{p \in A_{t+}} \mathrm{Normal}(y_{p,t}; n_{p,t}, \delta_y) \prod_{p \in A_t} \mathrm{Normal}(y^\star_{p,t}; m_{p,t}, \delta_y) d\alpha_{t-1}]_{q(-\alpha)}$$

$$(2)$$

where $A_{t+}$ is the set of open projects in $t$ in which $y_{p,t} > 0$ and $-\alpha$ is the set of all latent variables but $\alpha$. Marginalizing $\alpha_{t-1}$ we end up with the following quantities:

$$f(\boldsymbol{\alpha}_t) = \text{Normal}(\boldsymbol{\alpha}_t; \boldsymbol{\mu}_t, \boldsymbol{\Sigma}_t) \text{ where: } \boldsymbol{\Sigma}_{t-1}^{\star} = \left(\boldsymbol{\Sigma}_{t-1}^{-1} + \mathbf{A}'\mathbf{A}\right)^{-1}$$
$$\boldsymbol{\Sigma}_t = \left(\mathbf{S}_t + \mathbf{I} - \mathbf{A}\boldsymbol{\Sigma}_{t-1}^{\star}\mathbf{A}'\right)^{-1} \text{ and } \boldsymbol{\mu}_t = \boldsymbol{\Sigma}_t \left(\boldsymbol{b}_t + \mathbf{A}\boldsymbol{\Sigma}_{t-1}^{\star}\boldsymbol{\Sigma}_{t-1}^{-1}\boldsymbol{\mu}_{t-1}\right) \tag{3}$$

and the matrices $\mathbf{S}_t$ and $\boldsymbol{b}_t$ are time-dependent and are constructed as

$$\mathbf{S}_t = (\boldsymbol{\lambda}_y \otimes \boldsymbol{\lambda}_y) \odot \sum_{p \in A_{t+}} \text{E}[\boldsymbol{\theta}_p \otimes \boldsymbol{\theta}_p]_{q(\boldsymbol{\theta}_p)} + (\boldsymbol{\lambda}_{y^\star} \otimes \boldsymbol{\lambda}_{y^\star}) \odot \sum_{p \in A_t} \text{E}[\boldsymbol{\theta}_p \otimes \boldsymbol{\theta}_p]_{q(\boldsymbol{\theta}_p)} \text{ and}$$
$$\boldsymbol{b}_t = \boldsymbol{\lambda}_y \odot \sum_{p \in A_{t+}} \text{E}[\boldsymbol{\theta}_p](y_{p,t} - (\boldsymbol{\lambda}_c^\mathsf{T} c_{p,t} + \rho_y)) + \boldsymbol{\lambda}_{y^\star} \odot \sum_{p \in A_t} \text{E}[\boldsymbol{\theta}_p](\text{E}[y_{p,t}^\star] - (\boldsymbol{\lambda}_{c^\star}^\mathsf{T} c_{p,t} + \rho_{y^\star}))$$
$$\tag{4}$$

This is the usual derivation of the VBKM as seen in the literature [8] and the basic difference is that the expectations of topic proportions $\boldsymbol{\theta}$ and $y^\star$ elements are absorbed in the matrices $\mathbf{S}_t$ and vectors $\boldsymbol{b}_t$. As a special case, when $t = 1$, $\boldsymbol{\Sigma}_1 = (\mathbf{S}_1 + \mathbf{I})^{-1}$ and $\boldsymbol{\mu}_1 = \boldsymbol{\Sigma}_1 \boldsymbol{b}_1$. The backward messages procedure follows the same scheme as previous equations [8], where we can make use of these matrices once again. All the rest of the algorithm is similar to [8].

### 3.3  $q(y^\star)$ Derivation

The hurdle bit of the model we define in this work provides partial information about the states $y_{p,t}^\star$ given the observation of $y_{p,t}$. If $y_{p,t} = 0$, then $y_{p,t}^\star$ has got to be negative and it must be positive provided that $y_{p,t} > 0$. Having this in hand, we derive the variational distribution of these elements as:

$$q(y_{p,t}^\star) \approx \mathbb{1}_{sign(y_{p,t})=sign(y_{p,t}^\star)}\text{Normal}(y_{p,t}^\star, \text{E}[m_{p,t}], 1) = \begin{cases} rTN(y_{p,t}^\star, \text{E}[m_{p,t}], 1) \text{ if } y_{p,t} > 0 \\ lRN(y_{p,t}^\star, \text{E}[m_{p,t}], 1) \text{ if } y_{p,t} = 0 \end{cases} \tag{5}$$

where $\mathbb{1}$ is the indicator and $sign$ is the signal function and $rTN$ and $lTN$ stand for right-truncated and left-truncated Normal distributions [9] (Chap. 19), respectively. All of this is a direct derivation of Bayesian Probit Regression [10,11].

**M-Step.** The M-Step of the algorithm is standard and will not be discussed here. In order to estimate the parameters of the model, we need only the first and second moments of the existing latent variables which are easy to calculate. We substitute these expectations log likelihood of the model and perform gradient descent using the Limited-memory BFGS algorithm.

**Identifiability Issues.** Due to the latent nature of the topic heats, their usage in the Hurdle part of the model turns out to be unidentifiable, unless we enforce constraints into the parameters domain. We enforce the parameters $\boldsymbol{\lambda}_y$ and $\boldsymbol{\lambda}_{y^\star}$ to be $\geq 0$. By doing so we define that the "warmer" a topic is the more important it is to have a larger proportion of projects' definitions taken by that topic, and analogously, the "colder" a topic is at a given moment the less it is going to contribute for a project to obtain donations.

## 4 Experiments and Results

For our experiments, we scraped a first dataset from Kickstarter for which we used to construct the topics used in the modeling. We preprocessed the data and ended with 9086 different terms. These terms and these texts were used to construct the topics, which were then fed into the model and kept fixed. The second and most important dataset was obtained throughout 7 contiguous months, from April 2014 to November 2014, in which we collected data of approximately 45 K projects, which were collected regularly at every 12 hours to get snapshots of these projects. We collected only project-related features, such as *goal*, *duration*, *number of rewards* and textual description. We also constructed a time-varying feature which we call $\Delta_{p,t}$ that represents the scaling (unity-based $[0,1]$ normalization) of the duration of a project, e.g. a project $p$ which starts at time-point 31 and ends at time-point 60 will have features $\Delta_{p,45} = 0.5$, $\Delta_{p,60} = 1$ and so on. This feature is added twice in the covariate set, one time in a square form, to simulate the $U$-shape format of the donations to projects observed in [12]. Additionally to that, we included an autoregression component to every project history. We did so by adding three covariate variables: one indicator for the starting point of the projects, one indicator if the project has received donation in the previous time-point and the value of such donation in the log-scale.

We evaluated the proposed model by separating the projects according to the categories defined by Kickstarter and by learning the model making use of half of the time-points and performing all the estimations on the projects that were active at this time cut. We fixed the number of topics $K$ to 10 (picking the number of topics of a model is usually an ad-hoc task depending on the domain of the instances of the problem, although there are algorithms that automatically estimates an optimal number of topics [13]). For every combination of these elements, we made use of standard evaluation metrics. We explored the "topic heat" trajectories to visualize and analyze the overall behaviour of the market and additionally to this analysis, we used the estimated $\alpha$ states to compare the Mean Absolute Error (MAE) and Root Mean Square Error (RMSE) of Linear Regression in the regression task of estimating the next donation amount a project will receive.

### 4.1 Results

For each category, we present in Fig. 1 the relative normalized ($[-1, 1]$ scale) of the expected value of $\alpha$ given all donations (smoothing distribution) for every data point in the training dataset. We can interpret these graphical descriptions as follows: positive values for topic heats mean that projects containing a big

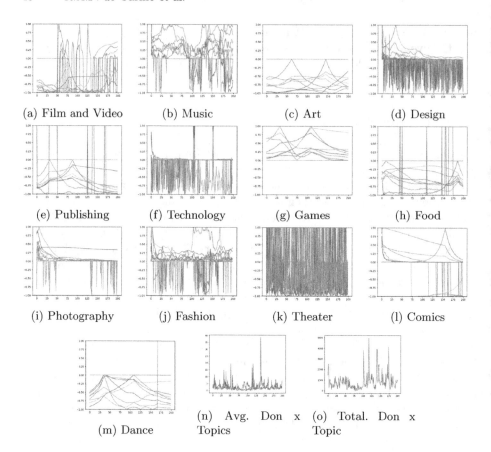

(a) Film and Video    (b) Music    (c) Art    (d) Design

(e) Publishing    (f) Technology    (g) Games    (h) Food

(i) Photography    (j) Fashion    (k) Theater    (l) Comics

(m) Dance

(n) Avg. Don x Topics

(o) Total. Don x Topic

**Fig. 1.** Normalized $[-1, 1]$ topic heat through time. (Best seen in color - Each color represents a specific topic) (Color figure online)

chunk of text referring these topics will likely get more donations, while negative values for topic heats imply having big chunks of the descriptions devoted to these topics will negatively influence the likelihood of getting more donations.

With this understanding in hand, we observe some interesting relations in this figure. First of all, we observe a difference in the heat of the topics for each different category, which is a natural observation due to the diverse nature of these categories. For some categories, such as Art, Technology, Games, and Photography, there is a clear tendency of some topics having consistent more importance and others, while in Music and Comics there is a variability and change in the most important topics. In a deeper view of the unveiled relations, let us pay attention to Fig. 1n where we observe the average of the donations times the topic proportions of topics 0 and 3 in projects of the category Music. First, we observe that there is, in general, more money related to topic 3, which is positive in the majority of the time. Around time-point 75 to 115 the heat of topic 3 decreases and becomes negative, so we observe a decrease in the money related to this topic in these time-points as well. On the other hand, in this very same period, the topic heat for topic 0 maintains a positive value and we observe more money connected to this topic in this period. Another observed effect can be seen comparing Fig. 1e to Fig. 1o where we present the sum of donations time topic 0 in the Publishing category. By comparing these elements we can justify the fast surge and decay of topic 0 in this category, which is related to a rapid surge in the donations related to this topic received by projects in this category. This is an effect of the latent variables trying to accommodate and smooth these surges.

For the regression task we use the covariates present in the settings with and without the $\boldsymbol{\theta} \odot \boldsymbol{\alpha}$ components, which are called *complete* (C) and *baseline* (B) respectively. For each new time in the testing set, we updated the Linear Model adding the new data in the time-point in the training set and predicting the data in the following time point. A summary of the results is presented in Table 1. As we can observe, when adding the information of the latent topic heats, the simple linear regression algorithm achieves better average results of RMSE and MAE in most of the categories. This provides empirical evidence that adding the topic heat information into black-box models may provide them valuable data to regression tasks.

**Table 1.** Average (over time) RMSE and MAE regression values for Linear Regression - Test set (White rows for baseline model and Grey rows for complete model)

| | Film and Video | Music | Art | Design | Publishing | Technology | Games |
|---|---|---|---|---|---|---|---|
| RMSE | 473.8124 | 222.58390 | 140.93211 | 562.3657 | 216.08103 | 368.8366 | 679.5713 |
| MAE | 116.4483 | 74.92392 | 44.03222 | 176.6190 | 48.39902 | 187.7912 | 227.7411 |
| **RMSE** | **472.6523** | **222.17493** | **143.09378** | **555.2242** | **216.13318** | **354.1419** | **672.7153** |
| **MAE** | **102.3413** | **69.29326** | **39.99157** | **124.7995** | **46.77517** | **129.2110** | **204.0611** |
| | Food | Photography | Fashion | Theater | Comics | Dance | |
| RMSE | 270.90690 | 98.02709 | 189.13682 | 216.70607 | 120.64323 | 220.32350 | |
| MAE | 62.69218 | 34.82984 | 62.73192 | 89.88659 | 59.32592 | 93.79268 | |
| **RMSE** | **272.85313** | **96.38986** | **188.72441** | **216.02921** | **132.40438** | **222.0048** | |
| **MAE** | **61.52649** | **26.34658** | **57.94165** | **84.96142** | **65.82996** | **93.5107** | |

## 5   Related Work

The idea of connecting textual data to numerical output has been studied recently in different scenarios. Some works connect the topic indicators of the words of a text to numeric values attached to the document (possibly a label) [14–16] but due to the enormous dimension of the vocabulary of the texts, these models suffer the curse of dimensionality when trying to connect these elements. On the other hand, other works deal directly with topic proportions in different ways: Labeled LDA [17] constructs a generative model for which a set of labels influence the topic proportion of texts, Associative Topic Model (ATM)[18] makes use of time-varying priors on topic proportions to predict a possibly multivariate time-series outcome related to documents that occur in different time-points.

Crowdfunding as an internet-based market is a relatively new subject and as such is the research on this topic. Several studies model the model both the likelihood of success of projects or the amount of donations that projects are going to receive. In general, these contributions select a set of project and social covariates, used as inputs to a black-box model. [19] makes use of kNN, auto-regressive and SVM models and a discretization scheme on the number of donations, along with social predictors to predict the likelihood of success of projects. [20,21] follow the same direction. [22] points the characteristic that donors make donations to projects in the same category as the previous projects they have donated to. Also on the point of topics and texts, [23] studies the textual characteristics of projects and their successes. Thes issues of retaining donors and recommending projects to possible donors are studied in [24] and [25].

## 6    Conclusions

We present a generative approach to model topic-sentiment variables. These variables are easy to visualize and give a clear picture of the time-varying sentiment attached to topics, in a topic model sense By doing so, we provided interesting insights to understand the dynamics of the important market of crowdfunding, while the constructed variables also improve the performance of regression algorithms and the proposed model can also be used and extended in different domains right out of the box.

**Acknowledgments.** Rafael Carmo was supported by Capes - Science Without Borders Programme (Process 99999.001034/2013-08) - Brazil.

## References

1. Mollick, E.R.: Containing multitudes: the many impacts of kickstarter funding. Available at SSRN 2808000 (2016)
2. Greenberg, M.D., Pardo, B., Hariharan, K., Gerber, E.: Crowdfunding support tools: predicting success & failure. In: CHI 2013 Extended Abstracts on Human Factors in Computing Systems. ACM (2013)
3. Blei, D., Ng, A., Jordan, M.: Latent dirichlet allocation. J. Mach. Learn. Res. **3**, 993–1022 (2003)
4. Durbin, J., Koopman, S.J.: Time Series Analysis by State Space Methods, vol. 38. OUP Oxford (2012)
5. Cragg, J.G.: Some statistical models for limited dependent variables with application to the demand for durable goods. Econometrica J. Econ. Soc. **39**, 829–844 (1971)
6. Jordan, M., Ghaharamani, Z., Jaakkola, T., Saul, L.: An introduction to variational methods for graphical models. In: Learning in Graphical Models, pp. 105–162 (1998)

7. Bernardo, J., Bayarri, M., Berger, J., Dawid, A., Heckerman, D., Smith, A., West, M., et al.: The variational bayesian em algorithm for incomplete data: with application to scoring graphical model structures. Bayesian Stat. **7**, 453–464 (2003)
8. Beal, M.J., Ghahramani, Z.: The variational kalman smoother. Gatsby Computational Neuroscience Unit, University College London, Technical report (2001)
9. Greene, W.H.: Econometric Analysis. Pearson Education India (2003)
10. Consonni, G., Marin, J.M.: Mean-field variational approximate bayesian inference for latent variable models. Comput. Stat. Data Anal. **52**(2), 790–798 (2007)
11. Jaakkola, T.S., Qi, Y.: Parameter expanded variational bayesian methods. In: Advances in Neural Information Processing Systems, pp. 1097–1104 (2006)
12. Kuppuswamy, V., Bayus, B.L.: Crowdfunding creative ideas: the dynamics of project backers in kickstarter. UNC Kenan-Flagler Research Paper (2015)
13. Wang, C., Paisley, J., Blei, D.: Online variational inference for the hierarchical dirichlet process. In: Proceedings of the Fourteenth International Conference on Artificial Intelligence and Statistics, pp. 752–760 (2011)
14. Zhang, C., Kjellström, H.: How to supervise topic models. In: Agapito, L., Bronstein, M.M., Rother, C. (eds.) ECCV 2014. LNCS, vol. 8926, pp. 500–515. Springer, Cham (2015). doi:10.1007/978-3-319-16181-5_39
15. Blei, D.M., McAuliffe, J.D.: Supervised Topic Models. In: NIPS - Advances in Neural Information Processing Systems, pp. 121–128 (2008)
16. Rabinovich, M., Blei, D.M.: The inverse regression topic model. In: ICML - International Conference on Machine Learning, pp. 199–207 (2014)
17. Ramage, D., Hall, D., Nallapati, R., Manning, C.D.: Labeled LDA: a supervised topic model for credit attribution in multi-labeled corpora. In: Proceedings of the 2009 Conference on Empirical Methods in Natural Language Processing (2009)
18. Park, S., Lee, W., Moon, I.C.: Supervised dynamic topic models for associative topic extraction with a numerical time series. In: CIKM - International Conference on Information and Knowledge Management, pp. 49–54 (2015)
19. Etter, V., Grossglauser, M., Thiran, P.: Launch hard or go home!: predicting the success of kickstarter campaigns. In: Proceedings of the First ACM Conference on Online Social Networks, pp. 177–182. ACM (2013)
20. Chen, K., Jones, B., Kim, I., Schlamp, B.: Kickpredict: Predicting kickstarter success (2013)
21. Kamath, R., Kamat, R.: Supervised learning model for kickstarter campaigns with rmining (2016)
22. Rakesh, V., Choo, J., Reddy, C.K.: Project recommendation using heterogeneous traits in crowdfunding. In: International AAAI Conference on Web and Social Media (2015)
23. Mitra, T., Gilbert, E.: The language that gets people to give: phrases that predict success on kickstarter. In: Proceedings of the 17th ACM Conference on Computer Supported Cooperative Work & Social Computing, pp. 49–61. ACM (2014)
24. Althoff, T., Leskovec, J.: Donor retention in online crowdfunding communities: a case study of donorschoose.org. In: Proceedings of the 24th International Conference on World Wide Web, pp. 34–44. ACM (2015)
25. An, J., Quercia, D., Crowcroft, J.: Recommending investors for crowdfunding projects. In: Proceedings of the 23rd International Conference on World Wide Web, pp. 261–270. ACM (2014)

# Regression Tree for Bandits Models
# in $A/B$ Testing

Emmanuelle Claeys[1,2,3]([✉]), Pierre Gançarski[1], Myriam Maumy-Bertrand[2],
and Hubert Wassner[3]

[1] Strasbourg University, CNRS, ICUBE, 300 Bd Sébastien Brant,
67400 Illkirch-Graffenstaden, France
claeys@unistra.fr
[2] Strasbourg University, CNRS, IRMA, 7 Rue René Descartes,
67000 Strasbourg, France
[3] AB TASTY, 3 Impasse de la Planchette, 75003 Paris, France

**Abstract.** In the context of Web $A/B$ testing, dynamic assignment of
traffic aims to promote the best variation ($A$ or $B$) as quickly as possible. However, dynamic assignment is difficult to use when the difference
between $A$ and $B$ affects the visitor differently according to his/her personal characteristics and his/her history (number of visits, navigation on
the website ...). In this paper, we propose a dynamic assignment strategy based on a visitor segmentation determined automatically from the
visitors navigation and characteristics.

**Keywords:** $A/B$ Test · best arm identification · bandit models · regression tree · time clustering · topic modeling

## 1 Introduction

The quality of a website (usability, iconography, customer experience or visitor's
navigation) is fundamental for an e-merchant (i.e., corresponding to business
based on the web). Developing and adapting the website to improve it is often
essential to maintain or increase sales. However, changes can negatively affect
sales. To avoid this negative effect, e-merchants must carefully observe the visitor
responses to the website changes to evaluate the relevance of these changes. On
a sample of e-visitors, an $A/B$ test consists of comparing several variations of
the same element. Usually two variants are available, which are denoted by $A$
(e.g., the original web-page) and $B$ (a variation to be tested). During an $A/B$
test, a ratio of visitors only sees a variation while the other visitors only see
the other one. The effectiveness of each variation is evaluated according to this
e-merchant's objective which is generally based on the conversion rate of the
variation (number of clicks on an element, add of items to basket, number of the
transactions...). After the test, for the *exploitation phase*, the e-merchant can
choose to implement the variation $B$, to keep the original $A$ or to implement a
custom strategy. To implement the observation phase, it is necessary to define the

© Springer International Publishing AG 2017
N. Adams et al. (Eds.): IDA 2017, LNCS 10584, pp. 52–62, 2017.
DOI: 10.1007/978-3-319-68765-0_5

strategy of assignment of a variation to a new visitor, i.e., the page the visitor will exclusively see. Indeed, during a given $A/B$ test, a variant is definitively affected to a visitor even if he/she comes back again. Consequently, a visitor only sees one of the variants.

As a consequence of this assignment, it is impossible to know the visitor behaviour if he/she have been assigned to the other variant. Thus, comparing the effectiveness of each of the two pages according to the visitors' behaviours is very hard because the tested populations are totally disjointed.

Strategies based on statistical tests to improve the assignment exist but they are static: the assignment ratio between $A$ and $B$ can not change during all the $A/B$ test. However, if identification of the best variation is crucial for the e-merchant, the priority remains the profitability of the website. Thus if, during the test, the new variation leads to very positive effect or, at the opposite, to very negative effect, the e-merchant would want to reduce the observation phase and increase the exploitation phase (for example, by using the best variation immediately). A way to circumvent this problem, is to change the ratio of assignment during the test: the better a variation, the higher the ratio of assignment to this variation, and reciprocally. During the exploration phase used to establish statistical confidences, a dynamic assignment strategy increases/decreases the ratio of visitors affected to the variations according to their performances until identifying the best variation. Then all visitors are assigned to the best variation and the test stops.

Almost all approaches which implement such strategies are based on the bandit model and give good results. Bandit methods automatically adapt the ratio of assignment and try to find the best variation as quickly as possible (exploration phase) in order to assign all the visitors to this best solution as soon as it is determined (exploitation phase). This approach tries to solve the exploration-exploitation dilemma introduced by Robbins et al. [11]. It can also be viewed as a reinforcement learning method [2]. Nevertheless, when the population is not homogeneous [14], the use of only one bandit is often not appropriate (see Sect. 2). In this paper, we present a new approach consisting of two steps (Sect. 3). The first one searches the most homogeneous subgroups into the visitors according to their navigation on the website (navigation, interest ...) and their own characteristics (e.g., localisation, navigator used...) using clustering algorithms and non-parametric regression trees. The second step uses a specific assignment strategy to each of them (i.e., a bandit algorithm for each group) to actually make the test. We also present the first experiments carried out on an existing fashion website (Sect. 4). Then, we conclude and present some perspectives (Sect. 5).

## 2   Dynamic Assignment Using Bandit Model

The problem is to assign a new visitor to a page variation. But at the beginning of the $A/B$ test, the a priori information about the conversion rate associated with each page is unknown. Thus, the conversion rates (namely, the expected reward) of the two pages must be determined empirically. To do that, the variation $A$ (resp. $B$) is associated to the arm $Arm_1$ (resp. $Arm_2$) of a bandit. Then, an iteration $i$ of the bandit algorithm consists of assigning the $i$-th visitor to a variation according to the respective expected conversion rate of the two arms. This expectation is then updated according to the visitor behaviour (conversion or not, amount of the transaction, ...). Initially, both expectations are unknown but it is supposed that the gain of each of the arms follows a probability distribution according to the normal law [1]. The aim of a bandit algorithm is then to find empirically these two distributions and to decide the best one while limiting regret (i.e., the cost of a bad choice) due to the discovery of the best arm (*exploration phase*). If one page is better than the other, it is better to assign visitors to it as quickly as possible (*exploitation phase*) while avoiding premature convergence. This problem is known as the "exploration vs exploitation dilemma" [11].

Let $\nu_i^j$ be the distribution of probability associated with $Arm^j$ at the iteration $i, i \in \{1, ...n\}$ and $\mu_i^j$ the a priori expectation associated with arm $j$ ($j \in \{1, 2\}$). Let $Z_i^j$ be the reward (conversion or not, amount of the transaction, ...) associated with the $i$-th visit obtained by $Arm_j$. We assume that $Z_i^j \in \mathcal{L}^1$.

Let $Arm^*$ be the best arm to be found and $\mu^*$ its expectation. By definition $\mu^* = max(\mu_n^1, \mu_n^2)$.

To identify $Arm^*$, the algorithm has to explore different options (exploration step).

The goal of bandits models is to minimize cumulative regret during testing. Cumulative regret grows each time the algorithm chooses a sub-optimal arm:

$$\mathcal{R}_i(\nu) = i\mu^* - \mathbb{E}_\nu[\sum_{k=1}^{i} Z_k]. \tag{1}$$

An effectiveness strategy limits regret as [7]:

$$\inf \lim_{t \to +\infty} \frac{\mathcal{R}_i(\nu)}{\log(t)} \geq \sum_{Arm:\mu^{Arm}<\mu^*} \frac{(\mu^* - \mu^{Arm})}{\mathbf{KL}(\nu^{Arm}, \nu^*)} \tag{2}$$

where $\mu^{Arm}$ is a sub-optimal arm expectation, $\nu^{Arm}$ is a sub-optimal arm distribution of probability and $\mathbf{KL}$ is the Kullback-Leibler divergence between two distributions: the bigger the difference between the optimal and sub-optimal distribution is, the quicker the process converges. Conversely, the distance $\mathbf{KL}$ is unusable if the two distributions are very similar.

In summary, bandits models can handle the "exploration-exploitation dilemma" thought dynamic assignment. Nevertheless, if too many visitors are not impacted by the test (i.e., impacted persons are too few to influence initial distribution), the bandit model will not be able to find the best variation.

To solve this problem, some variants of bandit model exits. For instance, the contextual bandits method uses a context vector with $d$ dimensions (visitor's context) to make the best decision.

For instance:

- **LinUCB** (Upper Confidence Bound)**:** assumes a linear dependency (to be identified) between the expected reward and the context vector [8]
- **UCBogram algorithm:** assumes a non-linear dependency (to be identified) between the expected reward and the context vector.
- **NeuralBandit algorithm:** a neural network approach where network learns the context and the associated reward.
- **KernelUCB algorithm:** a nonlinear version of **LinUCB** using a kernel matrix. Used for on-line learning.

(Note that this list is not exhaustive; the reader can find more detailed information in Galichet et al. [4]).

In fact, these methods require a prior knowledge on the context, on the relevance of the attributes and on the correlation between them. Unfortunately, in most cases, the e-merchant can not provide such knowledge making them difficult to use.

Our idea, and the main purpose of our work, is to find groups (also called segments) in which the difference between the two distributions is significant (i.e., $\mathbf{KL}(\nu^{Arm}, \nu^*) > 0$) and in which people have the same behaviour (i.e., the individual conversion rate is compatible with the both distributions). So, for each group, a dynamic assignment can be produced by a bandit as the conditions to use it are verified.

## 3     A New A/B Test Procedure

To solve issues introduced in the previous section, we propose a new A/B test approach which combines the two steps:

1. a *preliminary analysis* (offline) to identify visitors subgroups (or segments) according to their behaviours during their visits, and to determinate if these subgroups have been differently impacted by the modification of the page (see Sect. 3.1).
2. a *analysis step* (online) corresponding to the actual test which independently uses a bandit algorithm to each identified subgroup (see Sect. 3.2).

### 3.1     Step 1: Preliminary Analysis

To create subgroup of visitor which have been similarly impacted by the modification, we assume that visitors of a such subgroup had have similar navigation between the arrival of the visitor on the website and a potential conversion. As we dispose of historical data describing all the pages browsed by the visitors before the test during a given period (of generally two or three weeks), we can use this temporal information to extract such subgroups of visitor. The preliminary analysis we propose, is composed of three steps

1. creation of two clusterings of visitor using historic navigation data.
2. extraction of the topics of interest to improve the visitor profile.
3. patterns searching in visitor profiles to highlight common behaviour.

**Creation of Different Clusterings of Visitors.** To categorize the visitors, two clustering are created using browsing history. The first one is based on the vector $V_{presence}$ which indicates if a visitor visits the website within time intervals (usually, each day), the second on the vector $V_{pages}$ which gives the number of pages browsed by a visitor in each time interval.

Generally, a website is built according to the customer experience. So, it can integrate an implicit expected navigation from the home page to the tested page. Nevertheless, the visitor can start browsing from any of the pages on this path, or even from the tested one, for instance using a search engine. In fact, the browsing history can have different lengths, and can possibly be empty. Thus, in distance-based clustering algorithm, the use of Euclidean distance are not available.

As these two vectors can be seen as time series assigned to each visitor, we can use a classical time series clustering method based on the well-known partitioning algorithm K-MEANS and on the Dynamic Time Warping (D.T.W.) [10]. Centroids calculation are made by the D.T.W. Barycenter Averaging (D.B.A.) method [10], which is an iterative method to calculate clusters' average. This approach is known to be very effective although sensitive to noise and to extreme values. In our experiment, we use the a implementation provided by the **R** language [12].

**Taking Topic of Interest into Consideration.** In order to improve the quality of our visitors profiling, a topic of interest(s) is assigned to each visitor according to the historic navigation (1) by looking for the different key words of browsed pages, (2) by extracting topics from these keywords, (3) by associating a such topic to all the browsed page and (3) by associating all the topics associated to the pages seen by the visitor during his/her navigation. Note that a topic can be associated to several pages. The most used topic extraction method for discrete or symbolic data where topics are not correlated is L.D.A. (Latent Dirichlet Allocation) [5] which is a combination of Bayesian statistical models.

**Patterns Searching in Visitor Profiles.** The purpose of this step is to extract subgroups (or segments) of visitors, before the test (only related to the original variation, the variation B not yet being created) with similar conversion rates. Indeed it is assumed that (1) the test can only affect one or more subgroups and (2) visitors from a same subgroup have same behaviour on the both variations. So we can really compare the two distributions associated to the both variations because the homogeneously of the population.

We apply a non-parametric regression tree algorithm (e.g., conditional inference trees, *CTREE* algorithm, [6]) for discover these subgroups. This regression tree tests statistically significant difference in the conversion rate (Pearson chi-square test, with a 95% confidence level) for each visitor profile (composed of the cluster the visitor belongs to and the and his/her topics series). The built tree can be used to predict the conversion rate for a visitors from his/her profile (see Sect. 4).

The tree is recursively generated by dividing the population of a node into two subsets at each iteration. The node to divide is chosen through a test of co-variances [9] and the most discriminant variable is selected for splitting [13].

To identify the attributes (clusters vs topics) that better characterise the groups, the tree performs different tests such as the Spearman test, the Wilcoxon-Mann-Whitney test or the Kruskal-Wallis test but also permutation tests based on ANOVA statistics or correlation coefficients (in our experiments, with discrete explanatory classification variables, we use a Kruskal-Wallis test). We compared, using the Kruskal–Wallis non-parametric method, all the combinations of clusters to find the more discriminant ones (i.e., the combination which give the most different ratios of conversion). Note that in the **R** language implementation, the *CTREE* algorithm calculate multiple comparisons (with a Bonferroni correction due to multiple comparisons) [3] to choose the most appropriate test.

Recall that our objective is to use this subgroup to better variation assignment by associating a bandit to each subgroup corresponding to a leaf of this tree.

### 3.2 Step 2: Analysis (A/B Test)

To assign a new visitor during the analysis step (online), all information about his/her navigation from the webside homepage to the page to be tested are registered. As presented in Sect. 3.1, the two vectors corresponding to the visit are then calculated and the visitor is affected to the nearest cluster using D.T.W.

Efficient clustering incremental methods do not exist. So, it is difficult to improve the clusters during the test phase. We decide to build them before trough a training period.

The topic series is also affected to the profile. Then the visitor is also affected to a subgroup using the regression tree. As introduced previously, each group extracted by the regression tree contains visitors having similar behaviour. Then we can associate to each leaf of the tree, a bandit occurrence. Then the A/B test itself can start.

## 4 Experiments

### 4.1 The Data

We have applied our method on a dataset based on an $A/B^1$ test without dynamic assignment (60% of global traffic sent to $A$ and 40% to $B$) realised from 11/07/16 to 25/07/16 for a fashion e-commerce websites. We selected visitors arriving before 15/07/16 (midnight) to build our pre-analysis (*train_set*), and visitors arrived after 15/07/16 to test A/B itself (*exp_set*).

The reward was a binary value[2]. The reward function follows a Bernoulli distribution whose parameters are estimated over time.

---

[1] Variation B consisted of changing the link of a "return" button.
[2] 0: no purchase or 1: purchase.

## 4.2   The Test

According to the two step described in Sect. 2, we produced the clustering
*clustPage* (resp. *clustVisite*) based on the vector $V_{presence}$ (resp. $V_{pages}$). For
technical reasons, we had only used the topic (called *clust* in the following) corresponding at the last page browsed by the visitor rather all topic encountered
during the navigation.

Then, we generated a regression tree (Fig. 1) using exclusively the visitors
from *train_set* who have seen the original version *A* (Step 1). The preliminary
analysis identified 10 distinct groups

For instance, *"Node3"* corresponds to visitors belonging to *clustPage#3* and
to *clustVisite#2* or *clustVisite#3*.

For Step 2 corresponding to the A/B test itself, we affected a bandit at each
group corresponding to a leaf in the regression tree. Then we apply our dynamic
assignment strategy for all the "new" visitors from *exp_set*.

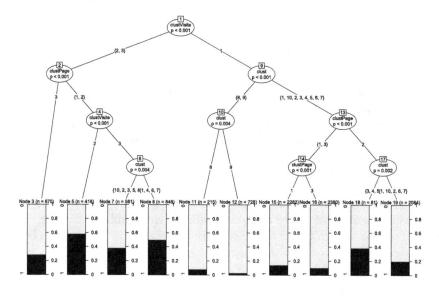

**Fig. 1.** Regression tree of *train_set*

## 4.3   Discussion

To evaluate performances of our method, we compare it with

- a same bandit model, but applied to all the visitors (without group) referred
  here as **Binomial Bandit**.
- a popular contextual bandit algorithm refereed here as **LinUCB**

  Two aspects are observed in particular:

1. Cumulative reward and ratio of cumulative rewards to visitors.
2. Probability to identify the best variation.

**Table 1.** Best variation and cumulative reward by group

| Bandit | Node 3 | Node 5 | Node 7 | Node 8 | Node 11 | Node 12 | Node 15 | Node 16 | Node 18 | Node 19 | Total |
|---|---|---|---|---|---|---|---|---|---|---|---|
| Winner | A | B | A | B | A | A | X | A | A | A | X |
| Rewards | 573 | 50 | 39 | 51 | 10 | 9 | 0 | 628 | 6 | 91 | 1457 |
| Visitors | 1878 | 99 | 131 | 181 | 310 | 475 | 0 | 7380 | 105 | 609 | 11168 |
| Proba | 0.66 | 0.74 | 0.88 | 0.63 | 0.52 | 0.72 | 0 | 0.96 | 0.73 | 0.91 | X |

**Cumulative Rewards.** With our approach, cumulative rewards is 1457 conversions while Binomial Bandit gets only 1252 ones. Table 1 details the best variation retained for each group, the cumulative reward obtained, the number of visitors associated to the group and finally the confidence on the choice of the optimal variation. Note that no visitor has been assigned to bandit #15.

**Confidence on the Best Variation.** Our method finds 6 optimal variations with a probability greater than 70% and generally before the 1000-th visitors while the Binomal Bandit requires a least 4000 visitors for the same confidence. Five bandits from them went into the exploitation phase before only one thousand of visitors.

To perform our experiment, we iteratively select a visitor. We apply the algorithm and if it decides to assign at this visitor to variation $(V_u)$ which differs to that given by the dataset, we search in the database, a visitor who presents similar characteristics but who is assigned to the variation $(V_u)$ in the dataset. If it fails, we randomly choose a visitor assigned to the variation $(V_u)$.

Moreover, as LinUCB is not totally deterministic, we have carried out an hundred runs on the algorithm. In the following we only report the mean of the cumulative rewards obtained by these runs.

Table 2 gives the performances of the three algorithms. It shows that LinUCB gets around 1519 (standard deviation of 49) cumulative rewards. Over 100 runs, it outperforms a Binomial Bandit approach but not ours, with which it compares well while being more complex.

Note that Binomial Bandit converges to the best variation (namely, B) but with a low confidence equals to 0.69.

**Detailled Analysis of Our Approach.** Figure 2 shows the relation between the groups previously defined and a *new* regression tree obtained from the *exp_set* dataset. It also includes the variations $A$ and $B$ as explanatory variables.

**Table 2.** Comparison of the three methods

| Approach | Our approach | Binomial bandit | LinUCB |
|---|---|---|---|
| Rewards/Visitors | 0.13 | 0.11 | 0.13 |
| Visitors | 11168 | 11168 | 11168 |

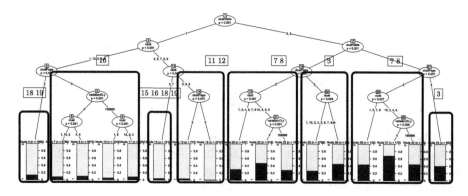

**Fig. 2.** Regression tree of *exp_set* with associated bandits

The idea is to evaluate the overlap between a tree built from the training set and the tree obtained with the validation set. For each visitor belonging to a leaf of the new tree, we search the subgroup which he/she belongs to in the first tree and thus, the bandit that would have been used to assign it to a variation. In the Fig. 2, each rectangle denotes the bandit's names for each subgroup of *exp_set* dataset. For instance, the leaf on the right only contains visitors associated to bandit #3 while the leaf at the left, contains visitors associated to bandit #18 and bandit #19. The second subgroup on the right contains to bandit #7 and bandit #8, and so on.

The idea is to compare if we can find visitors impacted by the test on the groups previously constructed for bandit assignment. We can observe if a visitor have been impacted by the modification of the page in the reality with discriminant variable of regression tree obtained from the *exp_set* dataset[3].

In reality, A was better than variation B only for small subgroups. According to the tree of Fig. 2, there are 3 profiles impacted by the test:

- Visitors belonging to *clust#1*, *clust#2*, *clust#4*, *clust#6* or *clust#10* and *clusPage#3* and *clusVisit#1*. These visitors have been assigned using the bandit #16. Our method chooses variation A.
- Visitors belonging to *clust#2*, *clust#6*, *clust#8* or *clust#10* and *clusPage#2* and *clustVisit#3*. These visitors have been assigned using the bandit #7 and #8. Our method chooses A (select bandit with highest probability).
- Visitors belonging to *clust#2*, *clust#4*, *clust#6* or *clust#10* and *clusPage#2* and *clustVisit#2* These visitors have been assigned using the bandit #7 and #8. Our method chooses A.

The Table 1 shows the detailed results of our method.
Our method adds the following benefits:

- Identification of the optimal variation for a particular group of visitors.
- Faster convergence towards the operational phase.

---

[3] For a technical reason, variation A is called 0 and variation B is called 166888.

– Limitation of convergence towards the wrong variation.

Our approach had a algorithmic complexity comparable to U.C.B one for experimental period. Moreover, the computation can be done in a parallel way. Nevertheless, we suppose that the actual reward probability distribution does not change over the time (hypothesis of stationary).

## 5   Conclusion and Future Work

This paper presents a new approach of dynamic assignments strategy based on a visitor segmentation determined automatically from the visitors navigation and characteristics. To produce this segmentation we have applied temporal and semantic clustering methods on visits database. To implement the dynamic allocation strategy we have used so-called "bandits" algorithm. To validate our approach, we have compared it with a dynamic allocation algorithm without a prior segmentation. Results are very promising and the concept has strong growth potential.

Currently we automatically determine the optimal number topic number and test our approach with other e-merchants. We also work on the quality of the clusters.

For future work, we plan to use more parameters related to the global traffic of the website in order to anticipate the "peaks" and "dips" of traffic interfere with analysis (differences between the variations can be reduced for a low traffic).

## References

1. Agrawal, R., Teneketzis, D., Anantharam, V.: Asymptotically efficient adaptive allocation schemes for controlled markov chains: finite parameter space. IEEE Trans. Autom. Control **34**(12), 1249–1259 (1989)
2. Auer, P., Cesa-Bianchi, N., Freund, Y., Schapire, R.E.: The nonstochastic multiarmed bandit problem. SIAM J. Comput. **32**(1), 48–77 (2003). doi:10.1137/S0097539701398375
3. Bonferroni, C.E.: Teoria statistica delle classi e calcolo delle probabilità. Pubblicazioni del R Istituto Superiore di Scienze Economiche e Commerciali di Firenze **8**, 3–62 (1936)
4. Galichet, N.: Contributions to Multi-Armed Bandits: Risk-Awareness and Sub-Sampling for Linear Contextual Bandits. Theses, Université Paris Sud - Paris XI, September 2015. https://tel.archives-ouvertes.fr/tel-01277170
5. Grün, B., Hornik, K.: Topicmodels: an R package for fitting topic models. J. Stat. Softw. **40**(13), 1–30 (2011). https://www.jstatsoft.org/v040/i13
6. Hothorn, T., Hornik, K., Zeileis, A.: Unbiased recursive partitioning: a conditional inference framework. J. Comput. Graph. Stat. **15**(3), 651–674 (2006)
7. Lai, T., Robbins, H.: Asymptotically efficient adaptive allocation rules. Adv. Appl. Math. **6**(1), 4–22 (1985). doi:10.1016/0196-8858(85)90002-8
8. Li, L., Chu, W., Langford, J., Schapire, R.E.: A contextual-bandit approach to personalized news article recommendation. In: Proceedings of the 19th International Conference on World Wide Web, WWW 2010, NY, USA, pp. 661–670 (2010). doi:10.1145/1772690.1772758

9. Molinaro, A.M., Dudoit, S., van der Laan, M.J.: Tree-based multivariate regression and density estimation with right-censored data. J. Multivar. Anal. **90**(1), 154–177 (2004). http://www.sciencedirect.com/science/article/pii/S0047259X04000296. Special Issue on Multivariate Methods in Genomic Data

10. Petitjean, F., Ketterlin, A., Gançarski, P.: A global averaging method for dynamic time warping, with applications to clustering. Patt. Recogn. **44**(3), 678–693 (2011). doi:10.1016/j.patcog.2010.09.013

11. Robbins, H.: Some aspects of the sequential design of experiments. Bull. Am. Math. Soc. **58**(5), 527–535 (1952). http://www.projecteuclid.org/DPubS/Repository/1.0/Disseminate?view=body&id=pdf_1&handle=euclid.bams/1183517370

12. Sarda-Espinosa, A.: Time series clustering along with optimizations for the dynamic time warping distance. dtwclust R package version 3.1.0 (2017). https://github.com/asardaes/dtwclust

13. Strasser, H., Weber, C.: On the asymptotic theory of permutation statistics. Report, Vienna University of Economics and Business Administration (1999). https://books.google.fr/books?id=pieBNAEACAAJ

14. Wald, A.: Sequential tests of statistical hypotheses. Ann. Math. Statist. **16**(2), 117–186 (1945). doi:10.1214/aoms/1177731118

# Seasonal Variation in Collective Mood via Twitter Content and Medical Purchases

Fabon Dzogang[1(✉)], James Goulding[2], Stafford Lightman[3], and Nello Cristianini[1(✉)]

[1] Intelligent Systems Laboratory, University of Bristol, Bristol, UK
{fabon.dzogang,nello.cristianini}@bristol.ac.uk
[2] N-LAB, University of Nottingham, Nottingham, UK
[3] Henry Wellcome Laboratories for Integrative Neuroscience and Endocrinology, School of Clinical Sciences, University of Bristol, Bristol, UK

**Abstract.** The analysis of sentiment contained in vast amounts of Twitter messages has reliably shown seasonal patterns of variation in multiple studies, a finding that can have great importance in the understanding of seasonal affective disorders, particularly if related with known seasonal variations in certain hormones. An important question, however, is that of directly linking the signals coming from Twitter with other sources of evidence about average mood changes. Specifically we compare Twitter signals relative to anxiety, sadness, anger, and fatigue with purchase of items related to anxiety, stress and fatigue at a major UK Health and Beauty retailer. Results show that all of these signals are highly correlated and strongly seasonal, being under-expressed in the summer and over-expressed in the other seasons, with interesting differences and similarities across them. Anxiety signals, extracted from both Twitter and from Health product purchases, peak in spring and autumn, and correlate also with the purchase of stress remedies, while Twitter sadness has a peak in the Winter, along with Twitter anger and remedies for fatigue. Surprisingly, purchase of remedies for fatigue do not match the Twitter fatigue, suggesting that perhaps the names we give to these indicators are only approximate indications of what they actually measure. This study contributes both to the clarification of the mood signals contained in social media, and more generally to our understanding of seasonal cycles in collective mood.

**Keywords:** Social media mining · Emotions · Human behaviour · Periodic patterns · Computational neuroscience

## 1 Introduction

The existence of seasonal structures in the contents of Twitter has been known for some time [4], and these are believed to reflect (also) patterns of seasonal variation in people's sentiment. In 2016 [2] comparable patterns of variation were also detected in the query log of Wikipedia, showing seasonal changes relative to seasonal affective disorder, panic disorder, acute stress disorder and several

© Springer International Publishing AG 2017
N. Adams et al. (Eds.): IDA 2017, LNCS 10584, pp. 63–74, 2017.
DOI: 10.1007/978-3-319-68765-0_6

other mental health issues. We add to this literature by analysing the purchasing behaviour of over-the-counter remedies directly linked to anxiety, stress and fatigue by over 10 million individuals. In total we analyse 3.87 million transactions, collected over 4 years in all UK branches of a leading Health and Beauty retailer[1] The signals extracted from these datasets are designed to be robust to accidental changes in one specific product or keyword: they are based on the average of many standardised time series, each reflecting the relative frequency of a product or a word, so that changes in our indicators reflect coordinated changes of multiple words or products.

Specifically we leverage the contents of UK Twitter to measure collective expressions of anxiety, sadness anger, and fatigue in the population. We also make use of Health product transactions to measure anxiety, stress, and fatigue across the country. As such the lists of keywords extracted from Twitter and the categories of products extracted from retail transactional logs were related - but there was no expectation that the resulting outputs should actually be measuring the same phenomena, as the lists were compiled for different purposes and the measures performed on two different streams of data.

From Twitter we have extracted keywords denoting each aspect of the negative sentiment in the Linguistic Inquiry and Word Count (LIWC) [12]. Based on these word lists we compiled a 4 years indicator of the negative moods on the social platform. We have also extracted an indicator of fatigue based on the PANAS psychological word list [14]. Examples of mood and fatigue keywords are provided in Appendix. A comparison was then made between the seasonal timing of those indicators and the purchase patterns of consumers at the health and beauty stores across the UK. The transactions that we considered for analysis correspond to the bestsellers products in the search results of the retailer's online web interface, for the queries *anxiety*, *stress*, and *fatigue*.

In analysis, all signals show a clear seasonal behaviour, with all being under expressed in the summers. Additionally, two fundamental types of signals are evident, perhaps pointing to different underlying mechanisms: 1. signals which peak in the Winter (e.g.: Twitter sadness); and 2. signals which peak in the middle of seasons (e.g.: remedies for anxiety and Twitter anxiety). More specifically, both Twitter and anxiety-related products signals peak in spring and autumn, with the indicator for stress-related health products also showing similar seasonal peaks. In contrast, Twitter sadness, Twitter anger and fatigue-related products all show a single mid-winter peak. Surprisingly, fatigue-related products do not match the Twitter fatigue signal. This suggests that perhaps the names we give to these indicators are only approximate indications of what they measure.

---

[1] The data for purchases of over-the-counter Health products comes from a major UK health and beauty retailer, and has been made available to the University of Nottingham (JG), under an NDA agreement that restricts naming of the retailer. The raw data cannot be published, however examples of health and fatigue related products for which data has been made available are detailed in the "Appendix - Examples of health and fatigue related products", but have been partly anonymized to respect this agreement.

The sales patterns of Health products relating to fatigue fit with a shorter winter daylength, showing a prolonged plateau from mid-December to April. Interestingly purchases for stress and anxiety seem to have a delayed onset around the beginning of February - suggesting the necessity for some longer term biological processes before they are triggered - although there is also a peak in the Autumn as there is for Twitter anxiety and fatigue. Intriguingly, anxiety peaks in the period where the rate of change in daylength is maximal.

## 2  Data

A summary of the two datasets used in this study is provided in Table 1. Below we first describe our collection pipeline that we have used to gather public microposts from Twitter UK every ten minutes, then we provide a short description of the UK health and beauty retailer's data.

### 2.1  Twitter Microposts

Using the Twitter API, we collected tweets in the period from January 2010 to November 2014, querying for tweets in the 54 largest towns and cities in the UK without specifying keywords or hashtags. For each tweet, we collected the anonymised textual content, a collection date and time, and information about the location of the tweet (within 10 Km of one of the 54 urban centres). We automatically removed messages containing standard holiday greetings as they contained mood-related words while not necessarily representing an expression of mood (see Appendix - Greeting messages). Due to collection problems we also removed from our collection the year 2012 and the months of November and December 2014. As a result, we have 800 M individual tweets covering the time intervals between January 2010 and November 2014, except for the year 2012. Each tweet was tokenized using a tool designed specifically for Twitter text [3]. Hyperlinks, mentions and hashtags were discarded, along with words containing only special characters (e.g. emoticons).

Table 1. Summary of data and sampling.

|  | Twitter content | Health products |
|---|---|---|
| Sampling interval | 4 years | 4 years |
| Sampling frequency | hourly | daily |
| Sampling locations | 54 largest urban centers in UK | $> 1,000$ postcodes in UK |
| Sampling starts | January 2010 | January 2012 |
| Sampling ends | November 2014 | November 2015 |
| Sampling issues | year 2012 was removed | N/A |
| Population size | N/A | $> 10M$ consumers |
| Volume | 800M tweets | 3.87M transactions |

## 2.2   Health Products Purchases

Transactional data was sourced from a major UK health and beauty retailer to extract purchase counts of over-the-counter health-products sold at stores across the UK. A total of 48 products were queried from the retailer's point-of-sale transactional logs, retrieving in each case a time series of daily transactions in the period between January 2012 and November 2015. These series' reflect the basket of over 10 million anonymous consumers and account for a total of 3.8 million individual purchases at stores across the UK, for the 48 products identified. In order to respect commercial confidentiality the raw count of sales of each product on a given day was divided by the total number of sales on that day, obtaining a relative frequency, at the source. Also any potentially identifying information relating to customers was strictly removed prior to any analysis being undertaken.

## 3   Detection of Collective Moods and Fatigue

From Twitter we have extracted keywords denoting each aspect of the negative sentiment in LIWC [12]. Based on these word lists we compiled a 4 years indicator between January 2010 and November 2014 formed by 414 words for sadness, 956 for anger, and 450 for anxiety. We have also extracted an indicator of fatigue based on the PANAS psychological word list [14], formed by just four words: *tired*, *sleepy*, *sluggish*, and *drowsy*.

We wished to compare the seasonal timing of these textual indicators sampled across the UK with the purchase patterns for health products at the health and beauty stores located across the country. To that aim, we first determined the top 30 bestseller products in each of the categories of anxiety, stress, and fatigue, from the retailer's online web interface. Our queries, one per category, were issued in May 2017 and yielded 14 products for anxiety, 12 products for fatigue, and 30 products for stress. After manually screening them, queries were issued to the health retailer's point-of-sale datasets. For each category we then

**Table 2.** Summary of the categories.

|  | Category description | Nb words/products | Nb series' with gap ≤ 1 year |
|---|---|---|---|
| Twitter anxiety | Liwc | 450 words | 450 words |
| Twitter anger | Liwc | 956 words | 947 words |
| Twitter sadness | Liwc | 414 words | 413 words |
| Twitter fatigue | Panas | 4 words | 4 words |
| Health-products anxiety | Bestseller products | 14 products | 13 products |
| Health-products fatigue | Bestseller products | 12 products | 4 products |
| Health-products stress | Bestseller products | 27 products | 18 products |

compiled a 4 years indicator that reflects the coordinated changes of each best-seller product purchases for a condition (e.g. stress, anxiety, or fatigue) in the interval between January 2012 and November 2015. Only three remedies occurring via the retailer's web interface were omitted from the indicators, with one being aimed at healing baby stress and two others not appearing within the anonymized transaction logs over the period examined. We also found that remedies for stress did overlap with remedies for anxiety: eight products were shared in the two categories. In this study we treat these two categories separately.

A description of the categories used in each data source is provided in Table 2. Examples of keywords used on the social platform and products on the retailer's online web interface are provided in Appendix to illustrate each category.

## 4  Methods

The method we used to extract expressions of emotion from twitter text is based on the LIWC methodology [10]: counting the relative frequency of each words in a list with separate lists for anxiety, sadness, and anger. To account for discrepancies in word usage [1] and word volumes [11] in the collective and anonymous contents sampled across the UK, we further developed the method by first computing the relative-frequency of each word in each sample, then by standardising each individual time series in the sampling interval (each received zero mean and unit variance in the 4 years). The series of relative frequencies is detrended before standardisation using a two years centered window to reduce the effect of long trends in the data.

The indicator for a given emotion is the average of many standardised time series of relative frequencies. In this way a periodic change in the indicator can only result from the coordinated change of several words in the respective LIWC list, reducing the risk deriving from homonymy [2]. For 'fatigue' we used a short list by PANAS [14]), formed by the four words: *tired*, *sleepy*, *sluggish*, and *drowsy*.

The seasonal pattern of an indicator is the 12 months time series of median values per each day of the year, smoothed using a three month centred window. Each indicator on the transaction logs data was extracted following the same steps on pre-compiled lists of products linked to a condition (e.g. stress, anxiety, or fatigue) instead of word lists. To emphasis unsolicited purchases in the population, the procedure was slightly modified: we clipped the individual standardised time series of each product at 3 standard deviations, thus accounting for potential sales or promotional days on individual items.

It is important to note that the word-lists and product-lists have been generated a priori and are neither data dependent, nor chosen a posteriori. This avoids the risk of multiple testing.

## 5  Results

All signals from Twitter and from the pharmaceutical data are summarized by their seasonal pattern shown in Fig. 1, aligned for comparison. Levels of expression above average are illustrated as plain semi-arcs on Fig. 3. All the indicators

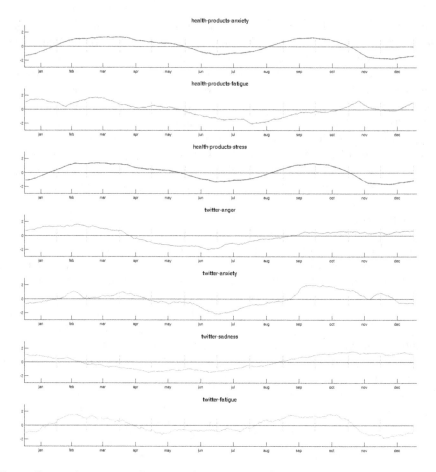

**Fig. 1.** Seasonal patterns relative to the negative moods evident in Twitter data and Remedy sales. Each pattern is aligned for comparison.

show a clear seasonal behaviour with all being under expressed in the summers. Some peaking in the Winter (e.g. Twitter sadness and anger), others peaking in the mid-seasons (e.g. health products anxiety and stress, Twitter anxiety, and the fatigue signals from both sources).

In more detail, we can observe from this data that health products purchases relating to fatigue fit with shorter winter daylength with a prolonged plateau starting in mid-December until April. Interestingly purchases for stress and anxiety products experience a delayed onset around February, suggesting the necessity for some longer term biological processes before they are triggered. Following the summer dip common to all moods and contiditions, both the patterns for stress and anxiety re-express in the Autumn, at a period when fatigue and overall negative sentiments over-express on Twitter. At this time of year expressions of anxious feelings reach remarkable levels on the social platform.

**Table 3.** Percentage of the indicator's variance that can be explained by the seasonal pattern in the 4 years interval. [*] indicates significance at 1% level, and [+] at 5% level; significance is assessed in a Monte Carlo simulation using N=100K random permutations of the indicator.

| Indicator | % Var. expl |
|---|---|
| Health-products stress | 33.5%[*] |
| Health-products anxiety | 28.6%[*] |
| Twitter anger | 03.0%[*] |
| Twitter sadness | 02.7%[*] |
| Health-products-fatigue | 02.5%[*] |
| Twitter anxiety | 01.2%[*] |
| Twitter fatigue | 00.8%[+] |

Assessing the similarity between each pattern along the year we find that purchases of remedies for anxiety and for stress correlate very strongly with expressions of fatigue on Twitter ($\rho = 0.80$ and $\rho = 0.83$ respectively). We also find a strong association with expressions of anxious feelings on the social platform ($\rho = 0.62$ and $\rho = 0.57$ respectively). More specifically we found that amongst the mood patterns measured on Twitter, anxiety was the most similar to purchases for stress and anxiety products; suggesting a strong agreement

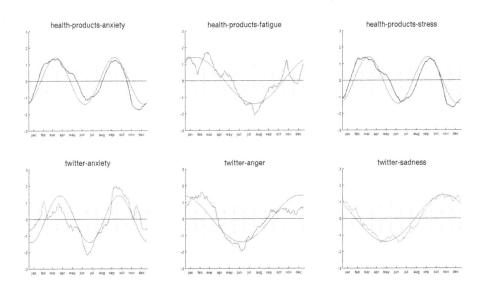

**Fig. 2.** Yearly and bi-annual patterns. Purchase patterns relative to anxiety, fatigue, and stress (first row) in the retailer's transaction logs; textual patterns relative to anger, anxiety, and sadness on the social platform (second row). An idealized sine wave of period 1 year (resp. 6 months) is overlaid on the yearly (resp. bi-annual) patterns.

between the textual contents published on the social platform and the purchase behaviours in the population. The agreement between the two sources can lend further support to the significant association found in [2] between periods of over-expressed anxiety on Twitter and periods of rain in the UK.

In spite of their late winter and late autumn overlap we found no significant association between purchase patterns for fatigue and expressions of fatigue on Twitter ($\rho = 0.09$), it is the dip in the three months period surrounding the end of year festivities (November/December/January), and the onset of Twitter fatigue early in the Autumn that differentiate them. Suggesting that perhaps the names we give to these indicators are only approximate indications of what they actually measure. Nonetheless a strong association was found between health products fatigue and expressions of anger on Twitter ($\rho = 0.69$). In addition to dark days and cold weather [2], this result shows that periods of fatigue (as measured by purchases of specific products) associate with expressions of angry moods (as measured by words used on Twitter).

The strength of each pattern's oscillation is summarized by the percentage of the variance it explains in the indicator across the 4 years series (see Table 3). With a maximum of 33.5% of variance explained for stress remedies,

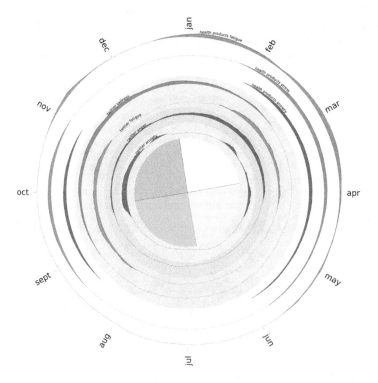

**Fig. 3.** Seasonal pattern of negative moods in Twitter, and patterns of purchasing behaviour for pharmaceutical remedies. Semi-arcs illustrate periods of over-expression in the smoothed series of median levels per each day of the year.

our indicators of purchase behaviour in the UK population are generally more seasonal than our indicators of collective moods on Twitter. In the 4 years intervals the social platform not only reacted to the seasons but it also responded to other significant events: for example in a recent study we found common change-points between the moods in Twitter and the gbp/euro rate in the hours following the vote for the EU membership referendum (brexit) [6].

Another striking difference that emerges from Fig. 1. Is that some patterns seem to experience two peaks in the middle-seasons, while other experience only one major resurgence in the Winter. We detailed these two types of behaviour by separating the patterns that experience a single peak in the year from those that over-express twice a year. For this purpose we overlaid on each pattern an idealized sine wave that maximally explains the variance in the indicator between a period of one year or six months (see Fig. 2). We found that health products anxiety/stress and Twitter anxiety experienced two peaks in the year, with the first occurring in March and the second in September; while fatigue related sales in the population as well as Twitter anger and sadness showed one single periodic resurgence concentrated in the winter months.

# 6    Conclusion

Seasonal affective disorders (SAD) is well described [8], mostly for the depressive symptoms occurring at the beginning of autumn and persisting until the end of winter. But seasonal changes also occur in healthy controls [15]. This is just one of the many seasonal variations that can affect mood and metabolism [9]. The typical symptoms of SAD include irritability, sadness, fatigue, decreased activity and libido, and changes in eating behaviour [7,13]. From a Neuroendocrinological perspective, median cortisol levels are lowest in the summer solstice quarter and over 8.5% higher in the winter solstice quarter [5].

In the present study we have compared two different classes of signals that are expected to capture variations in collective mood over the general UK population. One class of signals is formed by Twitter content collected over a period of 4 years, the other is formed by sales of over-the-counter Health Products in the UK, for items classed as related to anxiety, stress, fatigue.

We see broad agreement in their seasonal timing but also intriguing differences in the findings. Expressions of anxiety on the social platform associate strongly with purchase of anxiety and stress related health products in the population. Winter Remedies for fatigue also fit with expression of angry moods on Twitter and to a lower extent with the periodic resurgence of sad feelings in the winter months.

On the other hand, the fatigue that we extracted from Twitter could not be explained entirely with purchase of fatigue related medications. We found that the three months period between November and January was marked with under-expressed fatigue on the social platform, a behaviour that was not pronounced to the same extent in the transactions for fatigue. Note that similarly to health-product-fatigue the variations in twitter-fatigue were better explained by a sine wave of period one year peaking in the Winter months, however the pattern could not be well separated from the contribution of a six months cycle peaking in winter and in autumn. The difference between product-fatigue and twitter-fatigue still needs to be explained by further works. One of the future tests will be to focus the analysis on years that are available for both datasets, we will also increase the power of our tests by extending our sampling interval.

The key finding from the neurophysiological viewpoint is yet another confirmation of the strong circannual patterns followed by sentiment and mood, with more positive affect in the summers and more negative affect in the winters. Anxiety shows the same pattern both in the purchases and in the Twitter signals.

From the point of view of textual sentiment analysis, we believe that this is a step towards the validation of the LIWC methodology [10], of using word lists to detect changes of sentiment in a text. In the absence of ground truth data about mass mood changes during the year, we have shown that the sales of health-products could provide useful information, in the same way as Wikipedia logs were used in [2].

**Acknowledgments.** NC and FD are supported by the ERC advanced Grant ThinkBig. JG is supported by the EPSRC Neodemographics grant, EP/L021080/1.

# Appendix

*Greeting messages.* The signal about mood could be skewed by the presence of large amounts of standardised greeting messages in specific seasons, which make use of mood related words, while not expressing the mood of the writer. These standard greeting messages were removed from the data as follows: we ignored any Twitter post containing the word happy, merry, good, lovely, nice, great, or wonderful followed by either of christmas, halloween, valentine, easter, new year, mothers' day, fathers' day, and their variants (e.g. starting with a leading # or separated by a dash, a space or ending with 's when applicable) was not considered for analysis.

We verified that posts matching this pattern were indeed concentrated in very specific days (the expected ones for each holiday).

*Examples of mood and fatigue keywords.* Table 4 illustrates the most popular words on Twitter for each indicator of mood and for fatigue. The mood keywords are based on the LIWC word lists, those for fatigue are based on the PANAS word list.

*Examples of health and fatigue related product.* Fig. 4 gives an example of the most popular products for the searches *anxiety*, *stress*, and *fatigue* on the UK Health and Beauty retailer's online web interface, as queried on May 2017.

**Table 4.** Most popular mood words extracted from Twitter based on the LIWC word lists. Our indicator of fatigue on the social platform is based on the PANAS word list, formed by just four words.

|          | Mood words |
|----------|------------|
| Sadness  | *Miss lost sad missed cry missing alone lose crying hurt low fail broke Hurts losing tears loss sadly disappointed failed* |
| Anxiety  | *Crazy shame worry scared awkward horrible doubt scary confused alarm worried* *Fear upset emotional worries afraid pressure nervous avoid risk* |
| Anger    | *Shit fuck hate fucking hell stupid damn bitch hit* *Mad annoying cut kill fight crap jealous cunt piss fucked pissed* |
| Fatigue  | *Tired sleepy sluggish drowsy* |

| anxiety | stress | fatigue |
|---------|--------|---------|
| quiet life tablets | this works deep sleep pillow spray | clarins men anti-fatigue eye serum |
| karma st john's wort extract tablets | bach rescue remedy dropper | maybelline baby skin instant fatigue blur primer |
| schwabe karma mood tablets | kalms tablets | clinique anti-fatigue cooling eye gel |
| kalms tablets | rescue cream | embryolisse smooth radiant complexion immediate anti-fatigue |
| bach rescue remedy dropper 10ml | aromatherapy pure essential oil - 20ml lavender | collection reviving & anti-fatigue illuminating primer |
| pharmaton vitality capsules | aromatherapy pure essential oil - 10ml lavender | l'oreal paris nude magique cc cream anti- fatigue |
| bach rescue remedy spray | bach rescue remedy spray | nuxe creme prodigieuse eye contour - anti-fatigue moisturising eye cream |
| tisserand de-stress aromatherapy roller ball | bach rescue pastilles - blackcurrant with sweeteners | creme prodigieuse® enriched - anti-fatigue moisturising rich cream (dry skin) |
| "own-brand" stress relief tablets | bach rescue pastilles with sweeteners | creme prodigieuse - anti-fatigue moisturising cream (normal to combination skin) |
| vitano rhodiola rosea root extract film-coated tablets | "own-brand" st john's wort tablets | wellman anti-fatigue under eye serum |

**Fig. 4.** Top 10 most popular Health products presented for the searches *anxiety*, *stress*, and *fatigue* on the retailer's online web interface.

# References

1. Bamman, D., Eisenstein, J., Schnoebelen, T.: Gender identity and lexical variation in social media. J. Sociolinguistics **18**(2), 135–160 (2014)
2. Dzogang, F., Lansdall-Welfare, T., Cristianini, N.: Seasonal fluctuations in collective mood revealed by wikipedia searches and twitter posts. In: 2016 IEEE International Conference on Data Mining Workshop (SENTIRE) (2016)

3. Gimpel, K., Schneider, N., O'Connor, B., Das, D., Mills, D., Eisenstein, J., Heilman, M., Yogatama, D., Flanigan, J., Smith, N.A.: Part-of-speech tagging for twitter: annotation, features, and experiments. In: Proceedings of the 49th Annual Meeting of the Association for Computational Linguistics: Human Language Technologies: Short Papers, vol. 2, pp. 42–47 (2011)
4. Golder, S.A., Macy, M.W.: Diurnal and seasonal mood vary with work, sleep, and daylength across diverse cultures. Science **333**(6051), 1878–1881 (2011)
5. Hadlow, N.C., Brown, S., Wardrop, R., Henley, D.: The effects of season, daylight saving and time of sunrise on serum cortisol in a large population. Chronobiol. Int. **31**(2), 243–251 (2014)
6. Lansdall-Welfare, T., Dzogang, F., Cristianini, N.: Change-point analysis of the public mood in UK twitter during the brexit referendum. In: 2016 IEEE International Conference on Data Mining in Politics Workshop (DMIP) (2016)
7. Leonard, W., Levy, S., Tarskaia, L., Klimova, T., Fedorova, V., Baltakhinova, M., Krivoshapkin, V., Snodgrass, J.: Seasonal variation in basal metabolic rates among the yakut (sakha) of northeastern siberia. Am. J. Hum. Biol. **26**(4), 437–445 (2014)
8. Melrose, S.: Seasonal affective disorder: an overview of assessment and treatment approaches. Depression Res. Treat. (2015)
9. Migaud, M., Butrille, L., Batailler, M.: Seasonal regulation of structural plasticity and neurogenesis in the adult mammalian brain: focus on the sheep hypothalamus. Front. Neuroendocrinol. **37**, 146–157 (2015)
10. Pennebaker, J.W., Boyd, R.L., Jordan, K., Blackburn, K.: The development and psychometric properties of liwc2015. Technical report (2015)
11. Piantadosi, S.T.: Zipfs word frequency law in natural language: a critical review and future directions. Psychon. Bull. Rev. **21**(5), 1112–1130 (2014)
12. Tausczik, Y.R., Pennebaker, J.W.: The psychological meaning of words: liwc and computerized text analysis methods. J. Lang. Soc. Psychol. **29**(1), 24–54 (2010)
13. Walton, J.C., Weil, Z.M., Nelson, R.J.: Influence of photoperiod on hormones, behavior, and immune function. Front. Neuroendocrinol. **32**(3), 303–319 (2011)
14. Watson, D., Clark, L.A.: The panas-x: manual for the positive and negative affect schedule-expanded form (1999)
15. Winthorst, W.H., Roest, A.M., Bos, E.H., Meesters, Y., Penninx, B.W., Nolen, W.A., Jonge, P.: Self-attributed seasonality of mood and behavior: a report from the netherlands study of depression and anxiety. Depress. Anxiety **31**(6), 517–523 (2014)

# Skin Cancer Detection in Dermoscopy Images Using Sub-Region Features

Khalid Eltayef$^{(\boxtimes)}$, Yongmin Li$^{(\boxtimes)}$, Bashir I. Dodo$^{(\boxtimes)}$, and Xiaohui Liu$^{(\boxtimes)}$

Department of Computer Science, Brunel University London, London, UK
{Khalid.Eltayef,Yongmin.Li,Bashir.Dodo,XiaoHui.Liu}@brunel.ac.uk

**Abstract.** In the medical field, the identification of skin cancer (Malignant Melanoma) in dermoscopy images is still a challenging task for radiologists and researchers. Due to its rapid increase, the need for decision support systems to assist the radiologists to detect it in early stages becomes essential and necessary. Computer Aided Diagnosis (CAD) systems have significant potential to increase the accuracy of its early detection. Typically, CAD systems use various types of features to characterize skin lesions. The features are often concatenated into one vector (early fusion) to represent the image. In this paper, we present a novel method for melanoma detection from images. First the lesions are segmented by combining Particle Swarm Optimization and Markov Random Field methods. Then the K-means is applied on the segmented lesions to separate them into homogeneous clusters, from which important features are extracted. Finally, an Artificial Neural Network with Radial Basis Function is applied for the detection of melanoma. The method was tested on 200 dermoscopy images. The experimental results show that the proposed method achieved higher accuracy in terms of melanoma detection, compared to alternative methods.

**Keywords:** Melanoma detection · Skin cancer · Dermoscopy images · Skin lesion

## 1 Introduction

Malignant melanoma is considered one of the most deadly forms of skin cancer, and the mortality rate caused by it has increased significantly. It is the seventh most common malignancy in women, and the sixth most common in men. However, early detection of it is particularly important, since it can often be cured with a simple excision [23]. Therefore, with an early diagnosis of melanoma, the mortality rate can be reduced.

Dermoscopy is one of the major tools for the diagnosis of melanoma, it is widely used by dermatologists, due to its value in early stages for malignant melanoma detection. It provides better visualization of several pigmented structures such as streaks, dots, pigment networks and blue-white areas, which are invisible to the naked eye [24]. Due to the presence of the hair and several artifacts such as oil, air bubbles and gel drops on the images. In addition, the

© Springer International Publishing AG 2017
N. Adams et al. (Eds.): IDA 2017, LNCS 10584, pp. 75–86, 2017.
DOI: 10.1007/978-3-319-68765-0_7

borders of the lesions are not visible for the radiologists. Computer Aided Diagnosis (CAD) systems became very important and necessary to help the doctors to interpret the images clearly, and support their diagnosis.

In the last two decades, CAD systems of melanoma detection have reduced the gap between the medical and engineering knowledge, since these systems try to mimic the performance of dermatologists when diagnosing a skin lesion area. Thus, help them to differentiate between melanoma and benign lesions in less time [7,22].

In dermoscopy image analysis, one usually enhances the images first and segments the lesion areas (ROI). This is followed by extracting several features, which could be local or global, and using them to learn an appropriate classifier, in order to predict the lesion label (melanoma or non-melanoma). Each step of above process depends on the previous one. For instance, the classification stage depends on the performance of all previous steps. Therefore, to get a high classification rate and increase the accuracy of the diagnosis of skin cancer, all or most of the previous steps should be implemented with the best strategy. It is well-known that the skin lesion classification methods are usually based on the feature extraction. Therefore, the extraction of representative features of the lesions under analysis is a very important stage for efficient classification [19]. For this purpose, we focused on the best way to extract the optimal features from images, using an improved method to segment the lesion.

In this paper, four main steps have been implemented to build an automatic process for detection of malignant melanoma in dermoscopy images. First, a pre-processing step was applied on each image for the purpose of removing artifacts such as air bubbles, hairs and lightening reflection. Second, the Particle Swarm Optimization (PSO) [13] and the Markov Random Field (MRF) [17] methods are integrated to segment the lesion areas. Therefore, the k-means was applied on segmented image (lesion), in order to separate each homogeneous set of pixels in one group (cluster). Consequently, several features are extracted from the given lesion based on existing clusters. Thus, by using a trained classifier, the lesion is classified in one of two classes of benign and melanoma.

The paper is organized as follows. Section 2 provides an overview of previous works. The proposed approach is explained in detail in Sect. 3. The evaluation of the detection system and the results obtained are discussed in Sect. 4. Finally, conclusions are presented in Sect. 5.

## 2    Previous Work

In recent years, there has been an increasing interest in early detection of skin cancer using CAD systems. Most of these diagnosis techniques are based on the ABCD rule to analyse four parameters (Asymmetry, Border, Color and Diameter) [4,6,10,11,20]. In addition, the 7-point checklist criterion is also widely used for the same purpose [3]. Barata et al. [8] proposed a new approach to extract pigment networks from dermoscopy images using a bank of direction filters and many morphological operations. Two distinctive properties: region pigment network intensity and geometry were used, and several features were extracted. Then, an Adaboost algorithm was used to classify the given region as either normal or abnormal. Also, a Bag-of-Features (BoF) model for the classification of melanoma in dermoscopy images was implemented by Barata et al. [7]. The authors used two different types of local descriptors: color and texture, and their performance was evaluated separately, and then compared in order to assess their ability to describe the different dermoscopic features. The same research group presented a new method to classify skin cancer images as either melanoma or non-melanoma [5]. Color features and texture features were used based on color histograms in three different colors (HSV, L*a*b*, and Opponent) and gradient related histogram. All used features were extracted globally and locally from each image. Therefore, the authors investigated the best way to combine the features by applying two strategies (early and late fusion). A Random Forests classifier yielded the best results. Celebi et al. [10] proposed a machine learning method for automated quantification of clinically significant colors in dermoscopy images. The K-means clustering approach was used to cluster each image with an optimal K value, which was estimated separately using five commonly used cluster validity criteria. Eltayef et al. [15] proposed an automated skin cancer diagnosis system on dermoscopy images using pigment network structures. Five features were extracted from the segmented image (pigment network) and used to feed the Artificial Neural Network as classification stage. An automatic framework for detection of melanoma from dysplastic nevi was proposed by Rastgoo et al. [21]. They combined several extracted features such as colour, shape, size and texture features with well-known texture features such as local binary pattern, grey-level co-occurrence matrix, histogram of gradients and the Gabor filter. Support Vector Machines (SVM), gradient boosting and random forest methods were used to evaluate the performance of their work. A new methodology for color identification in dermoscopy images was introduced by Barata et al. [6]. The authors used the Gaussian mixtures model to learn a statistical model for five colors (black, dark brown, light brown, blue-gray and white). Therefore, the learned mixtures were used to assess the colors of a larger set of images. Alfed et al. [2] proposed a new method for melanoma diagnosis. They used a bank of direction filters to segment pigment networks from images, then extracted an few features from the segmented image and used Artificial Neural Network (ANN) as classifier. The same group introduced a new method for improving a bag-of-words approach by combining color histogram features and first order moments with the Histogram of Oriented Gradients (HOG) [1]. Three classifiers methods were used in their work K-Nearest Neighbors (KNN), SVM and AdaBoost, where the SVM achieved the best results.

## 3   The Proposed Method

The proposed method for skin cancer detection is described and discussed in this section. As an initial step, the quality of the image is improved by detecting and removing several artifacts, such as air bubbles, lightening reflection and hairs.

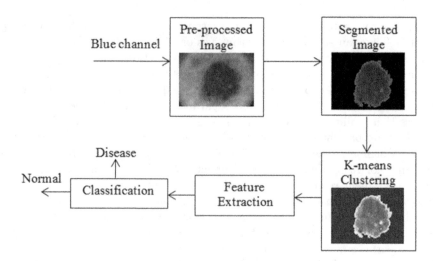

**Fig. 1.** Overview of the proposed approach for skin lesions classification.

Then, the skin lesion which is suspected to be a melanoma or non-melanoma is segmented from the surrounding healthy skin by applying PSO with MRF methods. Then, the k-means approach was applied on segmented image (lesion) to segment the lesion into sub region. Then, several features are extracted from each sub region. Finally, an Artificial Neural Network (ANN) classifier is trained to classify the skin lesion as either melanoma or normal skin. The overview of the proposed approach is illustrated in Fig. 1.

### 3.1   Pre-Processing of Dermoscopic Images

Usually, dermoscopic images do not have the expected quality to perform the diagnostic analysis. Thus, the step of image pre-processing is very important and necessary, in order to reduce the amount of artifacts and noise in the images. Skin lines, oil, air bubbles, lightening reflection and dark hair are present in almost every image. Therefore, to improve the quality of the images and prevent false positive detections, these kinds of noise must be detected and removed. According to [22], the blue channel of the color images is selected, as it has been experimentally shown to provide the best discrimination in most images.

**Reflection Detection.** In dermoscopy images, reflections appear as a result of presence of air bubbles caused by placing oil or jell before capturing the image. A sample thresholding method is implemented to classify each pixel as either a reflection artifact or background. The pixel can be classified as a reflection artifact if its intensity value is higher than threshold $T_{R1}$ as well as, if its intensity value minus the average intensity $I_{avg}(x, y)$ of its neighbours is higher than threshold $T_{R2}$, i.e.

$$\{I(x, y) > T_{R1}\} and \{(I(x, y) - I_{avg}(x, y)) > T_{R2}\} . \tag{1}$$

where $I$ is the image, $I_{avg}(x, y)$ is the average intensity value in surrounding neighbourhood with dimensions $11 \times 11$ and $T_{R1}$, $T_{R2}$ are set to 0.7 and 0.098 respectively.

**Hair Detection and Inpainting.** To improve the quality of the image and obtain correct diagnosis of melanoma skin cancer, the hair which is covering almost all lesions in dermoscopy images should be detected and removed, as a normal first stage of medical image processing. This step is required for effective segmentation and classification steps. To do so, the directional Gabor filters are implemented using a bank of 64 directional filters. Various parameters are used in the Gaussian filters at each phase. The images are filtered by each directional filter with various parameters. Therefore, the difference of Gaussians is carried out, followed by finding the local maximum at each image pixel location. Thus, the threshold approach is implemented to classify each pixel as either hair or background. Finally, the image inpainting method [12] is applied to fill in the gaps, which were occurred by multiplying the hair mask with the gray scale images. The details of the method can be found in [8,14,15].

## 3.2 Image Segmentation

The purpose of image segmentation is to separate the homogeneous lesions from the surrounding healthy skin. It is the most important phase for analysing images properly since it affects the subsequent steps accuracy. However, appropriate segmentation in dermoscopy images is a challenging task, because the lesions have large variations in size, shape and color, as well as, existence of low contrast between the lesions and surrounding healthy skin. In this work the image segmentation stage was implemented in two steps, the lesion area was segmented in the first step while the sub region inside the segmented lesion were segmented in the second.

**Skin Lesion Segmentation.** The automatic segmentation approach was implemented to extract the lesion area. The PSO and the MRF methods were combined, in order to minimize the energy function. The image segmentation is formulated as an optimization problem of the energy function with MRF theory. The POS method is used to perform the initial labeling based on the optimal threshold value, which was obtained by maximizing the fitness function. Then, an additional local search is performed for each segmented image by integrating it with MRF method. The purpose of that is to minimize the energy function or maximize the probability of pixel allocation to a cluster by using Maximum A Post Priority (MAP) [17]. More details of the approach can be seen in [16].

**Sub-Region Clustering.** Our aim is to divide the segmented lesion into a few clusters, with a more homogeneous distribution of pixels for each clusters. A simple K-means approach is used for this purpose. The outcomes (binary masks or segmented lesions) from the previous step are multiplied by gray scale images (bottom left in Fig. 2), in order to be able to separate the pixels whose located inside the lesion into several groups. The K-means clustering method is used since it is very simple and has low computational complexity. In addition, the number of clusters (k), usually could be determined easily. Hence the number of clusters has been experimentally obtained and set as $k = 5$.

The Euclidean distance is used to calculate the distance between the image pixels and the centroids of the clusters. Each single pixel was assigned to the appropriate cluster, based on its distance. Thus, the location of each cluster was updated and the pixels were re-assigned. This process continues until no more changes to cluster membership. The final result of this step is several homogeneous clusters, which can be used for the subsequent step of feature extraction. An example can be seen in Fig. 2.

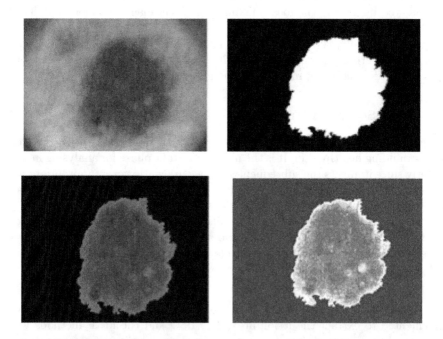

**Fig. 2.** Segmentation and clustering: Blue scale image (top left), segmented binary image (top right), segmented blue scale image, i.e. the lesion (bottom left) and result of clustering (bottom right).

### 3.3 Feature Extractions

The third phase of the proposed approach aims to extract several type of features from the segmented lesions. The density and the regular distribution of the blue color are the main properties, and can be used to identify the images. In this work, three color moments with color histogram are used as features, in order to determine the skin lesion type.

**Color Moments.** Color moments can be used to distinguish images based on their color distribution. Typically, probability distributions are characterized by the number of unique moments [1]. Consequently, they can be used as color features. The first color moment can be interpreted as the average color in each sub region inside the lesion, and can be calculated using the following equation:

$$E_i = \sum_{i=1}^{N} \frac{1}{N} Pi. \tag{2}$$

where $N$ is the total number of pixels inside the sub region and $Pi$ is the pixel value. The second color moment used as a feature is the standard deviation, which can be obtained by taking the square root of the variance of the color distribution.

$$\sigma_i = \sqrt{(\frac{1}{N} \sum_{i=1}^{N} (pi - E_i)^2)}. \tag{3}$$

where $E_i$ is the average value and $N$ is the total number of the pixels inside the sub region.

The third and last color moment used in our approach is the skewness, which means how asymmetric the color distribution is, and therefore, it gives useful information about the shape of the color distribution. Skewness can be calculated as:

$$S_i = \sqrt[3]{(\frac{1}{N} \sum_{i=1}^{N} (pi - E_i)^3)}. \tag{4}$$

**Color Histogram.** Color histogram is a way to represent the distribution of the composition of colors in images. It shows the number of pixels in each type of color in the image. The histogram associated with the blue color component $I_c, c \in \{3\}$ is given by:

$$h_c(i) = \frac{1}{N} \sum_{x,y} b_c(I_c(x,y))i = 1, ..., B_c. \tag{5}$$

where $N$ is the number of pixels inside the sub region, $i$ is histogram bin, $B_c$ in the number of bins and $b_c(.)$ is the characteristic function of $ith$

$$b_c(I_c(x,y)) = \left\{ \begin{array}{l} 1 \quad I_c(x,y) \quad \in \text{ ith color bin} \\ 0 \text{ otherwise} \end{array} \right\}. \tag{6}$$

The bins are defined by dividing the color component range into intervals with the same width. For all histograms, the number of bins is given by:

$$B_c \in \{5, 10, 15, 20\}, \tag{7}$$

and it was found that the best performance was achieved when the number of bins set as 10. Therefore, the elements of each sub region are sorted into 10 equally spaced bins between the minimum and maximum values of it.

The color moments and color histogram features are concatenated into one vector in order to represent the image. It is worth mentioning that all the extracted features are obtained from only the blue color moment, since it provides the best discriminatory performance. The same number of features was obtained from all images, three features related to the color moment and ten features from color histogram. These features are used as an image descriptor and the images are classified using machine learning techniques.

### 3.4    Skin Lesion Classification

The classification stage consists of interpretation and identification of skin lesion information based on the extracted features. Features from all images are stored in a database and fed to an Artificial Neural Network (ANN) as a classifier. The output of the classifier is (1) for melanoma and (0) for normal skin. Two-layer ANN (single hidden layer and output layer) are used with 100 neurons in the hidden layer. The Radial Basis Function (RBF) as an activation function given by [9] is used in the network. After training, the learned classifier can be used to classify new images as a melanoma or not.

## 4    Experimental Results

The proposed approach was evaluated on a public database PH2 [18] which provides 200 dermoscopy images. Each image in the database was classified by the radiologist as either normal or abnormal (the ground truth labels). The extracted features from all images were used as an input to the ANN classifier, with the ground truth labels as the output. The training and testing process was performed by using a 5-fold stratified cross-validation method. The images were split between five subsets, each one of them had approximately the same number of melanoma and non-melanoma. Several criteria have been used to evaluate the performance of the proposed approach.

To evaluate the performance of our method, we used 5 common evaluation criteria i.e. Sensitivity or True Positive Rate (TPR), Specificity or True Negative

**Table 1.** Results of lesion classification in dermoscopy images.

| Fold | SE (TPR) | SP (TNR) | AC | FNR | MSE |
|------|----------|----------|-----|-----|-----|
| 1 | 0.9874 | 0.9000 | 0.9698 | 0.01257 | 0.0470 |
| 2 | 1.0000 | 0.8750 | 0.9748 | 0.0000 | 0.0420 |
| 3 | 0.9748 | 0.8250 | 0.9447 | 0.0251 | 0.0592 |
| 4 | 0.9937 | 0.8250 | 0.9597 | 0.0062 | 0.0474 |
| 5 | 0.9811 | 0.8250 | 0.9497 | 0.0188 | 0.0514 |
| Mean | 98.74% | 85.00% | 95.97% | 0.0125% | 0.0494% |

Rate (TNR), Accuracy (AC), False Positive Rate (FPR), and Mean Square Error (MSE). The AC is defined as the sum of the true positives (images correctly classified as melanoma) and the true negatives (normal skin correctly identified as a non-melanoma), divided by the total number of images. The TPR is defined as the total number of true positives divided by the total number of images marked in the ground truth as melanoma. The TNR is defined as the total number of true negatives divided by the total number of images marked in the ground truth as normal skin. The FNR is calculated as the total number of false negatives divided by the number of images marked as non-melanoma in the ground truth image. It is worth mentioning that a perfect classification would have sensitivity (TPR) of 1 and FNR and MSE of 0.

The obtained results of our method can be seen in Table 1. It indicates that our approach by extracting several features from the sub region inside the segmented lesion achieved better results in terms of sensitivity, specificity and accuracy. In addition, the FPR and MSE are very desirable.

Quantitative comparison between various methods is difficult since different datasets and criteria have been used. However, we were able to evaluate the performance of our approach against Barata et al. [5,7,8], Eltayef et al. [15] and Alfed et al. [1] as they have the same objectives and they are based on the same database. Table 2 shows the comparison results.

To facilitate the performance comparison between our method and the alternate approaches, parameters such as sensitivity (SE), specificity (SP) and accuracy rates (AC) are computed for each method against the ground-truth. Table 2 shows the results of performance comparison. Out of the three criteria, the proposed method performed better than all other methods by SE and AC, but it is not as good as Eltayef [15] and Barata [5], in terms of SP. However, it is worth mentioning that the ground truth was used as the segmented images in both methods, while we did both segmentation and classification from the original images. Also, the work presented in [5] requires more computational time, because a number of image patches are used for feature extraction and this could be expensive when the patch size is large. In addition, the late fusion strategy was used in their work, which needs extra time to classify each type of features separately.

Table 2. Performance comparison with several methods.

| Method | SE | SP | AC |
|---|---|---|---|
| Barata [8] | 91.10% | 82.10% | 86.20% |
| Barata [7] | 93.00% | 85.00% | —— |
| Barata [5] | 98.00% | **90.00%** | —— |
| Alfed [1] | 91.00% | 85.00% | —— |
| Eltayef [15] | 92.30% | **95.00%** | 90.00 |
| Proposed method | **98.74%** | **85.00%** | **95.97%** |

## 5    Conclusions

In this paper, a comprehensive approach to melanoma detection in dermoscopy images was developed. The input images are first pre-processed by detecting and removing the noise. Then the lesions are segmented by applying PSO and MRF methods. As opposed to direct feature extraction from the segmented lesions, K-means is applied and the desired features, such as the color moments and color histogram, are extracted at the sub-region (cluster) level. These features are fed into an ANN with Radial Basis Function as an activation function for final melanoma classification. The proposed approach achieved approximately 96.0% accuracy, 99.00% sensitivity and 85.00% specificity on a dataset of 200 images. A comparison against several alternative methods shows that the proposed method achieved overall superior performance in terms of sensitivity and accuracy. Consequently, it has a great potential to detect melanoma in early stage and support the clinical diagnosis.

The main contributions of the work are as follows.

1. A comprehensive method including the whole process of image enhancing, segmentation of lesions and melanoma classification is developed.
2. A method for lesion segmentation is proposed by combining the PSO and MRF methods;
3. Feature extraction at the sub-region level is performed by separating the segmented lesions into homogeneous clusters.

## References

1. Alfed, N., Khelifi, F., Bouridane, A.: Improving a bag of words approach for skin cancer detection in dermoscopic images. In: 2016 International Conference on Control, Decision and Information Technologies (CoDIT), pp. 024–027. IEEE (2016)
2. Alfed, N., Khelifi, F., Bouridane, A., Seker, H.: Pigment network-based skin cancer detection. In: 2015 37th Annual International Conference of the IEEE Engineering in Medicine and Biology Society (EMBC), pp. 7214–7217. IEEE (2015)
3. Argenziano, G., Catricalà, C., Ardigo, M., Buccini, P., De Simone, P., Eibenschutz, L., Ferrari, A., Mariani, G., Silipo, V., Sperduti, I., et al.: Seven-point checklist of dermoscopy revisited. Br. J. Dermatol. **164**(4), 785–790 (2011)

4. Aribisala, B.S., Claridge, E.: A border irregularity measure using a modified conditional entropy method as a malignant melanoma predictor. In: Kamel, M., Campilho, A. (eds.) ICIAR 2005. LNCS, vol. 3656, pp. 914–921. Springer, Heidelberg (2005). doi:10.1007/11559573_111

5. Barata, C., Celebi, M.E., Marques, J.S.: Melanoma detection algorithm based on feature fusion. In: 2015 37th Annual International Conference of the IEEE Engineering in Medicine and Biology Society (EMBC), pp. 2653–2656. IEEE (2015)

6. Barata, C., Figueiredo, M.A., Celebi, M.E., Marques, J.S.: Color identification in dermoscopy images using gaussian mixture models. In: 2014 IEEE International Conference on Acoustics, Speech and Signal Processing (ICASSP), pp. 3611–3615. IEEE (2014)

7. Barata, C., Marques, J.S., Mendonça, T.: Bag-of-features classification model for the diagnose of melanoma in dermoscopy images using color and texture descriptors. In: Kamel, M., Campilho, A. (eds.) ICIAR 2013. LNCS, vol. 7950, pp. 547–555. Springer, Heidelberg (2013). doi:10.1007/978-3-642-39094-4_62

8. Barata, C., Marques, J.S., Rozeira, J.: A system for the detection of pigment network in dermoscopy images using directional filters. IEEE Trans. Biomed. Eng. **59**(10), 2744–2754 (2012)

9. Buhmann, M.D.: Radial Basis Functions: Theory and Implementations, vol. 12. Cambridge University Press, Cambridge (2003)

10. Celebi, M.E., Zornberg, A.: Automated quantification of clinically significant colors in dermoscopy images and its application to skin lesion classification. IEEE Syst. J. **8**(3), 980–984 (2014)

11. Clawson, K., Morrow, P., Scotney, B., McKenna, D., Dolan, O.: Computerised skin lesion surface analysis for pigment asymmetry quantification. In: 2007 International Machine Vision and Image Processing Conference, IMVIP 2007, pp. 75–82. IEEE (2007)

12. Criminisi, A., Perez, P., Toyama, K.: Object removal by exemplar-based inpainting. In: Proceedings of 2003 IEEE Computer Society Conference on Computer Vision and Pattern Recognition, vol. 2, p. II. IEEE (2003)

13. Eberhart, R., Kennedy, J.: A new optimizer using particle swarm theory. In: Proceedings of the Sixth International Symposium on Micro Machine and Human Science, MHS 1995, pp. 39–43. IEEE (1995)

14. Eltayef, K., Li, Y., Liu, X.: Detection of melanoma skin cancer in dermoscopy images. In: Journal of Physics: Conference Series, vol. 787, p. 012034. IOP Publishing (2017)

15. Eltayef, K., Li, Y., Liu, X.: Detection of pigment networks in dermoscopy images. In: Journal of Physics: Conference Series, vol. 787, p. 012033. IOP Publishing (2017)

16. Eltayef, K., Li, Y., Liu, X.: Lesion segmentation in dermoscopy images using particle swarm optimization and markov random field. In: IEEE International Symposium on Computer-Based Medical Systems (2017)

17. Geman, S., Geman, D.: Stochastic relaxation, gibbs distributions, and the bayesian restoration of images. IEEE Trans. Pattern Anal. Mach. Intell. **6**, 721–741 (1984)

18. Mendonça, T., Ferreira, P.M., Marques, J.S., Marcal, A.R., Rozeira, J.: Ph 2-a dermoscopic image database for research and benchmarking. In: 2013 35th Annual International Conference of the IEEE Engineering in Medicine and Biology Society (EMBC), pp. 5437–5440. IEEE (2013)

19. Oliveira, R.B., Papa, J.P., Pereira, A.S., Tavares, J.M.R.: Computational methods for pigmented skin lesion classification in images: review and future trends. Neural Comput. Applic., 1–24 (2016)

20. Pellacani, G., Grana, C., Seidenari, S.: Algorithmic reproduction of asymmetry and border cut-off parameters according to the abcd rule for dermoscopy. J. Eur. Acad. Dermatol. Venereol. **20**(10), 1214–1219 (2006)
21. Rastgoo, M., Garcia, R., Morel, O., Marzani, F.: Automatic differentiation of melanoma from dysplastic nevi. Comput. Med. Imaging Graph. **43**, 44–52 (2015)
22. Silveira, M., Nascimento, J.C., Marques, J.S., Marçal, A.R., Mendonça, T., Yamauchi, S., Maeda, J., Rozeira, J.: Comparison of segmentation methods for melanoma diagnosis in dermoscopy images. IEEE J. Sel. Top. Sign. Proces. **3**(1), 35–45 (2009)
23. Wighton, P., Lee, T.K., Lui, H., McLean, D.I., Atkins, M.S.: Generalizing common tasks in automated skin lesion diagnosis. IEEE Trans. Inf. Technol. Biomed. **15**(4), 622–629 (2011)
24. Zhou, H., Chen, M., Zou, L., Gass, R., Ferris, L., Drogowski, L., Rehg, J.M.: Spatially constrained segmentation of dermoscopy images. In: 5th IEEE International Symposium on Biomedical Imaging: From Nano to Macro, ISBI 2008, pp. 800–803. IEEE (2008)

# Bucket Selection: A Model-Independent Diverse Selection Strategy for Widening

Alexander Fillbrunn[1,2]([✉]), Leonard Wörteler[1], Michael Grossniklaus[1], and Michael R. Berthold[1,2,3]

[1] Department of Computer and Information Science, University of Konstanz, 78457 Konstanz, Germany
{alexander.fillbrunn,leonard.worteler,
michael.grossniklaus,michael.berthold}@uni-konstanz.de
[2] Konstanz Research School Chemical Biology (KoRS-CB), Konstanz, Germany
[3] KNIME AG, 8005 Zurich, Switzerland

**Abstract.** When using a greedy algorithm for finding a model, as is the case in many data mining algorithms, there is a risk of getting caught in local extrema, *i.e.*, suboptimal solutions. Widening is a technique for enhancing greedy algorithms by using parallel resources to broaden the search in the model space. The most important component of widening is the selector, a function that chooses the next models to refine. This selector ideally enforces diversity within the selected set of models in order to ensure that parallel workers explore sufficiently different parts of the model space and do not end up mimicking a simple beam search. Previous publications have shown that this works well for problems with a suitable distance measure for the models, but if no such measure is available, applying widening is challenging. In addition these approaches require extensive, sequential computations for diverse subset selection, making the entire process much slower than the original greedy algorithm. In this paper we propose the *bucket selector*, a model-independent randomized selection strategy. We find that (a) the bucket selector is a lot faster and not significantly worse when a diversity measure exists and (b) it performs better than existing selection strategies in cases without a diversity measure.

## 1 Introduction

An important component of many algorithms that need to traverse a large model space is greedy search. Data mining algorithms such as hierarchical clustering, rule set induction, and decision tree learning all rely on its ability to find good enough solution states quickly. In this setting, the increasingly available parallel computing resources can be leveraged in two ways. First, proposals have been made to use these resources to improve the *runtime performance* of greedy search algorithms, *i.e.*, to find the same results faster [2,6]. A second use of parallel computing resources that has been studied is to improve the *task-based performance* of greedy algorithms by addressing their shortcomings, namely their tendency to

© Springer International Publishing AG 2017
N. Adams et al. (Eds.): IDA 2017, LNCS 10584, pp. 87–98, 2017.
DOI: 10.1007/978-3-319-68765-0_8

get caught in local extrema. These techniques propose to search the model space in parallel, maintaining and improving multiple diverse (possibly incomplete) solutions until a better final model is found.

In this paper, we focus on this second class of greedy search algorithms and, in particular, on the technique of *widening* [1], which aims to maintain diverse models with little or, ideally, no communication between the parallel workers. In order to prevent workers from pursuing similar solutions because they show promise at first, widening relies on a distance measure and a selection strategy that together enforce diversity. However, deciding which distance measure and selection strategy to use for a given problem can be challenging itself.

On the one hand, designing a measure that selects good and diverse models from a large model space is a multi-objective optimization problem, which, in general, is difficult to solve. Nevertheless, widening has shown very promising results for problems that have a suitable model-dependent distance measure. Examples of such problems include the set cover problem [9], KRIMP [12], Bayesian networks [13], and hierarchical agglomerative clustering [7].

On the other hand, the state-of-the-art selection strategies either require a distance matrix for the models or the models to be sorted according to their quality. These preconditions put severe limits on the independence of individual parallel workers as they have to exchange their models frequently, resulting in a large increase in runtime over the sequential greedy algorithm.

In order to address these two challenges, we propose a model-independent diverse selection strategy for widening, the so-called *bucket selector*. We argue that our selection strategy is a powerful technique to enable widening for problems that have no straightforward distance measure or that require more independence of workers than state-of-the-art selection strategies can provide. As an example of such a greedy search problem, we use the set covering problem as a demonstrator and the ordering of joins during query optimization in relational database systems [14].

In summary, this paper makes the following contributions:

- We propose the bucket selector, a novel selection strategy that uses randomized partitioning in order to facilitate exploration of the model space while still exploiting promising areas (Sect. 4).
- In our evaluation, we demonstrate that (a) the performance of the bucket selector is not significantly worse for problems that have a well-known model-dependent distance measure, (b) it outperforms existing selection strategies for problems without such a measure and (c) in both cases, it finds results faster than other strategies (Sect. 5).

We begin in Sects. 2 and 3 by giving an overview of related work and an introduction to widening, respectively. Concluding remarks are given in Sect. 6.

## 2   Related Work

Work focusing on the diversity of models has been done mostly in the areas of ensemble learning and genetic algorithms, but interesting approaches can also be found in satisficing planning, where the objective is to find a *good enough* solution for a given problem.

A very popular and powerful method of utilizing parallel resources is the learning of multiple predictors concurrently, *i.e.*, ensemble learning, resulting in a group of models that leverages the *wisdom of the crowd* by combining the individual models' predictions, *e.g.*, by (weighted) majority vote. *Bagging* [4] is a method that is often used to ensure that the learners specialize in different areas of the feature space. In ensemble learning, promoting diversity among the individual members is an important part of building the joint classifier, as multiple very similar predictors would not contribute any additional knowledge. In general the diversity among members of an classifier ensemble is perceived to be high if the predictors' outputs differ. This makes sense for an ensemble, as a single model does not necessarily need to be good for every part of the input space, but overfit and focus on a fraction of it instead. When the goal is to select one generally good model in the end, this notion of diversity does not suffice.

Genetic algorithms provide a more fitting approach here, as they usually maintain a population, but the goal is to find one best solution. In order to avoid premature convergence of the population the selection of individuals to take over into the next iteration often consists of the fittest and some less fit individuals. Additionally, techniques such as fitness sharing [8] can be employed to enforce exploration of the model space.

Other approaches presented in previous widening publications use various distance metrics to enforce diversity. For set covering Ivanova *et al.* [9] use the Jaccard coefficient, Sampson *et al.* use the Frobenius Norm of the difference of the graphs' Laplacians to compare Bayesian networks [12] and an optimization based on $p$-dispersion-min-sum for KRIMP [13]. Fillbrunn *et al.* [7] compare incomplete hierarchical clustering trees by using the Robinson-Foulds metric. Most selection strategies presented in those publications are either computationally too expensive to be feasible for use in greedy algorithms due to them having to sort the models or build distance matrices, or they require extensive communication between workers.

## 3   Widening

Widening is an improvement for greedy algorithms that learn a model by iteratively refining it. Formally, such a sequential algorithm performs the following update on the model $m \in \mathcal{M}$ in each iteration:

$$m' = s(r(m)). \tag{1}$$

Here, $r$ is the refinement function that builds all possible successors for a model and $s$ is a selection function that selects a single model, usually based on some heuristic $\psi : \mathcal{M} \mapsto \mathbb{R}$ for the model's quality. The update rule is applied either until the model does not change anymore, cannot be refined further, or is deemed good enough. For a widened and parallel algorithm the update rule is extended for multiple models and the selection function is split into a thread/worker local and a global part:

$$\{m'_1, \ldots, m'_k\} = s_{global} \left( \bigcup_{i=1}^{k} s_{local}(r(m_i)) \right), \qquad (2)$$

where $k$ is the number of models held by parallel workers and refined individually. The selection function is applied locally to each worker and then globally to the union of these selected refinements to produce the next $k$ models. We call $k$ the *width* of the algorithm. Instead of a greedy search through the model space, this approach resembles a beam search, enabling broader exploration of the model space while still exploiting promising areas thoroughly. For $k = 1$ a widened algorithm behaves greedily and if $k$ is large enough the model space is searched exhaustively. If the selection operator simply selects models based on $\psi$, the solutions found by the individual workers are likely to either converge or be very similar, leading to inefficient use of compute resources and a lack of exploration. We call this simple and fast approach *top-k*. Widening aims to counteract this behavior by making the workers choose sufficiently different models $m'_1, \ldots, m'_k$. But at the same time, the models must be good according to the heuristic, so that the workers do not optimize solely based on diversity. The solution to this multi-objective optimization problem is the responsibility of the selection function $s$.

Ivanova *et al.* [9] introduced the diverse top-$k$ selector which selects the best $k$ models that are also sufficiently diverse from one another. While the top-$k$ selection strategy does not enforce diversity at all, diverse-top-$k$ cannot be run as efficiently because in each operation the models have to be sorted by quality. Additionally, it relies on a model-specific distance measure that is not always easy to find.

In the next section, we present the bucket selector, a randomized selection strategy that does not require a distance function, is easily parallelizable with little communication between the workers, and shows promising results in our initial performance study.

## 4   The Bucket Selector

The bucket selector enhances diversity by selecting not just the best models, but by partitioning the models into buckets and selecting the best model from each bucket. In which bucket a model is placed is determined by a hash function $h : \mathcal{M} \mapsto [0, k)$. The $i$th model selected by $s_{local}$ and $s_{global}$ is therefore:

$$m'_i = \max_{m:h(m)=i} \psi(m).$$

Because the same hash function is used locally and globally, $s_{local} = s_{global}$ as is also the case for the top-$k$ selector.

The choice for the hash function is obviously problem-dependent. In addition to the usual properties of hash functions (even distribution of hash values, same hash value for identical models) it would also be desirable to guarantee an even distribution of hash values *in each step* of the algorithm and possibly also equal hash values for *semantically* equivalent models. We plan to investigate these issues in subsequent work.

Using the bucket selector, we achieve diversity by using the hash function, while also ensuring that the same model is not selected twice. By fixing the bucket number for which a worker is responsible, we can further limit the necessary inter-worker communication to the transfer of the best model in a bucket to the respective worker node, avoiding the central collection of models altogether. For $k = 3$ this means that Worker 1 sends the best model in Bucket 2 to Worker 2 and the best model in Bucket 3 to Worker 3. Workers 2 and 3 do the same for Buckets 1 and 3 or 1 and 2, respectively. As a consequence, $k \cdot (k - 1)$ models have to be transferred in each iteration.

Given $k$ workers and $n$ models, the time complexity of the hash bucket selector is $\mathcal{O}(n + k^2)$: $n$ for finding the best model in each bucket locally and $k^2$ because each worker gets all models in its bucket and must find the best one.

Partitioning the models based on their hash value ensures that the same models produced by different workers still end up in the same bucket and the selection step is deterministic. A downside of the hashing is the fact that the models may not be evenly distributed across buckets or buckets may even stay empty if $k$ is large and the number of refinements is small. To ensure that approximately the same number of models competes for the top spot in each bucket, we can randomly shuffle the model list and assign a model $m_j$ at index $j$ to a bucket using the function $h \mapsto j \bmod k$. In contrast to hash partitioning the shuffle approach is sensitive to duplicate models: a good model that is sorted into multiple buckets may also be the best model in all of them and therefore reduce the diversity of the selected models. Adding explicit deduplication increases the time complexity to $\mathcal{O}(nk + k^2)$ because for each model that is put at the top spot of a bucket, the top spots of all other buckets have to be checked for a duplicate.

If determinism is important, it is also possible to hash the models into $l > k$ buckets and iteratively merge the smallest buckets until $k$ buckets are left. The complexity for this approach is $\mathcal{O}(n(l - k) + k^2)$ and its exploration behavior compared to that of top-$k$ is shown in Fig. 1. It should be noted that if the number of refinements is much larger than $k$, the probability of empty buckets is very small and bucket merging is not necessary.

## 5    Evaluation

In this section, we present the results of an initial study that we carried out by applying the bucket selector to the set cover problem and the join ordering problem for query optimization in relational database systems. The set cover problem serves as a baseline to evaluate the bucket selector on a problem with a known good distance measure, whereas the join ordering problem is used as an

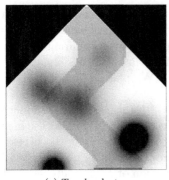    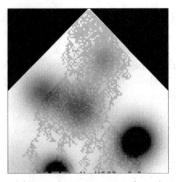

(a) Top-$k$ selector                    (b) Hash bucket selector ($l$=60)

**Fig. 1.** Exploration behavior of the top-$k$ selector and the hash bucket selector with bucket merging. The space is searched from top to bottom. One pixel denotes one model that is refined by moving to one of the 3 pixels below it. The darker a pixel, the lower is the model's cost.

example of a problem that has no straightforward distance measure. For both of these problems, we analyze and compare different selectors w.r.t. the quality of the solutions they find and their runtime performance with different grades of parallelization.

### 5.1   The Set Cover Problem

The set cover problem is an NP-hard problem from combinatorics [11, p. 414]. Given a collection of sets $\mathcal{S}$ whose union spans a universe $\mathcal{U}$, the goal is to find the smallest subset $\mathcal{C} \subseteq \mathcal{S}$ that still spans the whole universe.

An iterative algorithm for solving the problem greedily has been given by Johnson *et al.* [10]. The algorithm starts with an empty temporary cover $\hat{\mathcal{C}}$ and iteratively adds the set from $\mathcal{S}$ that covers the largest number of currently uncovered elements of $\mathcal{U}$. Given a model with temporary cover $\hat{\mathcal{C}}$ and the set of sets that are not yet taken $\mathcal{S}' = \{s | \forall s : s \in \mathcal{S}, s \notin \hat{\mathcal{C}}\}$, we can refine it and find set $\mathcal{R}$ of all possible new temporary covers with:

$$\mathcal{R} = \bigcup_{t \in \mathcal{S}'} \hat{\mathcal{C}} \cup t .$$

From this set we can then select the $k$ covers that cover the largest part of $\mathcal{U}$ for the next iteration. For a heuristic estimation of a cover's quality we can therefore simply use the size of the union of the sets contained within it, *i.e.* the size of the partial cover.

We compute the hash values of our models based on the bit set containing the indices of the sets contained in the cover. In our implementation we make use of the `java.util.BitSet#hashCode()` method.

## 5.2   Set Cover Results

We evaluated the set covering performance of the bucket selector on three bench-mark data sets: *rail507*, *rail516*, and *rail582* [3], all of which are from a real-life train-crew scheduling problem. We ran each experiment 200 times, shuffling the order in which we presented the data to the algorithm each time. When models are sorted by quality, ties between two models are broken based on the smallest set-index that is contained in the cover[1]. For the diverse top-$k$ selector we used a threshold based on the *jaccard index* as described by Ivanova *et al.* [9].

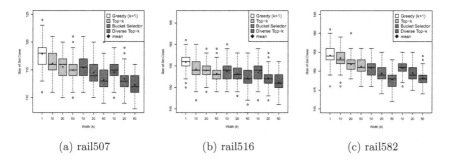

|  (a) rail507 | (b) rail516 | (c) rail582 |

**Fig. 2.** Results of solving the set cover problem with different selectors, data sets and $k$. The bucket selector uses a random partitioning.

As can be seen in Fig. 2, increasing the number of models maintained in parallel has a positive effect on the final size of the found set cover. Diverse Top-$k$ with $k = 10$ even achieves a slightly better mean set cover size than Top-$k$ with $k = 50$. We also see that the bucket selector performs better than the top-$k$ algorithm in all datasets, but worse than the diverse top-$k$ approach for *rail507* and *rail516*. While the improvements the bucket selector brings over Top-$k$ seem small, it has to be noted that this comes at no additional computational cost, as will be shown in Sect. 5.5.

## 5.3   The Join Ordering Problem

In contrast to the set cover problem, the join ordering problem for query opti-mization in a relational database system does not have a representation with a straightforward distance measure. We therefore demonstrate for this problem how join queries can be optimized using widening and the bucket selector.

The join ordering problem has the form $\langle \mathcal{T}, \mathcal{P} \rangle$, where $\mathcal{T}$ are the tables to be joined and $\mathcal{P}$ are predicates of the form $\langle A, B, s \rangle$, connecting tables $A$ and $B$ with

---

[1] Due to the nature of the set cover problem and the chosen heuristic, models with the same score occur frequently and other tie-breaking methods may be feasible. This is, however, out of scope of this work.

selectivity $s$. The selectivity denotes the fraction of rows retained from the cross-product $A \times B$, so that the resulting table has the cardinality $|A \bowtie B| = |A| \cdot |B| \cdot s$.

In order to find an efficient execution plan for a database query comprising multiple joins, the query optimizer needs to enumerate all orders in which the elements of $\mathcal{T}$ can be joined together into one join tree, *i.e.* all binary trees where the leaves are elements of $\mathcal{T}$ and the inner nodes are joins with zero or more predicates from $\mathcal{P}$. We disregard commutations (*i.e.*, $A \bowtie B \equiv B \bowtie A$), which reduces the number of candidates to inspect. For $N$ tables there are $1 \cdot 3 \cdot 5 \cdot \ldots \cdot (2N - 3) = (2N - 3)!! \in \Omega\left(4^N/N^{3/2}\right)$ distinct trees. For each of these join orders, the query optimizer estimates the cost of the corresponding query execution plan based on a cost model that represents CPU and I/O overhead. For a large number of tables ($N > 15$) the exhaustive enumeration of all possible table combinations is infeasible and query optimizers typically need to restrict the search space or resort to greedy or genetic algorithms to find good plans. As we focus exclusively on the ordering of joins in this paper, the sizes of the original and intermediate tables is the main factor that influences the overall cost of a query execution plan.

In this section, we show that widening with the bucket selector performs favorably compared to the top-$k$ selector and the simple greedy algorithm, suggesting that diversity is an important factor in the search for good join orders.

In the widening framework a model for a join optimization problem has the form $\langle \mathcal{R}, \mathcal{P} \rangle$ where $\mathcal{R}$ are relations (tables or joins) and $\mathcal{P}$ are the remaining predicates. In the initial model each element of $\mathcal{R}$ is a table ($\mathcal{R} = \mathcal{T}$). To refine an unfinished model $m = \langle \mathcal{R}, \mathcal{P} \rangle$ we take each predicate $P = \langle R_1, R_2, s \rangle$ with $R_1, R_2 \in \mathcal{R}$ from $\mathcal{P}$ and create a new model $m' = \langle R_1 \bowtie R_2 \cup \mathcal{R} \setminus R_1, R_2, \mathcal{P} \setminus P \rangle$. The cost of $m'$ is then: $cost(m) + |R_1| \cdot |R_2| \cdot s$.

In the selection step we can use this formula for our quality heuristic and select the $k$ models with the lowest cost estimation, *e.g.*, $\psi(m) = \frac{1}{cost(m)}$.

The hash function for join trees is computed by first assigning a unique random number to every table and calculating hash values of inner nodes in a bottom-up fashion using the formula $h(a \bowtie b) = 31 \cdot (31 \cdot j(a \bowtie b) + h(a)) + h(b)$, where $j(\cdot)$ returns the number of joins in a join (sub-)tree. The hash value of the whole tree is that of its root node.

## 5.4   Join Ordering Results

We evaluate our bucket selector on the join optimization problem with randomly generated tables and join predicates. For this evaluation, we use a random bucket selector that first puts one random model in each bucket and then assigns random buckets to the remaining models, *i.e.*, each bucket always has at least one model, given the number of models is greater or equal $k$. The hash bucket selector uses a hash code based on the join tree structure.

We limit our evaluation to joins in a snowflake schema, a topology that is common in data warehouses. A snowflake schema consists of a large *fact table*, which is connected to multiple *dimension tables*, which in turn have connections to further, smaller tables. We selected this use case because it is known to be challenging: in a snowflake schema, optimal plans often involve cross-products of the smaller tables, which are typically not enumerated by exact optimizers to reduce the search space.

In our experiments the fact table has a cardinality between 500,000 and 1,000,000 rows, the dimension tables have a cardinality between 500 and 1000 rows and the outer tables have a cardinality between 10 and 100 rows. The predicates for the joins between the fact and the dimension tables have a selectivity between 0.0001 and 0.0005 and the dimension tables are connected to the outer tables by predicates with a selectivity between 0.01 and 0.05. Each experiment is repeated 1000 times with random predicate selectivities and cardinalities within the given bounds. We also allow cross products, which are treated as joins with a predicate that has a selectivity of 1.0.

For a total of 16 tables arranged in a snowflake topology (1 fact table, 3 dimension tables and 12 outer tables) we can compare our results to the optimal plan. As can be seen in Fig. 3a both the random bucket and hash bucket selector outperform the top-$k$ selector in terms of plan costs. We also see that as $k$ increases the bucket approaches show greater improvement: while the top-$k$ selector's mean plan costs improve from 205% at $k = 1$ to 196% at $k = 100$, the random bucket selector improves from 165.3% to 139% and the hash bucket selector from 163.8% to 137.5%.

For a larger number of tables the calculation of the optimal plan becomes infeasible. In order to evaluate the bucket selector on such an example, we follow a similar approach as described by Bruno et al. [5] and compare the plans found by the bucket selector to the plans found by the top-$k$ selector. Figure 3b depicts the plan costs of the bucket selector relative to the top-$k$ selector. Both the random and hash bucket selector perform better, with a mean improvement of about 10%. The higher the width $k$ of the search is, the larger is the benefit the bucket selector has. With $k = 10$ the hash bucket selector produces on average plans with 89.9% of the top-$k$ selector's cost and with $k = 100$ this cost sinks to 83.9%.

In both experiments, we see that the hash bucket selector produces slightly better plans than the random bucket selector. This is due to the imperfect handling of duplicate models as explained in Sect. 4: a model that occurs multiple times in the refinements can be the top of multiple buckets if they are assigned randomly, thus resulting in less than $k$ selected models. As we can see in the next section, the random bucket selector in return offers a slightly better runtime without explicit deduplication.

## 5.5    Notes on Runtime

Apart from solution quality, the runtime of the search algorithm plays just as important a role. This especially concerns join plan generation, where the latency with which a database query optimizer can return a result to the user is the most crucial factor. In the following section, we evaluate the time the optimization algorithm needs to find a solution for different selectors. All benchmarks are

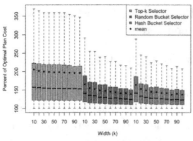
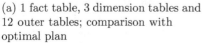
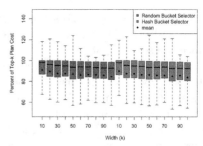

(a) 1 fact table, 3 dimension tables and 12 outer tables; comparison with optimal plan

(b) 1 fact table, 5 dimension tables and 25 outer tables; comparison with Top-$k$

**Fig. 3.** Plan costs achieved with different selectors on join order optimization with 16 (a) and 31 tables (b).

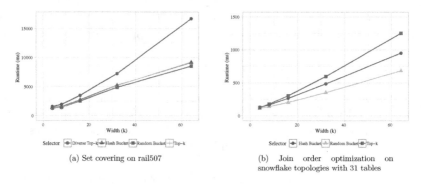

(a) Set covering on rail507

(b) Join order optimization on snowflake topologies with 31 tables

**Fig. 4.** Runtime for various values of $k$ in set covering and join order optimization.

performed on a machine with 64 GB of main memory and an Intel Core i7-5930K CPU with 6 physical and 6 virtual cores. For our experiments, we use the Java Microbenchmarking Harness (JMH)[2] to minimize the influence of Java's Just-in-Time compiler on the measurements. We run each experiment 200 times in batches of 20 iterations, creating a new virtual machine and performing five warm-up iterations for every batch.

For the set covering problem we run experiments with the *rail507* data set and compare the top-$k$, diverse top-$k$, hash bucket, and random bucket selectors. In Fig. 4a we see that diverse top-$k$ is considerably slower than the other selectors and that the random bucket selector slightly outperforms the hash bucket and the top-$k$ selectors in terms of runtime. That the hash bucket selector is as fast as the top-$k$ selector, despite theoretically better runtime complexity, is likely a result of the rather time consuming calculation of the hash function.

Our runtime evaluation of the join order optimization problem is conducted on randomly generated snowflake topologies with 1 fact table, 5 dimension tables

---

[2] http://openjdk.java.net/projects/code-tools/jmh/ (1/26/2017).

and 25 outer tables. We compare the top-$k$ selector and the hash and random bucket selectors. As depicted in Fig. 4b, both bucket selection strategies are faster than the top-$k$ approach, the random bucket selector finding a solution almost twice as fast for $k = 64$.

## 6 Conclusion

We have proposed the bucket selector—enforcing diversity with a random or hash-based partitioning of intermediate models—as a favorable alternative to simple top-$k$ selection, providing better results in terms of model quality and runtime. We also found that bucket selection cannot beat the solution quality of the diverse top-$k$ selection strategy in the set covering problem, where a good distance measure for the models exists. We thus see the bucket selector as a middle ground, being better than simple top-$k$ and providing the benefit of model independence and improved runtime over diverse top-$k$ selection.

As an example for a problem without a known good model distance measure, we presented our results for the heuristic optimization of query plans for snowflake joins. This problem seems to benefit from diversity enhancement during search, as we can quickly generate query plans that have on average 83.9% of the costs of the plans found by the top-$k$ approach. Our results motivate further work in the area of join order optimization concerning tests with other table topologies and the implementation of parallel query optimizers.

**Acknowledgements.** This work was partially funded by BMBF (grant 031A535C) and the Konstanz Research School Chemical Biology.

## References

1. Akbar, Z., Ivanova, V.N., Berthold, M.R.: Parallel data mining revisited. Better, not faster. In: Hollmén, J., Klawonn, F., Tucker, A. (eds.) IDA 2012. LNCS, vol. 7619, pp. 23–34. Springer, Heidelberg (2012). doi:10.1007/978-3-642-34156-4_4
2. Amado, N., Gama, J., Silva, F.: Parallel implementation of decision tree learning algorithms. In: Brazdil, P., Jorge, A. (eds.) EPIA 2001. LNCS, vol. 2258, pp. 6–13. Springer, Heidelberg (2001). doi:10.1007/3-540-45329-6_4
3. Beasley, J.E.: OR-Library: distributing test problems by electronic mail. J. Opl. Res. Soc. **41**(11), 1069–1072 (1990)
4. Breiman, L.: Bagging predictors. Mach. Learn. **24**(2), 123–140 (1996)
5. Bruno, N., Galindo-Legaria, C.A., Joshi, M.: Polynomial heuristics for query optimization. In: Proceedings of International Conference on Data Engineering (ICDE), pp. 589–600 (2010)
6. Zhihua, D., Lin, F.: A novel parallelization approach for hierarchical clustering. Parallel Comput. **31**(5), 523–527 (2005)
7. Fillbrunn, A., Berthold, M.R.: Diversity-driven widening of hierarchical agglomerative clustering. In: Fromont, E., De Bie, T., van Leeuwen, M. (eds.) IDA 2015. LNCS, vol. 9385, pp. 84–94. Springer, Cham (2015). doi:10.1007/978-3-319-24465-5_8

8. Goldberg, D.E., Richardson, J.T.: Genetic algorithms with sharing for multimodal function optimization. In: Proceedings of International Conference on Genetic Algorithms (ICGA), pp. 41–49 (1987)
9. Ivanova, V.N., Berthold, M.R.: Diversity-driven widening. In: Tucker, A., Höppner, F., Siebes, A., Swift, S. (eds.) IDA 2013. LNCS, vol. 8207, pp. 223–236. Springer, Heidelberg (2013). doi:10.1007/978-3-642-41398-8_20
10. Johnson, D.S.: Approximation algorithms for combinatorial problems. J. Comput. Syst. Sci. **9**(3), 256–278 (1974)
11. Korte, B., Vygen, J.: Combinatorial Optimization. Algorithms and Combinatorics. Springer, Heidelberg (2013)
12. Sampson, O., Berthold, M.R., Widened, K.: Better performance through diverse parallelism. In: Proceedings of International Symposium on Intelligent Data Analysis (IDA), pp. 276–285 (2014)
13. Sampson, O.R., Berthold, M.R.: Widened learning of Bayesian network classifiers. In: Boström, H., Knobbe, A., Soares, C., Papapetrou, P. (eds.) IDA 2016. LNCS, vol. 9897, pp. 215–225. Springer, Cham (2016). doi:10.1007/978-3-319-46349-0_19
14. Selinger, P.G., Astrahan, M.M., Chamberlin, D.D., Lorie, R.A., Price, T.G.: Access path selection in a relational database management system. In: Proceedings of International Conference on Management of Data (SIGMOD), pp. 23–34 (1979)

# Interactive Pattern Sampling
# for Characterizing Unlabeled Data

Arnaud Giacometti and Arnaud Soulet[(✉)]

Université François Rabelais Tours, LI EA 6300, Tours, France
{arnaud.giacometti,arnaud.soulet}@univ-tours.fr

**Abstract.** Many data exploration tasks require a target class. Unfortunately, the data is not always labeled with respect to this desired class. Rather than using unsupervised methods or a labeling pre-processing, this paper proposes an interactive system that discovers this target class and characterizes it at the same time. More precisely, we introduce a new interactive pattern mining method that learns which part of the dataset is really interesting for the user. By integrating user feedback about patterns, our method aims at sampling patterns with a probability proportional to their frequency in the interesting transactions. We demonstrate that it accurately identifies the target class if user feedback is consistent. Experiments also show this method has a good true and false positive rate enabling to present relevant patterns to the user.

**Keywords:** Pattern mining · Pattern sampling · Unlabeled data

## 1 Introduction

Many data exploration tasks are intended to characterize a part of data over another [9]. For instance, it is particularly the case to identify factors of a disease by comparing the data of ill patients to those of healthy ones, or to find fraudulent behaviors by comparing the data of scammers to others. Unfortunately, in practice, the collected data have not always the labels allowing to know what class an individual (healthy or sick) or a behavior (normal or fraudulent) belongs to. Of course, when the class label to characterize is absent, it is possible to use unsupervised analysis techniques (such as clustering, association rules or detection of outliers) to identify and characterize the target class. However, these techniques are often less effective because they focus on the majority trends taking into account all the data. To address this problem, an approach would consist in labeling data during the data preparation phase. Such a labeling process could be facilitated by an active learning method that can even be dedicated to an analysis approach [13]. However, labeling remains a particularly costly and tedious task, especially when the target class to study is really in minority. Furthermore, in many cases, the labeling can be difficult because experts have only an imperfect knowledge of the target class. Actually, this is another reason for the experts want the use of data mining tools. In other words,

© Springer International Publishing AG 2017
N. Adams et al. (Eds.): IDA 2017, LNCS 10584, pp. 99–111, 2017.
DOI: 10.1007/978-3-319-68765-0_9

we are facing a vicious circle: data analysis requires labeling which itself requires an analysis of data. Thus, the problem is how to label data to identify a target class while characterizing this target class with patterns.

In order to solve this problem, we propose to use the interactive pattern mining framework introduced in [7]. The central idea of this framework is to alternate between three steps. During the *mining step*, our system mines an initial batch of patterns using an adaptation of the two-step random procedure proposed in [3]. During the *interactive step*, the user provides feedback by evaluating whether the patterns of the batch are good descriptors or not of the target class. Then, during the *learning step*, the system updates a model of the target class using the user feedback. Thus, after each interaction with the user, we have a twofold challenge to overcome: (i) How can we update the model of the target class integrating the user feedback? and (ii) How can we draw patterns from the dataset taking into account the updated model of the target class?

In this paper, we propose a new interactive pattern sampling method to solve these two challenges at the same time. The outline of this paper is as follows. Section 2 reviews some work about active learning and interactive pattern mining. We state precisely our problem in Sect. 3. Our algorithm proposal is detailed in Sect. 4 where theoretical properties are presented (due to lack of space, proofs are omitted). Indeed, if the user feedback is consistent with the target class, we demonstrate that the transactions of the target class will be clearly identified and that the mined patterns will describe exactly these transactions. Finally, experiments in Sect. 5 show that the accuracy of the interactive system increases fairly quickly with the number of user feedback responses.

## 2   Related Work

To the best of our knowledge, there is no work on the mining of patterns characterizing a target class not known in advance. However, we benefit from the framework of interactive pattern mining [7]. Its primary goal is to present interesting patterns to the user. Even if user feedback is used for labeling data, this problem therefore differs from traditional active learning problems [12], the purpose of which is not to propose interesting queries to the user. This distinction is important for different reasons. First, the queries provided to the user are patterns, not transactions. In most active learning tasks, the feedback requested from the user is directly related to the objects to be classified and not a generalization of these objects (although there are few notable exceptions [1,10]). Second, the selection of the query presented to the user cannot only target the improvement of the classification model unlike conventional active learning. In order that the user continues to interact with the system, the latter has to mine patterns that are interesting for him/her (i.e., that describe the target class). Third, the query presented to the user at each iteration has to be computed in few seconds to maintain a satisfactory interaction. In traditional active learning, this constraint is not very strong because the query is selected from the dataset. It is much more difficult to mine the right pattern due to the huge search space.

In interactive pattern mining, one challenge is to select the relevant patterns while improving the learned model. In case of preference learning, the early methods [11,14] ignored the use of a criterion favoring the diversity of queries for acquiring a complete view of preferences. A recent approach [5] nevertheless showed interest to address this issue as done in active learning. It also showed the importance of randomization to promote good diversity. This randomization need justifies the use of pattern sampling [2]. In this paper, we also take advantage of the statistical properties of sampling to better learn the classification model and to better choose the query (mined patterns) as done in [6].

Another challenge is to mine new patterns at each iteration in few seconds to maintain a satisfactory interaction. This speed requirement is not satisfied by traditional methods of pattern mining. Thus, the first proposals [11,14] were based on a preliminary mining step and then, they re-ranked this preliminary collection of patterns according to the updated criterion stemming from the user preference model. This post-processing approach did not allow the discovery of new patterns. More recently, a beam search method [5] was proposed to extract at each iteration the new patterns that maximize the updated criterion (combining quality and diversity, in that case). Such an approach remains slow and it fails to find various patterns. In this context, pattern sampling [2] is an attractive technique because it gives a fast access to all the patterns, guaranteeing a very good diversity. In this paper, rather than using a stochastic method [2] or a SAT framework [4], we adopt the two-step procedure [3] that is linear with the database size.

## 3   Problem Statement

This section formulates the problem of characterizing a class from an unlabeled dataset, using pattern sampling and an interactive approach. Before, we remind basic definitions about pattern mining and we introduce the notion of oracle.

### 3.1   Basic Definitions

Let $\mathcal{I}$ be a set of distinct literals called items, an itemset (or a pattern) is a subset of $\mathcal{I}$ and the language of itemsets $\mathcal{L}$ is $2^{\mathcal{I}}$ (where $2^S$ denotes the powerset of $S$). A transactional dataset $\mathcal{D}$ is a multi-set of itemsets of $\mathcal{L}$. Each itemset, usually called transaction, is a data observation. For instance, Table 1 gives a transactional dataset with 4 transactions $t_1, \ldots, t_4$ described by 5 items $A$, $B$, $C$, $D$ and $E$. $\Delta$ denotes the set of all datasets.

Pattern discovery takes advantage of interestingness measures to evaluate the relevancy of a pattern. More precisely, an interestingness measure for a pattern language $\mathcal{L}$ is a function defined from $\mathcal{L} \times \Delta$ to $\Re$. For instance, the support of an itemset $X$ in a dataset $\mathcal{D}$, denoted $supp(X, \mathcal{D})$, is the proportion of transactions containing $X$: $supp(X, \mathcal{D}) = |\{t \in \mathcal{D} : X \subseteq t\}|/|\mathcal{D}|$. Pattern sampling aims at accessing the pattern space $\mathcal{L}$ by a sampling procedure simulating a distribution $p : \mathcal{L} \to [0, 1]$ that is defined with respect to an interestingness measure $m$:

**Table 1.** A toy dataset $\mathcal{D}$

| Trans | Items | | | | | Class |
|-------|---|---|---|---|---|-------|
| $t_1$ | A | B | | | E | + |
| $t_2$ | A | B | | | | + |
| $t_3$ | | B | C | D | | − |
| $t_4$ | | B | C | | | − |
| Known in advance | | | | | | Unknown |

**Table 2.** Evolution of weights with feedback

| Trans | Init | $B$ (−) | $BE$ (+) | $BD$ (−) |
|-------|------|---------|----------|----------|
| $t_1$ | $0.50 \pm 0.5$ | $\mathbf{0.27 \pm 0.5}$ | $\mathbf{0.51 \pm 0.5}$ | $0.51 \pm 0.5$ |
| $t_2$ | $0.50 \pm 0.5$ | $\mathbf{0.27 \pm 0.5}$ | $0.27 \pm 0.5$ | $0.27 \pm 0.5$ |
| $t_3$ | $0.50 \pm 0.5$ | $\mathbf{0.27 \pm 0.5}$ | $0.27 \pm 0.5$ | $\mathbf{0.13 \pm 0.3}$ |
| $t_4$ | $0.50 \pm 0.5$ | $\mathbf{0.27 \pm 0.5}$ | $0.27 \pm 0.5$ | $0.27 \pm 0.5$ |

$p(.) = m(.)/Z$ where $Z$ is a normalizing constant. In this way, with no parameter (except possibly the sample size), the user has a fast and direct access to the entire pattern language.

Assume now that the dataset $\mathcal{D}$ is partitioned into two subsets, denoted by $\mathcal{D}^+$ and $\mathcal{D}^-$, such that $\mathcal{D} = \mathcal{D}^+ \cup \mathcal{D}^-$ and $\mathcal{D}^+ \cap \mathcal{D}^- = \emptyset$. We say that the sub-dataset $\mathcal{D}^+$ contains the set of *positive* transactions, whereas the sub-dataset $\mathcal{D}^-$ contains the set of *negative* transactions. In our toy example (see Table 1), $t_1$ and $t_2$ are positive transactions, whereas $t_3$ and $t_4$ are negative ones. In our approach, we assume that the sub-datasets $\mathcal{D}^+$ and $\mathcal{D}^-$ are not known in advance, whereas the user want to discover patterns that characterize the subset $\mathcal{D}^+$ of positive transactions. In our toy example (see Table 1), because $supp(BE, \mathcal{D}^+) = 0.5$ and $supp(BE, \mathcal{D}^-) = 0$, the user is definitely interested by pattern $BE$. But, he/she is less interested by pattern $B$ since $supp(B, \mathcal{D}^+) = supp(B, \mathcal{D}^-) = 1$.

In that context, we assume that an oracle $\mathcal{O} : \mathcal{L} \to \{-, +\}$ models the user feedback. It means that $\mathcal{O}(X) = +$ (resp. $-$) iff the oracle gives a positive (resp. negative) feedback response for the pattern $X$. In Table 2, three patterns are drawn ($B$, $BE$ and $BD$) and the user feedback is indicated in parentheses. Since the user feedback about the same pattern $X$ may change during the process (the user may have an imperfect knowledge of the set $\mathcal{D}^+$ of positive transactions), we consider that $\mathcal{O}$ is a random variable. Thereby, $\mathbf{P}(+|X)$ will denote the probability of having a positive feedback given $X$ when the oracle is consulted. For instance, in our toy example, because $supp(BE, \mathcal{D}^+) = 0.5$ and $supp(BE, \mathcal{D}^-) = 0$, we could assume that $\mathbf{P}(+|BE) = 1$, meaning that the oracle always gives a positive feedback for $BE$. On the other hand, because $supp(B, \mathcal{D}^+) = supp(B, \mathcal{D}^-) = 1$, we could assume that $\mathbf{P}(+|B) = 0.5$, meaning that the user could evaluate pattern $B$ positively or negatively according to the objective of the user (i.e., discrimination or characterization of the target class).

### 3.2   Problem Formulation

In our context, since the user is not interested in all transactions in $\mathcal{D}$, but only in positive transactions in $\mathcal{D}^+$, we do not want to sample the pattern space according to the interestingness measure $m$ evaluated on $\mathcal{D}$, but on $\mathcal{D}^+$. Indeed, the interestingness measure $m$ evaluated on $\mathcal{D}^+$ is better suited because it enables

---

**Algorithm 1.** Interactive pattern sampling

---

**Input:** A dataset $\mathcal{D}$ and an oracle $\mathcal{O}$
1: Let $F$ be an empty sequence
2: Let $\omega_F(t) := 0.5$ for all $t \in \mathcal{D}$
3: **repeat**
4:    Draw a pattern $X$ from $\mathcal{D}$ according to its weighted support $supp_\omega$
5:    Add the user feedback to the sequence $F$
6:    Update the weight vector $\omega_F$ using $F$
7: **until** The user stops the process

---

us to focus on the patterns describing the set of positive transactions. Unfortunately, the set of positive transactions in $\mathcal{D}^+$ is not known in advance. Therefore, our problem can be formalized as follows:

**Problem 1** *Given a dataset $\mathcal{D}$ containing an unknown set of positive transactions $\mathcal{D}^+$ and an oracle $\mathcal{O}$, our problem consists in building a sequence of patterns $\langle X_1, \ldots, X_k \rangle$ such that the probability to draw a pattern $X_i$ at step $i$ tends to $supp(X_i, \mathcal{D}^+)/Z$ when $i$ tends to $+\infty$ where $Z$ is a normalizing constant.*

Note that at each step $i$, the oracle $\mathcal{O}$ will be used to evaluate the interestingness of the pattern $X_i$ presented to the user. The next sections show how to choose these patterns $X_i$ and how the user feedback $\mathcal{O}(X_i)$ can be used by the system to improve its knowledge of $\mathcal{D}^+$.

## 4    Interactive Pattern Sampling Algorithm

### 4.1    General Principles of the Approach

For addressing the problem formalized in Sect. 3.2, Algorithm 1 provides a sketch of our interactive system. Its key idea is to associate a weight $\omega_F(t)$ to each transaction $t \in \mathcal{D}$ that maintains an estimation of the class conditional probability $\mathbf{P}(+|t)$ (the probability that a transaction $t$ belongs to $\mathcal{D}^+$). Of course, all these weights $\omega_F(t)$ are initialized to 0.5 because the class is unknown at the beginning (line 2)[1], as shown in the second column in Table 2. But, at the end, the goal is to have $\omega_F(t) = 1$ iff $t \in \mathcal{D}^+$ (0 otherwise). For this purpose, our system alternates between three steps as proposed in [7]:

- **Mining step (line 4):** This step provides patterns by favoring those which are frequent in transactions with high weights $\omega_F$. More precisely, a pattern $X$ is sampled according to a weighted support $supp_\omega$. Typically, after the positive feedback on $BE$ (see Table 2), $AB$ will be more likely to be drawn than $BC$ because the total weight of $t_1$ and $t_2$ becomes higher than that of $t_3$ and $t_4$.

---

[1] It is also possible to set weights to 0 or 1 if the labels of some transactions are already known.

---

**Algorithm 2.** Weighted Support-based Sampling

---

**Input:** A dataset $\mathcal{D}$ and a weight vector $\omega$
**Output:** A random itemset $X \sim supp_\omega(\mathcal{L}, \mathcal{D})$
 1: Let weight vector $\omega'$ be defined by $\omega'(t) := 2^{|t|} \times \omega(t)$ for all $t \in \mathcal{D}$
 2: Draw a transaction $t \sim \omega'(\mathcal{D})$
 3: **return** an itemset $X \sim u(2^t)$

---

- **Interactive step (line 5):** During this step, the user evaluates whether the pattern $X$ is a good descriptor or not of the unknown sub-dataset $\mathcal{D}^+$ of positive transactions.
- **Learning step (line 6):** The system updates the weight $\omega_F(t)$ of each transaction $t$ containing $X$. Basically, if the user feedback is positive, the weight $\omega_F(t)$ is increased otherwise it is decreased (see Sect. 4.3 for more details). For instance, in Table 2, the weight of $t_1$ is increased after the draw of $BE$ while that of $t_3$ is decreased after the draw of $BD$.

In order that our system works, it is necessary to link the user feedback given on patterns (i.e., $\mathbf{P}(+|X)$) to the class conditional probabilities on transactions (i.e., $\mathbf{P}(+|t)$). Our approach is based on this central result which is independent of the mining and learning steps:

*Property 1 (Class Conditional Probability).* Given a transaction $t$ in $\mathcal{D}$ and a pattern langage $\mathcal{L}$, we have: $\mathbf{P}(+|t) = \sum_{X \in \mathcal{L}} \mathbf{P}(X|t) \times \mathbf{P}(+|X)$.

It is impossible to calculate the exact class conditional probability of a transaction because its calculation depends on the entire pattern language $\mathcal{L}$. Using Property 1, we show in Sect. 4.3 how we can estimate $\mathbf{P}(+|t)$ given a sequence of user feedback responses. Previously, while $\mathbf{P}(+|X)$ is straightforwardly provided by the oracle, the method used to a draw sequence of patterns $X$ is necessary to further detail $\mathbf{P}(X|t)$. This method is presented in the following Sect. 4.2.

### 4.2   Pattern Sampling According to the Weighted Support

In [3], the authors show how to sample patterns following a distribution proportional to their support. In our approach, we propose to sample patterns following a distribution proportional to their *weighted* support. More formally, given a dataset $\mathcal{D}$ and a weight vector $w$, the weighted support of a pattern $X$ in $\mathcal{D}$, denoted $supp_\omega(X, \mathcal{D})$, is defined by: $supp_\omega(X, \mathcal{D}) = \sum_{t \in \mathcal{D}, X \subseteq t} \omega(t) / \left( \sum_{t \in \mathcal{D}} \omega(t) \right)$.

Algorithm 2 adapts the two-step random procedure [3] to sample patterns according to their weighted supports. Using this algorithm, the weighted support is similar to the usual support at the beginning (when all weights are equal to 0.5). More interestingly, it is easy to see that $supp_\omega(X, \mathcal{D}) = supp(X, \mathcal{D}^+)$ (which solves Problem 1) if all positive transactions in $\mathcal{D}^+$ have 1 as weight and other transactions have 0 as weight after a long sequence of interactions with the user. However, we still have to show how we can learn the weights of the transactions, which is the goal of the following section.

---

**Algorithm 3.** Learning the weights

---

**Input:** A sequence $F = \{(X_1, f_1, s_1), ..., (X_k, f_k, s_k)\}$ of $k$ user feedback responses
**Output:** A updated set of weights $\omega(t)$

1: **for all** $t \in \mathcal{D}$ **do**

2: $\quad \bar{\omega}_F(t) := \dfrac{\sum_{(X_j, f_j, s_j) \in F, X_j \subseteq t} f_j / s_j}{\sum_{(X_j, f_j, s_j) \in F, X_j \subseteq t} 1 / s_j}$

3: $\quad \omega_F(t) := \dfrac{inf_F(t) + sup_F(t)}{2}$

4: $\quad$ **if** $inf_F(t) > 0.5$ **then** $\omega_F(t) := 1$

5: $\quad$ **if** $sup_F(t) < 0.5$ **then** $\omega_F(t) := 0$

6: **end for**

---

### 4.3 Learning the Weights of Transactions

In this section, we show how we can update the weights of the transactions from the user feedback. Assuming that patterns are sampled using Algorithm 2, given a transaction $t \in \mathcal{D}$, we know that $\mathbf{P}(X|t) = 0$ if $X \not\subseteq t$ and $\mathbf{P}(X|t) = \frac{1}{|2^t|}$ if $X \subseteq t$. Thus, using Property 1, we finally have:

$$\mathbf{P}(+|t) = \sum_{X \in \mathcal{L}} \mathbf{P}(X|t) \times \mathbf{P}(+|X) = \frac{1}{2^{|t|}} \sum_{X \subseteq t} \mathbf{P}(+|X) \tag{1}$$

Using this equation, Algorithm 3 shows how the probabilities $\mathbf{P}(+|t)$ can be estimated from a sequence of user feedback responses, and how these estimations can be used to update the weights of the transactions. Let $F = \{(X_1, f_1, s_1), ..., (X_k, f_k, s_k)\}$ be a sequence of $k$ user feedback responses, where $X_k$ is the pattern drawn at step $k$ in Algorithm 1, $f_k = 1$ if $\mathcal{O}(X_k) = +$ (0 otherwise), and $s_k = supp_\omega(X_k, \mathcal{D})$. At step 2 of Algorithm 3, we start to compute a first estimation $\bar{\omega}_F(t)$ of $\mathbf{P}(+|t)$ using a weighted arithmetic mean. The following property shows that $\bar{\omega}_F(t)$ tends to $\mathbf{P}(+|t)$ when the number of user feedback responses tends to infinity.

*Property 2 (Probability Estimations).* Given a dataset $\mathcal{D}$, for every transaction $t \in \mathcal{D}$, the weight $\bar{\omega}_F(t)$ converges to $\mathbf{P}(+|t)$ when the number of user feedback responses $|F|$ tends to infinity.

In practice, this property means that the addition of new feedback responses tends to improve the estimation of the probability $\mathbf{P}(+|t)$. In order to evaluate the estimation error, we benefit from a statistical result known as Bennett's inequality which is true irrespective of the probability distribution [8]. After $k$ independent observations of a real-valued random variable $r$ with range $[0, 1]$, Bennett's inequality ensures that, with a confidence $1 - \delta$, the true mean of $r$ is at least $\bar{r} - \epsilon$ where $\bar{r}$ and $\bar{\sigma}$ are respectively the observed mean and standard deviation of the samples and $\epsilon = \sqrt{\frac{2\bar{\sigma}^2 \ln(1/\delta)}{k}} + \frac{\ln(1/\delta)}{3k}$. We use this statistical result to bound the true value of $\mathbf{P}(+|t)$ from a sequence of user feedback responses $F$:

*Property 3 (Bounds).* Given a dataset $\mathcal{D}$, a sequence of user feedback responses $F$ and a confidence $1 - \delta$, the probability $\mathbf{P}(+|t)$ for a transaction $t$ is bounded as follows:

$$\underbrace{\max\{0, \bar{\omega}_F(t) - \epsilon\}}_{inf_F(t)} \leq \mathbf{P}(+|t) \leq \underbrace{\min\{\bar{\omega}_F(t) + \epsilon, 1\}}_{sup_F(t)}$$

with $\epsilon = \sqrt{2\bar{\sigma}^2 \ln(1/\delta)/k} + \ln(1/\delta)/3k$ where $\bar{\sigma} = \sqrt{\bar{\omega}_F(t) - \bar{\omega}_F(t)^2}$ is the empirical standard deviation of $\bar{\omega}_F(t)$.

This property is important because it gives information about the error of the estimation $\bar{\omega}_F$. In Algorithm 3, we use this property to compute $\omega_F(t) = \frac{inf_F(t) + sup_F(t)}{2}$, i.e. a corrected estimation of $\mathbf{P}(+|t)$. Since both bounds tend to $\mathbf{P}(+|t)$, it is easy to see that the corrected estimation $\omega_F(t)$ also tends to $\mathbf{P}(+|t)$ when the number of feedback responses increases. Finally, at lines 4 and 5 of Algorithm 3, we force the weight $\omega_F(t)$ to tend to 1 (resp. 0) when it is certain (with respect to the confidence level) that the probability $\mathbf{P}(+|t)$ is higher than 0.5 (resp. $\mathbf{P}(+|t) < 0.5$). For instance, after the evaluation of $BD$ in Table 2, the final weight $\omega_F(t_3)$ will be zero because $0.13 + 0.3 = 0.43$ is below 0.5.

### 4.4   Convergence and Complexity

It may be that we do not properly learn the set of positive transactions from the user feedback on the patterns if his/her feedback is not consistent. For instance, if $\mathbf{P}(+|X) = 0$ for all patterns $X \subseteq t$, then we compute $\mathbf{P}(+|t) = 0$ even if $t$ is truly a positive transaction. Therefore, we introduce the notion of consistent oracle:

**Definition 1 (Consistency).** *Given a set $\mathcal{D}^+ \subseteq \mathcal{D}$ of positive transactions, an oracle $\mathcal{O}$ is consistent with $\mathcal{D}^+$ iff for all transaction $t \in \mathcal{D}$, we have $\mathbf{P}(+|t) > 0.5$ if $t \in \mathcal{D}^+$, and $\mathbf{P}(+|t) < 0.5$ otherwise.*

Using this definition of consistency, and Property 3, it is possible to conclude on the good convergence of Algorithm 1:

**Theorem 1 (Convergence).** *Given $\mathcal{D}$ with $\mathcal{D}^+ \subseteq \mathcal{D}$ and an oracle $\mathcal{O}$ consistent with $\mathcal{D}^+$, for each transaction $t \in \mathcal{D}$, the weight $\omega_F(t)$ converges to 1 iff $t \in \mathcal{D}^+$ (otherwise to 0) when the number of user feedback responses $|F|$ tends to infinity. Consequently, the weighted support tends to the support in $\mathcal{D}^+$.*

Under the assumption of consistency, Algorithm 1 clearly solves the problem stated in Sect. 3.2. Interestingly, the time complexity of this approach in $O(k|\mathcal{D}||\mathcal{I}|)$ (where $k$ is the number of mined patterns) is excellent. Finally, as the weights can be calculated without keeping the details of all user feedback, the space complexity of the algorithm is linear with the size of the dataset.

# 5    Experimental Study

This section has the twofold objective of evaluating the quality of the class learning through user feedback and the quality of the patterns presented to the user. Note that the Java source code of the implementation used for this study is available at www.info.univ-tours.fr/~soulet/prototype/ida17/.

| $\mathcal{D}$ | $|\mathcal{D}|$ | $|\mathcal{I}|$ | $|\mathcal{D}^+_{min}|$ | $|\mathcal{D}^+_{min}|/|\mathcal{D}|$ | $\mathcal{D}$ | $|\mathcal{D}|$ | $|\mathcal{I}|$ | $|\mathcal{D}^+_{min}|$ | $|\mathcal{D}^+_{min}|/|\mathcal{D}|$ |
|---|---|---|---|---|---|---|---|---|---|
| abalone | 4,177 | 28 | 1,307 | 0.31 | mushroom | 8,124 | 119 | 3,916 | 0.48 |
| chess | 3,196 | 75 | 1,527 | 0.48 | page | 941 | 35 | 9 | 0.01 |
| cmc | 1,473 | 28 | 469 | 0.32 | sick | 2,800 | 58 | 171 | 0.06 |
| german | 1,000 | 76 | 300 | 0.30 | vehicle | 846 | 58 | 199 | 0.24 |
| hypo | 3,163 | 47 | 151 | 0.05 | | | | | |

**Fig. 1.** Features of UCI benchmarks

**Protocol** We report the experimental evaluations conducted on 9 datasets coming from the UCI Machine Learning Repository (archive.ics.uci.edu/ml). Table 1 provides the main features of each dataset. For each dataset $\mathcal{D}$, the minority class of $\mathcal{D}$ corresponds to the set of positive transactions. The cardinality of this minority class, denoted $\mathcal{D}^+_{min}$, is indicated in the last column of Table 1. We first perform experiments using a deterministic oracle (in the sense that its answer is constant for a given pattern). Given a set of positive transactions $\mathcal{D}^+ \subseteq \mathcal{D}$, this deterministic oracle is defined as $\mathcal{O}(X) = +$ if $supp(X, \mathcal{D}^+) > supp(X, \mathcal{D})$ ($-$ otherwise). Intuitively, a user is interested in a pattern if its support in the set of positive transactions $\mathcal{D}^+$ is higher than its support in the dataset $\mathcal{D}$.

First, we evaluate the quality of the *mining step* by considering the number of interesting patterns, i.e. patterns positively rated by the user. More precisely, we compute the ratio of positive feedback responses over the last 50 patterns provided to the user i.e., $\mathbf{P}(+) = \sum_{i=k-49}^{k} f_k/50$ given a sequence of user feedback responses $F = \langle (X_1, f_1, s_1), \ldots, (X_k, f_k, s_k) \rangle$ with $k \geq 50$.

Second, a confusion matrix is used for evaluating the quality of the *learning step*. More precisely, we consider that a transaction is classified in the positive class (resp. negative class) if its weight considering the margin of error is greater than 0.5 (resp. < 0.5). Thus, we introduce two sets of transactions defined by: $\mathcal{P}^+ = \{t \in \mathcal{D} \mid inf_F(t) > 0.5\}$ and $\mathcal{P}^- = \{t \in \mathcal{D} \mid sup_F(t) < 0.5\}$. In the very first iterations, it is clear that no class is predicted, i.e. $\mathcal{P}^+ = \mathcal{P}^- = \emptyset$. Then, as the interactions progress, the proportion of classified transactions, defined by $Completeness = \frac{|\mathcal{P}^+ \cup \mathcal{P}^-|}{|\mathcal{D}|}$, increases. In order to evaluate the quality of the learning step, we also use the True Positive Rate (*TPR*) and the False Positive Rate (*FPR*) measures defined respectively by: $TPR = \frac{|\mathcal{D}^+ \cap \mathcal{P}^+|}{|\mathcal{D}^+|}$ and $FPR = \frac{|\mathcal{D}^- \cap \mathcal{P}^+|}{|\mathcal{D}^-|}$.

All experiments were repeated 100 times and the arithmetic mean is used for averaging the values coming from those 100 measurements. Finally, the confidence level $1 - \delta$ is set to 0.8.

**Convergence** The left part of Fig. 2 gives the proportion of positive feedback responses with respect to the number of iterations (i.e., the number of patterns presented to the user). As expected, this quality measure increases as the number of user feedback responses increases, which means that more relevant patterns are presented to the user. Note that in the first iterations, the 4 datasets having the largest ratio $|\mathcal{D}^+_{min}|/|\mathcal{D}|$ (i.e., `abalone`, `chess`, `cmc`, `mushroom`) are also those having the best proportions of positive feedback responses. Indeed, it is easier to find patterns that characterize an important class than a small class as it is the case for `page`. However, after a sufficient number of iterations, the system is efficient to propose relevant patterns even for small positive classes. Furthermore, we can see that the proportion of positive feedback responses does not converge towards 1. This observation can be explained by the nature of the oracle used in the experiments. Indeed, an oracle based on a contrast measure is unfavorable to our sampling method based on a description measure (i.e., support). It can also be explained by the nature of the dataset. Indeed, the set of items of the

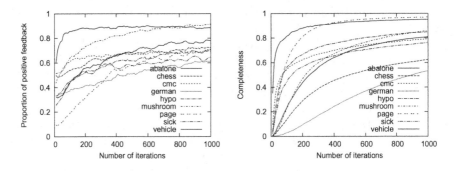

**Fig. 2.** Proportion of positive feedback responses and completeness

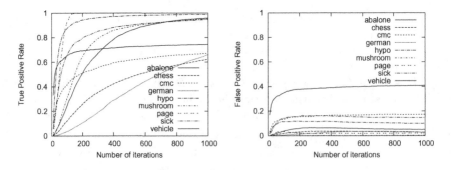

**Fig. 3.** *TPR* and *FPR* with the number of iterations

dataset is in general not adequate to perfectly characterize the target class, i.e. the class of positive transactions.

The right part of Fig. 2 gives the completeness (proportion of classified transactions) with the number of iterations. As expected, the completeness converges to 1 meaning that the method will arrive at classifying all transactions. Importantly, we observe that the method quickly learns the class of a majority of transactions on most datasets. Indeed, after 300 patterns, the completeness is greater than 0.5 for all datasets except chess and german (in this case, the oracle does not discriminate the two classes well). In order to evaluate more precisely the quality of the learning step, Fig. 3 plots the *TPR* and *FPR* with the number of iterations. Except for chess and german datasets, we observe that the *TPR* (proportion of positive transactions that are correctly classified in $\mathcal{P}^+$) increases and converges to their maximal value very fast. In particular, we can emphasize that it is the case for the datasets hypo, page and sick, even though the set of positive transactions for these datasets is very small (less than 6% of the whole dataset). Concerning the *FPR* (proportion of negative transactions incorrectly classified), we finally observe that it stabilizes to a low value (less than 20%) very fast (in less than 200 iterations) except for abalone.

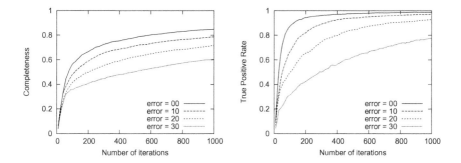

**Fig. 4.** Completeness and *TPR* according to an oracle with error on hypo

**Non-deterministic oracle** We now evaluate the impact of non-deterministic oracle by introducing an error component to the oracle. Experiments are carried out on hypo with 4 different error probabilities 30%, 20%, 10% and 0% (it means that the oracle gives an opposite feedback in $x$% of its answers). By observing the *Completeness* and *TPR* in Fig. 4, we observe that the convergence is guaranteed, but the required time increases with the error rate. Importantly, it is easy to see that the approach is robust because the error probability has no significant impact on the final value of the *TPR*, meaning that the set of positive transactions is correctly identified whatever the error probability.

# 6    Conclusion

This paper presents a new method of interactive pattern mining by benefiting from pattern sampling. Beyond its practical efficiency, this technique offers statistical guarantees on the learned class model and therefore, on the convergence of the interactive process. Experiments highlight this good convergence on several benchmarks. The number of classified transactions increases rapidly while the true and false positive rates remain satisfactory even if the target class consists in only few transactions. Besides, even if an end user can only make a limited number of feedback responses, the good convergence of the system is interesting because it is possible to envisage such a system in a context of crowdsourcing. We would intend to generalize this method to other interestingness measures more sophisticated than support, including measures for identifying contrasts between $\mathcal{D}^+$ and $\mathcal{D}^-$.

**Acknowledgements.** This work has been partially supported by the Decade project, Mastodons 2017, CNRS.

# References

1. Bessiere, C., Coletta, R., Hebrard, E., Katsirelos, G., Lazaar, N., Narodytska, N., Quimper, C.G., Walsh, T.: Constraint acquisition via partial queries. In: Proceedings of the 23rd IJCAI, pp. 475–481 (2013)
2. Bhuiyan, M., Mukhopadhyay, S., Hasan, M.A.: Interactive pattern mining on hidden data: a sampling-based solution. In: Proceedings of ACM CIKM, pp. 95–104 (2012)
3. Boley, M., Lucchese, C., Paurat, D., Gärtner, T.: Direct local pattern sampling by efficient two-step random procedures. In: Proceedings of the 17th ACM SIGKDD, pp. 582–590 (2011)
4. Dzyuba, V., van Leeuwen, M., De Raedt, L.: Flexible constrained sampling with guarantees for pattern mining. Data Min. Knowl. Disc. **31**, 1–28 (2017)
5. Dzyuba, V., Leeuwen, M.v., Nijssen, S., De Raedt, L.: Interactive learning of pattern rankings. Int. J. Artif. Intell. Tools **23**(06), 32 p. (2014)
6. Giacometti, A., Soulet, A.: Anytime algorithm for frequent pattern outlier detection. Int. J. Data Sci. Analytics **2**(3–4), 119–130 (2016)
7. Leeuwen, M.: Interactive Data Exploration Using Pattern Mining. In: Holzinger, A., Jurisica, I. (eds.) Interactive Knowledge Discovery and Data Mining in Biomedical Informatics. LNCS, vol. 8401, pp. 169–182. Springer, Heidelberg (2014). doi:10.1007/978-3-662-43968-5_9
8. Maurer, A., Pontil, M.: Empirical Bernstein bounds and sample variance penalization. arXiv preprint (2009). arXiv:0907.3740
9. Novak, P.K., Lavrač, N., Webb, G.I.: Supervised descriptive rule discovery: a unifying survey of contrast set, emerging pattern and subgroup mining. J. Mach. Learn. Res. **10**(Feb), 377–403 (2009)
10. Rashidi, P., Cook, D.J.: Ask me better questions: active learning queries based on rule induction. In: Proceedings of the 17th ACM SIGKDD 2011, pp. 904–912 (2011)

11. Rueping, S.: Ranking interesting subgroups. In: Proceedings of the 26th Annual International Conference on Machine Learning, pp. 913–920. ACM (2009)
12. Settles, B.: A practical test for univariate and multivariate normality. Computer sciences Technical report 1648, University of Wisconsin, Madison (2010)
13. Tong, S., Koller, D.: Support vector machine active learning with applications to text classification. J. Mach. Learn. Res. **2**, 45–66 (2001)
14. Xin, D., Shen, X., Mei, Q., Han, J.: Discovering interesting patterns through user's interactive feedback. In: Proceedings of the 12th ACM SIGKDD 2006, pp. 773–778 (2006)

# Searching for Spatio-Temporal-Keyword Patterns in Semantic Trajectories

Fragkiskos Gryllakis[1]([⊠]), Nikos Pelekis[2], Christos Doulkeridis[3],
Stylianos Sideridis[1], and Yannis Theodoridis[1]

[1] Department of Informatics, University of Piraeus, Piraeus, Greece
{fgryllakis,ssider,ytheod}@unipi.gr
[2] Department of Statistics and Ins. Science, University of Piraeus,
Piraeus, Greece
npelekis@unipi.gr
[3] Department of Digital Systems, University of Piraeus, Piraeus, Greece
cdoulk@unipi.gr

**Abstract.** Location-based social network users typically publish information about their location and activity (in the form of keywords) along time, thus providing the mobility data management research community with complex and voluminous data. In this work, we handle this kind of data as sequences in the Spatio-Temporal-Keyword (STK) domain. This modeling is coherent with the concept of semantic trajectories that has recently attracted the interest of this community. Our paper focuses on the efficient processing of pattern queries over the STK domain, hence called Spatio-Temporal-Keyword Pattern (STKP) queries. Our approach is based on efficient index structures that take into account the triple nature of these patterns and is developed in a NoSQL graph database. Through an extensive experimental study over real-life datasets, we demonstrate the effectiveness and efficiency of our proposal.

**Keywords:** Spatio-temporal-keyword patterns · Semantic trajectories · Graph DB

## 1 Introduction

The increasing use of mobile devices with location-sensing capabilities poses new challenges related to the efficient management and effective mining of large volumes of spatio-temporal data. Recently, due to the wide spread of location-based social networking (LBSN) services, such datasets are widely available, usually accompanied by textual annotations (consider Twitter, Foursquare etc.). Even in the cases where user-generated annotations are not provided, there are plenty of suitable methods that are able to extract relevant contextual information from open linked data sources. As such, the so-called semantic trajectories [10, 12, 13] have recently gained the interest of the mobility data management community.

In this context, we formulate and address the problem of Spatio-Temporal-Keyword Pattern (STKP) search over semantic trajectories. Typically, a semantic trajectory is defined as a sequence of sub-trajectories (coined episodes), usually representing 'move'

© Springer International Publishing AG 2017
N. Adams et al. (Eds.): IDA 2017, LNCS 10584, pp. 112–124, 2017.
DOI: 10.1007/978-3-319-68765-0_10

and/or 'stop' activities; each episode consists of spatial, temporal, and textual information about user's activity. Hence, a STKP query is defined as a sequence of (spatial, temporal, textual) constraints over episodes. The constraints are formulated by regular expressions, thus offering high expressiveness and flexibility in query formulation.

Our work is motivated by applications that require advanced searching and mining operations on top of semantic trajectories. As discussed in [12], searching a semantic trajectory database (STD) for *"people crossing the city center on their way from work back to home"* or *"people driving more than 20 km on their way from home to work"* or *"people spending more than 1 h daily for bring-get activities of their children at school"*, is a quite challenging task that can be efficiently supported by the STKP query formalism discussed in this paper. It is important to be mentioned that, for the purposes of this paper, textual information is considered to be a set of keywords (e.g. 'home', 'work', 'bring-get' etc.) an abstraction quite popular in the literature.

To support efficient processing of STKP queries, we follow the spatial-textual indexing [3] paradigm that tightly integrates spatial with textual information in a single indexing structure. The main challenge is to integrate temporal with spatial-textual information, also taking into account the sequential nature of trajectory data [14]. To this end, we propose two alternative indexing structures, called *TSR-tree* and *ESR-tree*, built at the trajectory and episode-level, respectively. We advocate the use of a graph DBMS that naturally supports connected objects, which, in our concept, are the episodes of a semantic trajectory.

In a nutshell, the contributions of this work are outlined as follows: (i) we formally define STKP queries on semantic trajectory databases (Sect. 2); (ii) we propose hybrid spatio-temporal-keyword index structures to support this type of search (Sect. 3.1) and, capitalizing on them, we design efficient STKP query processing algorithms for STKP search (Sect. 3.2) boosted by a selectivity estimation model (Sect. 3.3); and (iii) we realize our solutions over the NoSQL Neo4j graph DBMS [9] performing an experimental study over a real-life dataset from Foursquare (Sect. 4).

In addition, Sect. 5 presents related work in comparison with our proposal, and Sect. 6 concludes the paper, also discussing ideas for future work.

## 2   Problem Formulation

In this section, we first provide a definition about semantic trajectories as sequences of 'stop' episodes [17]. The definitions are formulated in the context of a graph database that provides an intuitive model in our case.

**Definition 1 (Episode, Semantic Trajectory, Semantic Trajectory Database).** A *semantic trajectory database* (STD) is a set of semantic trajectories. In turn, a *semantic trajectory st* is defined as a path $<v_1, \ldots, v_n>$ of successive nodes, where node $v_j$ corresponds to episode $ep_j$ in the semantic trajectory. In turn, *episode ep* is a structure of the form (*o-id, st-id, ep-id, MBB, tags, next*), where *o-id* is a unique identifier of the moving object, *st-id* is the identifier of the semantic trajectory of the object, *ep-id* is the identifier of the sub-trajectory this episode represents, *MBB* is the spatio-temporal range (spatial projection in 2D plane along with the temporal projection) where this episode takes place, *tags* is bag of keywords including information related to the semantics of the episode and *next* is the identifier of the episode that follows (if any).

**Definition 2 (Episode abstraction, matching episodes, STKP query).** Let $E_i$ be an *episode abstraction*, which is defined as a potentially incomplete episode (i.e. an episode where some of its spatial, temporal, textual properties may be missing). A STKP query over a STD is defined as a sequence Q: = $<E_1, ..., E_k>$ of episode abstractions, and returns the semantic trajectories that match Q. A semantic trajectory matches a sequence of episode abstractions when it includes *matching episodes* in the same sequence; episode $ep_i$ matches episode abstraction $E_i$ when $ep_i.MBB$ is covered by $E_i.MBB$ and $ep_i.tags \subset E_i.tags$.

**Example 1.** In Fig. 1, we depict a STD consisting of 3 semantic trajectories; each trajectory consists of four episodes. An example STKP query Q is also illustrated at the bottom right corner. In particular, Q consists of a number of episode abstractions; with notation $E_i$* corresponding to a number of zero or more episode abstractions of the form $E_i$ (for clarity of presentation the episode abstractions in Q distinguish the temporal from the spatial information, which is not the case in Defn. 1 where both are organized together in a MBB). Actually, Q searches for trajectories starting with zero or more episodes of any kind (see notation (*, *, *)* in Q), followed by an episode in a spatial [35, 35, 50, 50] and temporal [$t_{18}$, $t_{20}$] range with keyword 'RESTAURANT' and ending with an episode in a spatial [40, 40, 55, 55] and temporal [$t_{21}$, $t_{23}$] range with keyword 'DESSERT'. The output set includes semantic trajectory 1, which fulfills the above constraints.

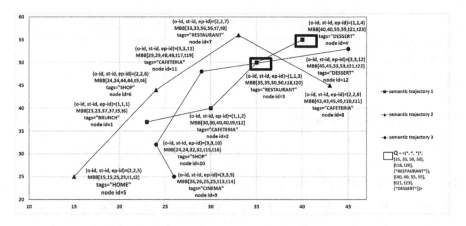

**Fig. 1.** Graph representation of an STD consisting of 3 trajectories along with a STKP query.

# 3   Indexing, Query Processing, and Optimization

In order to support STKP query processing, we first present two alternative structures for indexing STDs, and then we provide the respective search algorithms.

## 3.1   Indexing STDs

We envisage hybrid structures that tightly combine spatio-temporal with text indexes (3D R-trees and Inverted Files - IF, respectively), in order for both types of information to be used as first-class citizens when pruning the search space. The main distinction between the two indexes is whether they are built at the trajectory-or the episode- level, hence we call them *Trajectory-based Semantic R-tree (TSR-tree)* and *Episode-based Semantic R-tree (ESR-tree)*, respectively.

In principle, the construction mechanism of both structures includes two phases. In phase 1, a 3D R-tree is created by taking into account the spatio-temporal information (either at the entire trajectory or the episode level). Then, in phase 2, the respective IFs are built in a bottom-up fashion, and for each internal node, an IF is created for indexing the keywords corresponding to the entries of its child nodes. For each internal node, there exists a pointer to an IF that organizes all the tags of its child nodes. Even though our indexes are inspired by the IR-tree [16], we emphasize two notable differences: (a) our indexes preserve the sequential nature of the semantic trajectories at the leaf level (also known as trajectory preservation property) which has shown its advantages in trajectory-oriented queries [14], and (b) the temporal dimension has been smoothly incorporated and is integrated with the spatial-textual information. In the following paragraphs we provide more details for each of the proposed indexes.

**Trajectory-based Semantic R-Tree (TSR-tree).** In this index, a semantic trajectory is considered as the individual unit for the tree construction. As such, for each trajectory we compute its MBB and a list of the bags of tags that include the tags of the episodes, sorted by time. At the end, the tags in the list are concatenated to a single string in order to create a pseudo-word with all keywords. In order to exploit the graph database where our index resides, the leaf nodes including the afore-mentioned abstracted semantic trajectories become the starting nodes of the sequence of the episodes of the actual semantic trajectories. Then, we create IFs for all the internal nodes of the tree upon the pseudo-words. In Fig. 2, we illustrate the TSR-tree that indexes the STD of Example 1. The maximum number of children per node (fanout) is 2. In this figure, the nodes with ids 13-15 correspond to the approximated semantic trajectories, coined 'pseudo-trajectories', pointing to the first episode of the actual semantic trajectories.

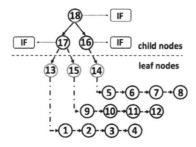

**Fig. 2.** The TSR-tree built upon the STD of Example 1.

**Episode-based Semantic R-tree (ESR-tree).** The ESR-tree considers the episodes of the semantic trajectories as its structural unit. In Fig. 3 (up), we illustrate the ESR-tree for the STD of Example 2. In particular, nodes with ids 1-12 (leaf nodes) represent the episodes of the three semantic trajectories whereas nodes with ids 13-16 represent non-leaf nodes. Solid directed lines represent ESR-tree internal relationships, dashed directed lines represent ESR-tree leaf relationships and dotted bi-directed lines capture the sequential nature of episodes in a trajectory. Figure 3 (down) illustrates the IF for each internal node.

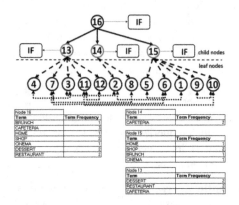

**Fig. 3.** The ESR-tree of Example 1 (up); the IF of each internal node (down).

Due to space constraints, we do not present detailed index construction algorithms, which are rather straightforward.

### 3.2   Processing STKP Queries

In this section, we present the respective algorithms for processing STKP queries using the afore-described indexes. All algorithms are illustrated in Fig. 4.

**STKP over TSR-trees:** Algorithm 1 is a typical breadth-first traversal algorithm taking into account that TSR-tree is built using entire semantic trajectories as building blocks. After the necessary initializations (lines 1-3), the algorithm finds which nodes satisfy the requirements of the query at each level of the tree (lines 4-5). When the traversal reaches the leaf level (line 6) the algorithm detects the semantic trajectories that satisfy the query requirements and adds them in the results set (line 10). This is performed by VerifyEpisode (Algorithm 2) that operates at the data level by scanning the sequence of the episode nodes, starting from the first episode (i.e. the one pointed by the «pseudo-trajectory» leaf node (line 7). As regards the VerifyEpisode routine, it checks simultaneously the spatio-temporal constraints of the episode abstraction of Q by using, on the one hand, the MBB of the 3D-Rtree node and, on the other hand, the keyword constraints in the IF.

| ALGORITHM 1.   TSR-tree-based STKP |
| --- |
| **Input:** The root of a TSR-tree, a sequence $Q$ $:=<E_1, ..., E_k>$ of episode abstractions. |
| **Output:** The set of semantic trajectories that match Q. |
| 1.    result = Ø, depthNodes = Ø, treeDepth = getTreeDepth (root) |
| 2.    depthNodes (0).add (root) |
| 3.    int i = 0 |
| 4.    **while** (i<treeDepth) |
| 5.      **for** each node in depthNodes (i) |
| 6.        **if** (i = treeDepth-1) **then** |
| 7.          **if** VerifyEpisode (node.next, Q) **then** |
| 8.            result.add (getSemTraj (node)) |
| 9.          **end if** |
| 10.        **else** |
| 11.          **if** VerifyEpisode (node, Q) **then** |
| 12.            depthNodes (i + 1).add (node) |
| 13.          **end if** |
| 14.        **end if** |
| 15.      **end for** |
| 16.      i = i + 1 |
| 17.    **end while** |
| 18.    **return** result |
| End of Algorithm |

| ALGORITHM 2.   VerifyEpisode |
| --- |
| **Input:** an episode $ep$, a sequence $Q := <E_1, ..., E_k>$ of episode abstractions |
| **Output:** TRUE if the semantic trajectory that $ep$ belongs to matches $Q$, FALSE otherwise. |
| 1.    int $count$ = 0; int $i$ = 0; int $stop$ = 0; |
| 2.    **for** each ($E_i$ in $Q$) |
| 3.      $j = stop$; |
| 4.      **while** (ep.next is not null) |
| 5.        **if** verify($E_i$, $ep_j$) **then** |
| 6.          $count = count + 1$; |
| 7.          $stop = j + 1$; |
| 8.          break; |
| 9.        **end if** |
| 10.      **end while** |
| 11.    **end for** |
| 12.    **if** $count$ = $Q$.length **then** |
| 13.      **return** TRUE; |
| 14.    **end if** |
| 15.    **return** FALSE; |
| End of Algorithm |

| ALGORITHM 3.   ESR-tree-based STKP |
| --- |
| **Input:** The root of a ESR-tree, a sequence $Q$ $:=<E_1, ..., E_k>$ of episode abstractions. |
| **Output:** The set of semantic trajectories that satisfy the requirements set by $Q$. |
| 1.    $result$ = Ø |
| 2.    $candEpisodes$= STKRangeQuery($E_1$) |
| 3.    **for** each $ep$ in $candEpisodes$ |
| 4.      **if** VerifyEpisode($ep$, $Q$) **then** |
| 5.        $result$.add(getSemTraj($ep$)) |
| 6.      **end if** |
| 7.    **end for** |
| 8.    **return** $result$ |
| End of Algorithm |

**Fig. 4.**  STKP query processing algorithms.

**STKP over ESR-trees:** Algorithm 3 follows the filter-and-refine principle. At the filtering step, it descends the tree in a breadth-first traversal, during which it simultaneously verifies each node with respect to the spatio-temporal and textual criteria of a single query episode abstraction, until it reaches the leaf level (see line 2 where the first episode abstraction, $E_1$, is the chosen one). The refine step applies the constraints at the episode (data) level of the graph. VerifyEpisode is used again after the initial verification of the selected episode abstraction. As already mentioned, this routine begins the verification of the remaining abstract episodes from the episode node that resulted from the tree traversal. Note that one can select an episode abstraction other than the first one, resulting in two invocations of VerifyEpisode: forward and backward scanning. The challenge here is how to make the best use of the index to filter the search space.

This will be discussed thoroughly in the next section, where a selectivity estimation model designed for this purpose is proposed.

### 3.3    Optimizing STKP Queries

Given a STKP query Q: = $<E_1, \ldots, E_k>$, our goal is to design a selectivity estimation model that identifies the most selective abstract episode $E_*$ in Q, in order to start query execution from this episode, thereby pruning candidate results the earliest possible.

Our key observation is that such a model can be computed by decomposing the computation of selectivity $S_i$ of an abstract episode $E_i$ (ST, K) in two parts, one computing selectivity $S^{ST}$ of the spatio-temporal component ST and another computing selectivity $S^K$ of the textual component K. Obviously, for the overall selectivity $S_i$ it holds that $S_i \leq \min(S^{ST}, S^K)$. Having selectivity estimations for each abstract episode of Q, the index traversal could start at the one having the minimum $S_i$. Consequently:

$$E_i = \left\{ E_j, 1 \leq j \leq k | \mathrm{argmin}_{1 \leq j \leq k}(S_j) \right\} \tag{1}$$

The challenge is to define an effective selectivity model for each of the two domains, i.e. effectively estimate $S^{ST}$ and $S^K$. Our proposal follows a histogram-based approach for domain ST and an inverted file (IF-based) approach for domain K.

In the spatio-temporal domain, given N sampling episodes $ep_j$ and a (user-defined) number of buckets B, the spatio-temporal histogram $H^{ST}$ that summarizes these episodes is defined as a vector:

$$H^{ST} = \left[ <R^{ST_1}, P^{ST_1}>, <R^{ST_2}, P^{ST_2}>, \ldots, <R^{ST_B}, P^{ST_B}> \right] \tag{2}$$

where $<R^{ST_1}, R^{ST_2}, \ldots, R^{ST_B}>$ corresponds to a uniform 3D (spatio-temporal) grid of size $\sqrt[3]{B} \times \sqrt[3]{B} \times \sqrt[3]{B}$ and $<P^{ST_1}, P^{ST_2}, \ldots, P^{ST_B}>$ denotes the ratios of the number of episodes that overlap the respective cells of the grid. Formally:

$$P^{ST_b} = \sum_{ep_j \mathrm{overlap} R^{ST_b}} \frac{1}{N}, 1 \leq b \leq B \tag{3}$$

Based on the above, we define spatio-temporal selectivity $S^{ST}$ of an abstract episode $E_i$ (ST, K) as:

$$S^{ST} = \begin{cases} \sum_{R^{ST_b} \mathrm{overlap} ST} P^{ST_b}, ST \neq {}'*' \\ N, ST = {}'*' \end{cases} \tag{4}$$

In the textual (keyword set) domain, given N sampling episodes $ep_j$ and assuming that their textual components are organized in an IF, we take advantage of the IF postings to estimate textual selectivity $S^K$. More formally, given an IF as a relation of tuples of the form $<\kappa, p_k, st_{ids}, ep_{ids}>$, where $\kappa$ is the atomic tag indexed by the IF, $p_k$ is the length of the postings list, namely the number of episodes that include keyword $\kappa$ in their tags' list, $st_{ids}$ is the $p_k$-length list of the identifiers of the semantic trajectories that include $\kappa$, and $ep_{ids}$ is the $p_k$-length list of offsets that represent the ordering of the episode in the sequence of episodes of the corresponding semantic trajectory, we define textual selectivity $S^K$ of an abstract episode $E_i$ (ST, K) as:

$$S^K = \begin{cases} min\{p_\kappa, \kappa \in K\}, K \neq {}^{'}*{}^{'} \\ N, K = {}^{'}*{}^{'} \end{cases} \tag{5}$$

The rationale behind Eq. 5 is that the selectivity of the textual domain is dominated by the keyword $\kappa \in K$ that minimizes the length $p_k$ of postings, in other words, the least frequent keyword in K, since this is the minimum number of episodes that we need to reach at the leaf level in order to check if they match with the given query pattern Q. Note that this estimation can be refined by restricting the postings only to the number of episodes that belong to distinct semantic trajectories, as in the general case a tag in a trajectory may appear in several episodes, thus counted multiple times.

Combining the component selectivity formulas, (an upper bound of) the overall selectivity S of an abstract episode $E_i$ (ST, K) is defined as:

$$S \leq min\{S^{ST}, S^K\} \tag{6}$$

# 4   Experimental Evaluation

In this section, we present our experimental study. All experiments were performed in a PC equipped with Intel Core i7-7700 CPU with 4 cores, at 3.6 GHz and 16 GB RAM. The proposed indexes and search algorithms are implemented as an extension of the Neo4j Spatial library and its R-Tree index. The inverted files are Apache Lucene indexes [1]. For our experimentation, we used the Foursquare New York dataset [4], which includes long-term (about 10 months - from Apr. 12, 2012 to Feb. 16, 2013) check-in data (227, 428 check-ins) in New York city collected from Foursquare social network.

Regarding parameter experimentation, we study the performance of our algorithms when varying different parameters, including: (a) the index fanout, (b) the grid granularity for the query optimizer, (c) the query length as number of abstract episodes, and (d) the size of STD. We generate queries of increasing query length, ranging from 2 to 10 abstract episodes. In particular, we generate query sets by randomly picking a semantic trajectory from the dataset and considering the location of the object as the query location at the finest spatio-temporal extent. Afterwards, we randomly choose a number of words (2, 4, 6, 8, 10) from the object as the query keywords and gradually increase the spatio-temporal extent of the chosen location in order for the query always to return at least one result. We also insert wildcards in between episodes abstractions in a random way.

In the following, we first report results related to index construction, focusing on time to build the index and its disk-size. Then, we study the performance of query processing for the proposed algorithms and indexes, also demonstrating the advantages in terms of performance when using the query optimization.

**Experiments on index construction.** The experiments on index construction demonstrate the time and space required to build the two alternative indexing structures with varying fanout (100, 200); note that as our implementation is on a graph DBMS, the fanout is not related with the block size. We also split the datasets into 4 datasets of different size, with each having the 25%, 50%, 75%, and 100% of the original size. Figure 5 (up) shows the differences in the sizes between the datasets regarding the spatio-temporal (i.e. 3D-Rtree) and text (i.e. IF) indexes; we notice that in all cases the former has considerably larger size than the latter. TSR-tree needs less space for storing the textual information in comparison with the ESR-tree. We also illustrate the total index size. As expected, ESR-tree needs more space than TSR-tree. This is due to the higher number of nodes and relationships generated in this index.

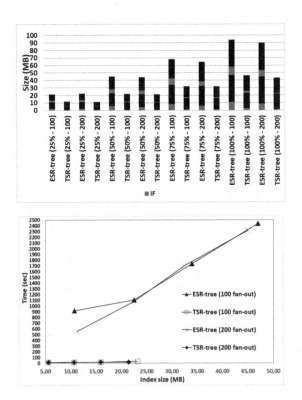

**Fig. 5.** Index size for the TSR-tree and ESR-tree w.r.t. fanout (up); Creation time for the ESR-tree and TSR-tree index (down).

Figure 5 (down) illustrates the results regarding the time for inserting the various fractions of the dataset to Neo4j. We notice that the creation time of the ESR-tree is larger compared to TSR-tree. This is rational as ESR-tree includes a higher number of nodes and relationships. Both cases exhibit a linear behavior.

**Experiments on STKP query performance.** Figure 6 (up) depicts the average performance (in terms of execution time) of the query processing algorithms for 5 different queries, each of which has the same query length (i.e. number of episode abstractions). For the ESR-tree-based algorithm we experimented with two different sizes of the regular grid that is required by our selectivity model, namely a $30 \times 30 \times 30$ and a $100 \times 100 \times 100$ grid. We also present the results for fanout equal to 100. From the results we notice that the ESR-tree-based STKP algorithm that uses our selectivity model has the best performance, followed by the algorithm ESR-tree-based STKP without the selectivity model. Moreover, the finer the grid the better performance is. Due to space constraints, we do not present results for fanout equal to 200, which however lead to similar conclusions.

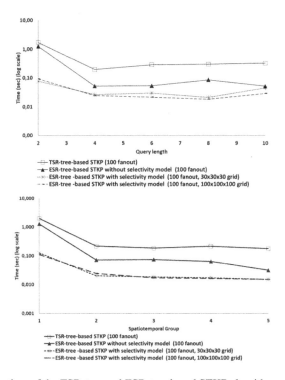

**Fig. 6.** Execution time of the TSR-tree and ESR-tree based STKP algorithms w.r.t. query length (up) and queries of different spatio-temporal extent (down).

On the other hand, in order to compare the query performance of the two algorithmic approaches with respect to the spatio-temporal extent of the queries, we used the same queries as in the previous experiment and we defined 5 different groups where each group is created by gradually increasing the spatio-temporal extent of the episode abstractions. This time each group has queries of varying length, namely 2, 4, 6, 8 and 10 abstract episodes. Figure 6 (down) depicts the average performance of the different groups. We notice that regardless of the spatio-temporal extent, it is the ESR-tree-based algorithm with the selectivity model, the smallest fanout and the finest grid that has the best performance.

# 5 Related Work

In this section, we briefly present the relevant domains of this work that include, on the one hand, spatial-keyword indexes and query processing algorithms and, on the other hand, pattern matching techniques in mobility data.

Several types of spatial-keyword queries have been discussed in the literature, including Boolean range, Boolean kNN, top-k NN and spatio-textual similarity joins [2, 3, 5, 7, 11]. Moreover, according to [3], geo-textual indexes that can support the efficient resolution of the above queries can be categorized in different categories, with respect to the spatial indexing scheme, the text index employed, and the hybrid manner that the spatial and text index are combined. For instance, the IR-tree [16] utilizes the well-known R-tree spatial index. Each node of the IR-tree has a pointer to an inverted file that represents all the textual data of the objects that are children of this node. The leaf nodes of the tree also have IFs for their textual data. Each non-leaf node has also a pseudo-document that represents with a proportional weight all the documents contained in the subtree rooted at this node.

Regarding pattern matching in mobility data, [8] represents a trajectory as a sequence of locations (defined as zones) and a trajectory pattern is defined as a sequence of symbols from a specific scale level. The pattern search follows an algorithm that is based on non-deterministic finite automata (NFA). Contrary to our approach, this research does not support temporal and topological constraints on the zones of a pattern. In [15], flexible patterns are used for searching regular expressions in trajectories with spatio-temporal criteria using a predetermined separation of the spatial area that is defined as a spatial alphabet. Differently from our approach, flexible patterns do not support textual (i.e. keywords) constraints.

In [6], the authors introduce the concept of symbolic trajectories, which in their simplest form are defined as a function of time over string labels. Symbolic trajectories focus mainly on queries with textual and temporal criteria, while spatial criteria are resolved in a post-processing step at the data object level. In our case, we propose hybrid index structures where space, time and textual information is of equal importance, since an efficient resolution of the STKP requires repetitive invocation of spatio-temporal-keyword matching queries.

In [12], we introduced a general SQL framework for querying semantic trajectory databases, where one of the provided functionalities was the discovery STKP. More specifically, we introduced a loosely connected index that is actually a spatio-temporal index (e.g. a 3D R-tree or a TB-tree) and an additional IF, connected only at the leaves of the spatio-temporal index. This approach can be considered as a solution when employing separate off-the-shelf indexes but, by default, it is weaker than the tightly integrated approach, which is proposed in this paper.

# 6 Conclusions

In this paper, we proposed hybrid spatio-temporal-textual indexes and query processing algorithms for the efficient STKP search in semantic trajectory databases. Between the two proposals (the trajectory-level-based TSR-tree and the episode-level-based ESR-tree), we conclude that regardless of the query length the search based on the ESR-tree, boosted by an appropriately designed selectivity model, is the winner, with the penalty of the higher index creation time and size compared with the TSR-index. In the future we plan to (a) investigate distributed and parallel algorithms to support real-time discovery of STKP over big datasets and (b) exploit STKP search in supporting expensive data mining operations over semantic trajectory databases, such as frequent motion patterns, semantic-based clusters and outliers.

**Acknowledgments.** This work has been partly supported by the University of Piraeus Research Center.

# References

1. Apache Lucene, http://lucene.apache.org/
2. Bouros, P., Ge, S., Mamoulis, N.: Spatio-Textual Similarity Joins. PVLDB **6**(1) (2012). doi:10.14778/2428536.2428537
3. Chen, L., Cong, G., Jensen, C.S., Wu, D.: Spatial keyword query processing: an exprerimental evaluation. PVLDB (2013). doi:10.14778/2535569.2448955
4. Dingqi, Y., Daqing, Z., Vincent, W.Z., Zhiyong, Y.: Modeling user activity preference by leveraging user spatial temporal characteristics in LBSNs. TSMC, **45**(1) (2015). doi:10.1109/TSMC.2014.2327053
5. Frentzos, E., Gratsias, K., Pelekis, N., Theodoridis, Y.: Algorithms for nearest neighbor search on moving object trajectories. Geoinformatica **11** (2007). doi:10.1007/s10707-006-0007-7
6. Guting, R.H., Valdes, F., Damiamni, M.L.: Symbolic Trajectories. ACM Trans. Spat. Algorithms Syst. **1**(2) (2015). doi:10.1145/2786756
7. Hariharan, R., Hore, B., Li, C., Mehrotra, S.: Processing spatial-keyword (SK) queries in geographic information retrieval (GIR) systems. In: Proceedings of SSDBM (2007). doi:10.1109/SSDBM.2007.22
8. du Mouza, C., Rigaux, P.: Mobility patterns. GeoInformatica **9**(4), 297–319 (2005). doi:10.1007/s10707-005-4574-9
9. Neo4j, Graph Database, http://www.neo4j.org/
10. Parent, C., Spaccapietra, S., Renso, C., Andrienko, G., Andrienko, N., Bogorny, V., Damiani, M.L., Gkoulalas-Divanis, A., Macedo, J., Pelekis, N., Theodoridis, Y., Yan, Z.: Semantic trajectories modeling and analysis. ACM Comput. Surv. **45**(4) (2013). doi:10.1145/2501654.2501656
11. Pelekis, N., Andrienko, G., Andrienko, N., Kopanakis, I., Marketos, G., Theodoridis, Y.: Visually exploring movement data via similarity-based analysis. JIIS **38**(2) (2012). doi:10.1007/s10844-011-0159-2
12. Pelekis, N., Sideridis, S., Theodoridis, Y.: Hermes$^{sem}$: a Semantic-aware framework for the management and analysis of our LifeSteps. In: Proceedings of DSAA (2015). doi:10.1109/DSAA.2015.7344849

13. Pelekis, N., Theodoridis, Y., Janssens, D.: On the management and analysis of our lifesteps. SIGKDD Explor. **15**(1), 23–32 (2013). doi:10.1145/2594473.2594478

14. Pfoser, D., Jensen, C.S., Theodoridis, Y.: Novel approaches to the indexing of moving object trajectories. In: Proceedings of VLDB (2000)

15. Vieira, M.R., Bakalov, P., Tsotras, V.J.: Querying trajectories using flexible patterns. In: Proceedings of EDBT (2010). doi:10.1145/1739041.1739091

16. Wu, D., Cong, G., Jensen, C.S.: A framework for efficient spatial web object retrieval. VLDBJ **21**(6), 797–822 (2012). doi:10.1007/s00778-012-0271-0

17. Zhang, C., Han, J., Shou, L., Lu, J., Porta, T.L.: Splitter: mining fine-grained sequential patterns in semantic trajectories. PVLDB **7**(9) (2014). doi:10.14778/2732939.2732949

# Natural Language Analysis of Online Health Forums

Abul Hasan[(✉)], Mark Levene, and David J. Weston

Department of Computer Science and Information Systems Birkbeck,
University of London, London WC1E 7HX, UK
{abulhasan,mlevene,dweston}@dcs.bbk.ac.uk

**Abstract.** Despite advances in concept extraction from free text, finding meaningful health related information from online patient forums still poses a significant challenge. Here we demonstrate how structured information can be extracted from posts found in such online health related forums by forming relationships between a drug/treatment and a symptom or side effect, including the polarity/sentiment of the patient. In particular, a rule-based natural language processing (NLP) system is deployed, where information in sentences is linked together though anaphora resolution. Our NLP relationship extraction system provides a strong baseline, achieving an $F_1$ score of over 80% in discovering the said relationships that are present in the posts we analysed.

**Keywords:** Natural language processing · Health related forums · Rule-based system

## 1  Introduction

Health related posts in medical forums often contain factual information regarding drug usage, the effectiveness of a drug when used to treat a symptom, and the experience of any adverse effects from the drug. In order to extract this type of information from free text in medical forums, natural language analysis of the text is required [5]. As has been pointed out by Karimi et al. [8], detecting, from unstructured text, the disease, treatment and symptom entities, their attributes and existing relationships, is a major research issue in the Natural Language Processing (NLP) domain. Although progress has been made in extracting entities, there is still the challenge of extracting specific relationships between these entities [7]. In particular, here we are interested in extracting relationships of the form (treatment, polarity/sentiment, symptoms/side effects), which represent relationships that provide us with information on the effectiveness or otherwise of various treatments, especially medication. When aggregated over many forum posts, such triples could inform practitioners and/or patients on the effectiveness of treatments beyond the information gathered from studies published in medical journals and by the pharmaceutical companies.

We report on an initial proof of concept of extracting such triples from about 1000 posts related to Parkinson's disease from the PatientsLikeMe website [14],

© Springer International Publishing AG 2017
N. Adams et al. (Eds.): IDA 2017, LNCS 10584, pp. 125–137, 2017.
DOI: 10.1007/978-3-319-68765-0_11

providing details of the algorithm we deployed and the results from a comprehensive evaluation of the algorithm. It is important to note that patients' comments in a forum, such as the one we are analysing, will contain slang and verbose, informal, descriptions of treatments and side effects (for example, using "body shaking" instead of the more formal "tremor"). Such informal terms are not normally present in standard ontologies such as the *Unified Medical Language System* (UMLS) [1]. As a result of this difficulty, much of the previous research in this area has focused on extracting formal medical terms from the free text. For example, Gupta et al. [7] attempted to extract drugs and treatments (DTs), and symptoms and conditions (SCs) terms present in the forum text. On the other hand, Nikfarajam et al. [12] and Sampathkumar et al. [16] extracted adverse drug reactions (ADRs) from Twitter and the DailyStrength forum [4]. Their research shows that social media is a valuable source for finding information related to drugs, symptoms and side effects.

We now give a brief summary of our NLP relationship extraction system, whose aim is to build triples, which can be aggregated to provide useful statistical medical information relating to the effectiveness of various treatments. Our model makes use of the following *concepts*:

1. Drugs (X), or more generally, treatments.
2. Symptoms (Y), which the drug is meant to treat.
3. Side effects (Z), which are caused by the drugs.
4. Polarity (P), which indicates how positive a treatment is or how negative a side effect is from the patient's point of view.

The system first extracts different health related concepts from the forum posts, and then creates structured information by forming a relationship between a drug and a symptom or side effect, through polarity analysis of the text. We termed such a relationship formally as a *disease triple* henceforth simply a *triple*.

Despite machine learning techniques such as *conditional random fields* (CRFs) [10] being very effective in NLP information extraction, we have chosen to build a *rule-based system* [10]. The reasons for this choice are:

1. To the best of our knowledge, this is the first attempt to extract, from social media, relationships in the form of disease triples, which include patient sentiment. There is no baseline system for such work and in order to attain deep knowledge of the use of natural language, specialised rules are often needed.
2. Dictionaries, lexicons and ontologies in the medical domain are built for extraction tasks from documents written by experts [7]. However, patients are, in most cases, not familiar with this terminology, so they tend to use commonly understood terms. As a result matching to such pre-built dictionaries often results in poor performance. In order to build common domain knowledge, it is first necessary to manually analyse a significant number of posts, and extend publicly available lexicons and gazetteers using a specifically designed set of generalised rules for extracting structured information from the free text.

3. Once the baseline is established it is possible to export the set of designed rules and resulting extended gazetteers to be used in a more sophisticated machine learning technique such as word2vec [11], to attain transferability and scalability when extracting triples from other forums.

Our overall contribution is the summarisation of health related forum posts by identifying relationship between concepts in the form of disease triples to provide a coherent structure, which can be used to extract meaningful medical statistical information.

## 1.1   A Motivating Example

Let us examine the following example post to motivate the research:
*"I take 600 mg of gabapentin at bedtime, helps me shake and kick less; and a donepezil 10 mg, settles me down allowing sleep. Clonazapam works great but I can't take the groggy, foggy head the next day."*
After various pre-processing steps and the application of linguistic rules, we create the disease triples as shown in Table 1. Most of the existing work concentrates on processing a single sentence. However, our method is a formal approach that uses information from the whole post, which may contain several sentences, by making use of *anaphoric relations* [6] present in the sentences. At this stage we do not identify dosages or temporal information present in the text; these will be tackled in future work.

**Table 1.** +, -, symp, side, drug, list, con and intens, denote positive polarity, negative polarity, symptom, side effects, drug/treatment, list of nouns, conjunction and intensifier, respectively.

| Sentence or sentence segment | Disease triple |
| --- | --- |
| I take 600 mg of gabapentin$_{drug}$ at bedtime, helps$_+$ me shake$_{symp}$ and$_{list}$ kick$_{symp}$ less$_-$; | (gabapentin, +, shake) <br> (gabapentin, +, kick) |
| and$_{con}$ a donepezil$_{drug}$ 10 mg, settles$_+$ me down allowing sleep$_{symp}$ | (donepezil, +, sleep) |
| Clonazapam$_{drug}$ works$_+$ great$_{intens}$ | (Clonazapam, +, ?) |
| but$_{con}$ I can't take$_-$ the groggy$_{side}$, foggy head$_{side}$ the next day. | (Clonazapam$_{drug\ anaphora}$, -, groggy) <br> (Clonazapam$_{drug\ anaphora}$, -, foggy head) |

The rest of the paper is organised as follows. Section 2 describes prior research related to our work. Section 3 discusses our text processing architecture and describes the details of our algorithm. The data set and its annotation are discussed in Sect. 4. In Sect. 5 we provide our findings and results from the experiment on the posts from PatientsLikeMe. Finally, in Sect. 6 we give our concluding remarks and discuss future work.

## 2    Related Work

A comprehensive review of NLP systems used for text processing in the medical domain can be found in [8]. Prior research in finding useful health related information from social media, mainly focused on *named entity extraction* (NER) tasks such as discovering ADRs in relation to a drug or treatment as reported in [8]. For example, Nikfarjam et al. [12] built a system, called ADRMine, using CRFs for recognising ADR mentions from social media. Features such as contextual, lexical and semantic parts-of-speech (POS) tags features were added to an existing CRF classifier. ADRMine also added word embedding features, created from word2vec [11], trained on Twitter and the DailyStrength [4] corpora. Moreover, Nikfarjam et al. [12] added clusters of words formed from word2vec to their supervised model as an additional feature. In a recent paper, Korkontzelos et al. [9] analysed the effect of sentiment analysis features in ADR classification, which made use of rules such as *"negation"* to improve the performance of their system. Dai et al. [3] also investigated features to use for finding ADR in Twitter posts.

On the other hand, Sampathkumar et al. [16] mined ADRs from health forums using a supervised *hidden Markov model* (HMM) [10]. The HMM provides statistical structure for the forum messages, where drug and side effects keywords representing the causal relation between the drug, side effects and other words, were encoded as hidden states. Concepts were extracted the from messages using existing medical lexicons. The model was trained with the positive samples of ADRs, and learnt the association between drugs and side effects through the presence of keywords. After training the forum messages, hidden states offered predictions from the preprocessed observed messages. The authors conducted various experiments by varying different components of the system. One of their findings was that the F-score of the supervised classification model is significantly lowered as the size of the dictionaries is reduced.

Comparing the results from the above models, it seems that the model using HMMs achieved better performance than that using CRFs. The reason for CRFs not performing as well, can be attributed partially to the pre-built dictionaries used for labelling the text, containing noisy data [7]. However, the lack of a common data set in this domain makes the judgement in comparing different models a difficult task.

Closely related to our work, is a semi-supervised algorithm deployed by Gupta et al. [7], where they extracted SCs and DTs from social media using lexico-syntactic patterns. At first, they labelled the concepts using dictionaries constructed from publicly available sources. Then, flexible patterns were created by looking at two to four words before and after the labelled tokens. Patterns were scored by a frequency measure, i.e. the top-$k$ most occurred patterns were chosen. They applied these patterns to all the sentences and extracted the matched phrases. These learned phrases were added to the dictionary, and the process was repeated until convergence. This method resulted in an improvement of the $F_1$-score by approximately 5% over the baseline lexicon-based approach. The authors also reported on the discovery of new DTs and SCs terms that were not present in the seed dictionaries. Also closely related is the recent work of Pain et al. [13]. They investigated classification methods for identifying drug/effect relations and their corresponding polarities for tweets.

We note that extracting entities from individual sentences in a post, is an important step towards analysing natural language in medical forums. However, linking the sentences in a post would allow a more comprehensive analysis of the text. Disambiguation of semantic relations between two expressions (sentences in our case) is known as *anaphora resolution*, where a later expression (the *anaphor*) has some semantic relation to an earlier expression (the *antecedent*). The rule-based system described in [6] is a knowledge-centric and pattern-based approach for disambiguating anaphoric references in clinical records. For our work, we took a slightly different approach by considering protagonist theory [6], which suggests that narrative events are centered on one or more key actors. Our analysis of the posts reveals that a drug mention found in a sentence is often referred to in subsequent sentences. Thus, in the case of drug mentions, we seek to find the antecedents of anaphoric expressions.

## 3  Methodology

The overall methodology is schematically shown in Fig. 1 and the corresponding pseudo-code of the algorithm is presented in Algorithm 1 below. In the following subsections we describe this methodology in more detail.

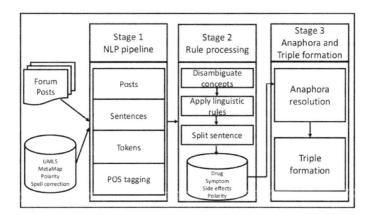

**Fig. 1.** Text processing architecture

### 3.1  NLP Pipeline

The text processing pipeline, constructed using GATE [2], splits the posts into sentences, tokenises the text and labels the tokens with their POS tags. At this stage, our system recognises drugs, symptoms, side effects and opinions present in the text by using different lexicons and plug-ins to construct the text processing pipeline. Apart from using some publicly available resources, we constructed gazetteers using domain knowledge and extended these by augmenting the terms during the training phase of the system.

### 3.2 Rule Processing

We disambiguate multiple concepts recognised in the previous stage by applying different linguistic rules, then split the sentences and compute the polarity score of each resulting segment.

**Disambiguation of Concepts.** We extracted four types of concepts from each sentence, i.e. drugs, symptoms, side effects and polarities. We now briefly discuss the concept extraction and disambiguation process.

- **Drug extraction.** We used a subset of the RxNorm terminology [17] as our drug gazetteer, where the vocabulary consisted of drugs and treatments used for Parkinson's disease; RxNorm is a normalised naming system for generic and branded drugs. For our purpose, it is very important to distinguish between the mentions of generic and branded drugs to avoid extracting the same type of drug mention multiple times. As such we have constructed two dictionaries for generic and branded drugs. A feature is added to the drug token according to the drug gazetteer it is extracted from. If a sentence contains both prescription and generic drug mentions, the generic drug mention is subsumed. Spelling mistakes in drug names are corrected using a normalised edit distance of two based on *Levenshtein's edit distance* [2].

- **Symptom extraction.** We used the MetaMap [1] annotation plug-in to annotate symptoms (sign or symptom semantic types) in sentences. We also constructed a separate symptom gazetteer using domain knowledge, and terms such as "voice", "smell" and "restless" were also added to the gazetteer. By default the polarity feature of symptoms recognised by the MetaMap program are set to negative. The symptom gazetteer contains explicit polarity for each symptom. Not all the symptoms present in the gazetteer are negative, for example concepts such as "sleep" and "energy" are marked as positive. The polarity feature is extracted from the gazetteer for each symptom.

- **Side effect extraction.** For side effects, a gazetteer using COSTART (Common Standard Thesaurus of Adverse Reaction Terms) [17] is constructed. We extended the dictionary by adding new terms during the training phase.

- **Polarity extraction.** We have collected the polarity terms from the MPQA Subjectivity Lexicon [18], which contains more than 8000 words annotated manually by the authors as positive, negative or neutral. The lexicon also includes the POS information for the terms. Symptom and side effect terms present in the dictionary are not labelled as polarity terms. The prior polarity score for positive and negative terms are set to $+1$ and $-1$ respectively, and neutral words are given as score 0. Our system matches a token with the lexicon if the POS information present in the lexicon is the same as the POS category of the token in the sentence.

**Linguistic Rules.** Matching polarity words is not enough in order to extract opinion from a sentence, and it often produces wrong result. A description of how valence of a lexical item is modified by the presence of different lexical items such as "negation", "modifiers" and "presuppositionals" can be found in [15]. Similar to [15] we applied following heuristic linguistic rules:

- **Negation.** If a negation word such as "not" and "don't" precedes in 0 to 3 tokens of a polarity, symptom or side effect concept, then the polarity of each concept is reversed and will generate the feature `negation_concept` (for example, "doesn't_work").
- **Modifiers.** Modifiers such as intensifying adverbs (for example, 'very", "strongly"), diminishing adverbs (for example, 'little", "kind of") increases or decreases the sentiment value of a concept and generate a feature `modifier_concept` (for example, "very_tired"). This rule modifies polarity of the first matched concept at a distance of 0 to 3 tokens.
- **Presuppositional.** The polarity of presuppositional items such as "barely" is multiplied with that of the concepts and polarity is flipped as a result. A feature such as `pre_concept` (for example, "barely_tremor") is generated after applying this rule. As for modifiers, this rule also changes the polarity of the first matched concept at a distance of 0 to 3 tokens.

**Split Sentence.** In this step, we first split a sentence and then calculate the final polarity score. These two steps are described below:

- **Sentence segmentation.** We used common conjunctions ( "and", "but") and "until" to segment a sentence. Our analysis revealed that "and" is sometimes used to connect two or more words to form a list rather than connecting two different parts of a sentence. We constructed a rule to find such conjunctions. For example if POS tags of the two tokens in either end of "and" are same, then it is denoted as a list of tokens and we do not segment sentence in such cases.
- **Final polarity score.** We add the polarity scores of all the opinion concepts in a sentence or segment of a sentence. The polarity of a triple is positive if total score is more than 0, negative if it is less than 0. If the score is 0, then polarity of the triple will be that of the symptom or side effect.

### 3.3   Triple Formation

At this stage, we first perform anaphora resolution and then form triples. These two steps are described in the following two paragraphs.

**Anaphora Resolution.** Messages in a forum contain sentences referring to the concepts mentioned earlier in the text. Our rule for finding anaphoric references for drugs is: if the current sentence has a drug mention, then the drug is carried forward to the next sentence in the text. Using this rule, if a triple has a drug

```
Data: P is a list of Posts
1  foreach p in P do
2  |   Split p into a list of sentences, S;
3  |   foreach s in S do
4  |   |   Tokenise s into a list of tokens, T;
5  |   |   foreach t in T do
6  |   |   |   Append the POS information;
7  |   |   |   Identify the concept class, C matching with the gazetteers;
8  |   |   |   Let, C = {X, Y, Z, P} where X, Y, Z, P are drug, symptom, side
   |   |   |   effect, and polarity, respectively;
9  |   |   |   Disambiguate concepts t;
10 |   |   |   Calculate the polarity, p by applying linguistic rules for t in P;
11 |   |   end
12 |   |   Split s into a list of segments, G using conjunctions, and, but, until;
13 |   |   foreach g in G do
14 |   |   |   Compute polarity score, SC;
15 |   |   |   Create a list of triples, L;
16 |   |   |   A triple is either (X, Y, SC) or (X, Z, SC);
17 |   |   |   Where, "?" is the placeholder for a missing concept;
18 |   |   |   foreach l in L do
19 |   |   |   |   Perform anaphora resolution for missing X;
20 |   |   |   end
21 |   |   end
22 |   end
23 end
```

**Algorithm 1.** Text processing algorithm

mention and a subsequent triple contains the default drug concept ("?"), then we replace the default with the drug found, and repeat the same process for all the sentences in text until we find a new drug mention. If we find multiple mentions of drugs then multiple triples are created containing each drug mention. We also look for the patterns such as "from X to Y", which indicates that the person actually stopped using drug "X" and moved on to using "Y". In such case, we add a feature to the drug ("X") token indicating that the algorithm should stop creating triple for this drug in subsequent sentences.

**Disease Triples.** We create a list of concepts by ordering them according to their offset from the beginning of a sentence. Triples are formed using the following format:

1. *Triple 1:* Drug, Polarity, Symptom
2. *Triple 2:* Drug, Polarity, Side effect.

The algorithm iteratively finds drug, polarity and symptom or side effect concepts using the order shown in the formation of triple. A triple is formed by taking three consecutive concepts of a different kind. In our algorithm, consecutive concepts (for example, drugs) of same kind, signals the starting point of a new triple. The algorithm places a default concept which is "?" in case of missing concepts.

# 4  Experimental Setup and Evaluation

The following subsections describe the data we collected and the procedure we used for verifying our annotation of the unstructured text. We then describe our experiment and its results.

## 4.1  Data Set

PatientsLikeMe [14] is an online health discussion forum, where patients with chronic health conditions can share their experiences living with disease. For our study, we extracted user comments from discussion threads related to Parkinson's disease. After registering with this website, pages were automatically scraped and the posts were anonymised by removing user IDs. A total of 1058 posts were collected from the period of April, 2016 to June, 2016.

500 posts were used for training and 400 for testing the system. The remaining posts were used for the annotation validation, described in the next section, where 58 posts were used to train the annotators and the remaining 100 posts for cross validation of the annotations.

## 4.2  Annotation Validation

The annotation for the dataset was carried out by the first author. In order to verify the fidelity of these annotations an experiment was conducted using a small subset of the data, where the level of agreement between the annotator and other annotators was measured.

Ten researchers from Birkbeck's department of Computer Science and Information Systems volunteered for the validation experiment. Annotators were trained by showing annotated posts (20 posts were chosen from the annotator training set of 58) and explaining each concept and triple types. The remaining 100 posts were divided randomly into five sets of 20. Each of the ten annotators were randomly assigned two sets such that each set would get two annotations from different annotators. The agreement between the actual annotator and volunteers were calculated using both Cohen's kappa statistic [19] and accuracy. A very high level of agreement in recognising drug, symptom, positive and negative strings was achieved. However, agreement and accuracy in recognising triples and side effects were somewhat lower (72.09% and 76.18% respectively). It was subsequently determined that two of the annotators had not fully understood the task. Table 2 shows the results with and without these two 'outlier' annotators. It can be seen that there is a better agreement in identifying triples, however agreement in identifying side-effects has been reduced. This is due to the small number of side-effect concepts in the validation set (10 in total).

**Table 2.** Annotation validation result. Kappa-O and Accuracy-O are the results after discounting the 2 'outlier' annotators.

| Concept | Kappa | Accuracy | Kappa-O | Accuracy-O |
|---|---|---|---|---|
| Drug and treatment | 87.37% | 94.40% | 86.84% | 94.16% |
| Symptom | 91.49% | 96.90% | 92.73% | 97.32% |
| Side effects | 76.18% | 99.40% | 71.05% | 99.23% |
| Positive polarity | 89.71% | 96.26% | 89.06% | 95.97% |
| Negative polarity | 90.72% | 97.15% | 89.77% | 97.02% |
| Triples | 72.09% | 88.79% | 76.18% | 90.72% |

### 4.3   Evaluation

To evaluate our proposed approach the standard measures of accuracy, precision, recall and $F_1$ [19] were used. Each post was split into segmented sentences as described in Sect. 3.2. Triples formed from a segmented sentence are then merged with those from other segments and subsumed in case of repetition. A sentence can contain zero of more concepts and consequently zero or more disease triples, as shown in Table 3.

**Table 3.** Test set summary

| Posts | Sentences | Zero triples | One triple | More than one triple |
|---|---|---|---|---|
| 400 | 2564 | 544 | 1447 | 573 |

## 5   Training and Test Results

Training of the system was conducted incrementally over five iterations. The training data was split into 5 sets of 100 posts each. At the beginning of the first iteration, we analysed the posts from the first set, i.e. annotated them with the concepts, generated rules and extended dictionaries, and then evaluated the system's output with the actual annotation. After we achieved a satisfactory performance, which meant a precision of approximately 80% or above, we moved on to the next iteration and followed the same procedure. The evaluation was carried out at each iteration by cumulatively adding a new set of posts to the posts from previous iterations.

It is interesting to see the overall performance from the training data, Table 4. This was achieved in a principled manner, as described in Sect. 3.2, involving the development of as few rules as possible. It should be noted that without anaphora resolution very few triples would have been successfully identified. In addition, dictionaries were extended when necessary. Though we are very successful in

**Table 4.** Training and test results

| Dataset | Training | | | | Test | | | |
|---|---|---|---|---|---|---|---|---|
| Concept | Accuracy | Precision | Recall | $F_1$ | Accuracy | Precision | Recall | $F_1$ |
| Drug | 95.06% | 99.63% | 99.12% | 97.29% | 90.71% | 88.29% | 95.14% | 91.59% |
| Symptom | 95.71% | 85.64% | 99.06% | 91.86 | 94.26% | 84.08% | 87.36% | 85.69% |
| Side effects | 99.06% | 87.95% | 98.50% | 92.92% | 98.42% | 80.25% | 93.53% | 86.38% |
| + polarity | 90.19% | 80.98% | 96.15% | 87.92% | 86.44% | 72.68% | 94.42% | 82.13% |
| - polarity | 90.61% | 79.60% | 94.04% | 86.22% | 87.08% | 73.57% | 88.52% | 80.35% |
| Triple 1 | 83.06% | 81.01% | 95.23% | 87.54 % | 73.93% | 71.11% | 96.02% | 81.71% |
| Triple 2 | 84.76% | 82.28 % | 94.96% | 88.16% | 74.47% | 71.31% | 96.81% | 82.13% |

recognising concepts, the system makes a few mistakes in disambiguating polarity terms. As a result, the performances at triple level are lower, which resembles that of recognising positive and negative polarity terms. This result is in line with our hypothesis (see Sect. 1) that we can establish a relation between drug and symptom and drug and side effects through the polarity of a sentence. Although the polarity dictionary [18] has been extended by incorporating common phrases and is also supported by set of generalised rules, there is still room for improvement.

For testing the system was run over the remaining 400 post test dataset, without any modification to the system. The results are shown in Table 4. We can see that the system has generalised well. In general the disease triple identification has had the greatest fall in performance, since recognising these relationships is dependent on accurately identifying the concepts from which they are comprised.

# 6    Concluding Remarks

We have proposed summarising potential useful medical information in free-form unstructured text, with disease triples. We have developed a strong baseline system, achieving an $F_1$ score of over 80%, in identifying these disease triples using traditional NLP methods, and have demonstrated that this approach can generalise successfully. One current limitation of our system, which is left as future work, is that we are not yet recording useful temporal/quantitative data such as dosages or frequency of recurrence of side effects.

Our next goal is to transfer the knowledge gained in our system to discover triples in other patient forums. In order to achieve this we anticipate the use of machine learning methods, such as word2vec [11], which can adapt to different usage of language than the PatientsLikeMe forum we have concentrated on.

# References

1. Bodenreider, O.: The unified medical language system (UMLS): integrating biomedical terminology. Nucleic Acids Res. **32**(suppl 1), D267–D270 (2004)
2. Cunningham, H., Maynard, D., Bontcheva, K., Tablan, V., Aswani, N., Roberts, I., Gorrell, G., Funk, A., Roberts, A., Damljanovic, D., et al.: Developing language processing components with gate version 6 (a user guide). University of Sheffield, Department of Computer Science (2011)
3. Dai, H.J., Touray, M., Jonnagaddala, J., Syed-Abdul, S.: Feature engineering for recognizing adverse drug reactions from twitter posts. Information **7**(2), 27 (2016)
4. DailyStrength: https://www.dailystrength.org/. Accessed 04 May 2017
5. Denecke, K., Deng, Y.: Sentiment analysis in medical settings: new opportunities and challenges. Artif. Intell. Med. **64**(1), 17–27 (2015)
6. Gooch, P., Roudsari, A.: Lexical patterns, features and knowledge resources for coreference resolution in clinical notes. J. Biomed. Inform. **45**(5), 901–912 (2012)
7. Gupta, S., MacLean, D.L., Heer, J., Manning, C.D.: Induced lexico-syntactic patterns improve information extraction from online medical forums. J. Am. Med. Inf. Assoc. **21**(5), 902–909 (2014)
8. Karimi, S., Wang, C., Metke-Jimenez, A., Gaire, R., Paris, C.: Text and data mining techniques in adverse drug reaction detection. ACM Comput. Surv. (CSUR) **47**(4), 56 (2015)
9. Korkontzelos, I., Nikfarjam, A., Shardlow, M., Sarker, A., Ananiadou, S., Gonzalez, G.H.: Analysis of the effect of sentiment analysis on extracting adverse drug reactions from tweets and forum posts. J. Biomed. Inform. **62**, 148–158 (2016)
10. Manning, C.D., Schütze, H., et al.: Foundations of Statistical Natural Language Processing, vol. 999. MIT Press, Cambridge (1999)
11. Mikolov, T., Sutskever, I., Chen, K., Corrado, G.S., Dean, J.: Distributed representations of words and phrases and their compositionality. In: Advances in Neural Information Processing Systems, pp. 3111–3119 (2013)
12. Nikfarjam, A., Sarker, A., O'Connor, K., Ginn, R., Gonzalez, G.: Pharmacovigilance from social media: mining adverse drug reaction mentions using sequence labeling with word embedding cluster features. J. Am. Med. Inform. Assoc. **22**, 1–11 (2015)
13. Pain, J., Levacher, J., Quinqunel, A., Belz, A.: Analysis of twitter data for postmarketing surveillance in pharmacovigilance. In: Proceedings of the 2nd Workshop on Noisy User-generated Text, pp. 94–101 (2016)
14. PatientsLikeMe: https://www.patientslikeme.com/. Accessed 21 Apr 2017
15. Polanyi, L., Zaenen, A.: Contextual valence shifters. In: Shanahan, J.G., Qu, Y., Wiebe, J. (eds.) Computing Attitude and Affect in Text: Theory and Applications. The Information Retrieval Series, vol. 20, pp. 1–10. Springer, Dordrecht (2006). doi:10.1007/1-4020-4102-0_1
16. Sampathkumar, H., Chen, X.W., Luo, B.: Mining adverse drug reactions from online healthcare forums using hidden markov model. BMC Med. Inform. Decis. Making **14**(1), 91 (2014)
17. U.S. National Library of Medicine: https://www.nlm.nih.gov/. Accessed 21 Jun 2016

18. Wilson, T., Wiebe, J., Hoffmann, P.: Recognizing contextual polarity in phrase-level sentiment analysis. In: Proceedings of Human Language Technology Conference and Conference on Empirical Methods in Natural Language Processing (HLT/EMNLP), pp. 347–354. Association for Computational Linguistics (2005)
19. Witten, I.H., Frank, E., Hall, M.A., Pal, C.J.: Data Mining: Practical Machine Learning Tools and Techniques. Morgan Kaufmann, San Francisco (2005)

# Freudian Slips: Analysing the Internal Representations of a Neural Network from Its Mistakes

Sen Jia, Thomas Lansdall-Welfare, and Nello Cristianini[✉]

Intelligent Systems Laboratory, University of Bristol, Bristol, UK
nello.cristianini@bristol.ac.uk

**Abstract.** The use of deep networks has improved the state of the art in various domains of AI, making practical applications possible. At the same time, there are increasing calls to make learning systems more transparent and explainable, due to concerns that they might develop biases in their internal representations that might lead to unintended discrimination, when applied to sensitive personal decisions. The use of vast subsymbolic distributed representations has made this task very difficult. We suggest that we can learn a lot about the biases and the internal representations of a deep network without having to unravel its connections, but by adopting the old psychological approach of analysing its "slips of the tongue". We demonstrate in a practical example that an analysis of the confusion matrix can reveal that a CNN has represented a biological task in a way that reflects our understanding of taxonomy, inferring more structure than it was requested to by the training algorithm. In particular, we show how a CNN trained to recognise animal families, contains also higher order information about taxa such as the superfamily, parvorder, suborder and order for example. We speculate that various forms of psycho-metric testing for neural networks might provide us insight about their inner workings.

**Keywords:** Deep learning · Taxonomy · Computer vision · Explainable AI · Blackbox testing

## 1 Introduction

Deep neural networks deliver state of the art performance in different areas of AI [14,21,23], particularly in computer vision, and promise to be deployed in many further domains [19]. However, for all their convenience, they do attract the criticism that they operate as black boxes [1]: that they can only pick up correlations, with no regard for causality or other theoretical frameworks that humans would consider more explainable. This criticism is often summarised as "correlation trumps causation", based on the observation that there is no clear way to interpret the internal configurations of weights learnt by the network, and that they are trained to perform a specific prediction task, and not directly

© Springer International Publishing AG 2017
N. Adams et al. (Eds.): IDA 2017, LNCS 10584, pp. 138–148, 2017.
DOI: 10.1007/978-3-319-68765-0_12

rewarded for developing a higher level understanding of the problem at hand. This criticism is not necessarily true though, as there are many reasons to believe that our own theoretical frameworks respond (also) to criteria of economy, and therefore that black box machine learning algorithms might find it useful to represent data in the same way [22].

Indeed, it has been known for a long time [5] that there is no real reason for a neural network to prefer an elegant representation of reality that can capture some higher level understanding of the problem, being trained only to perform correct predictions. Yet we, as humans, tend to prefer simpler and structured representations, often invoking Occam's razor.

This drives at a problem that has been considered for many years, of making AI explainable, but which has recently found new urgency, amid concerns that machines are going to soon be making decisions about us that we will unable to understand and for which they can offer no explanation [8,10,11]. This problem goes to the heart of the old question of how can we interpret the inner representations of reality that are inferred by a neural network, as a way to understand if it contains any unintended biases?

The direct reaction of the engineer has always been that of opening up the network and tackling the mess of connections and neurons [18], much like a surgeon performing brain surgery (a phrase sometimes applied to deep neural networks [12]). However, there is another way to reveal information about the internal structure of these networks, loosely analogous to that of Sigmund Freud, who was trying to understand the internal world view of a black box not by its successes (which after all are what it was trained for) but by its errors and mistakes. Far from being random, these errors resulting from machine learning tasks, much like slips of the tongue, might reveal hidden biases and aspects of how the neural network represents the world internally. A systematic study of the errors made by deep–networks might reveal useful information about their hidden biases and assumptions without the need to unravel the role of each internal connection.

We suggest these "Freudian slips" may offer a promising alternative to other black-box approaches [3,16] to analysing the internal organisation of a network without surgery, and perform a first experiment to demonstrate the method. Our analysis of errors shows that a deep convolutional neural network can learn more structure than it is specifically asked to.

As a testing ground, we selected the domain of biology, where there is already a fundamental framework, that of evolution and therefore of taxonomy, in which we can assess whether the network is learning a similar representation of the world to humans. For instance, since animal species are not uncorrelated but have evolved over time from one another, there could be benefits for a neural network to learn more than just its minimal task of recognising different taxa, but to internally represent information about the phylogenetic taxonomic structure. For instance, this can be seen in the way that phylogenetic taxonomies are inferred from data using maximum parsimony [7].

We start with the task of teaching a network to assign images of mammals curated from the web to one of 54 animal families, where we have 540,000 images of individual animals, belonging to 3585 species, organised in 54 distinct families, themselves organised into 27 orders, all within the Mammalian class. We train a deep convolutional neural network to recognise the 54 families, an interesting task in and of itself [2,9], and perform an analysis on the errors it makes in order to learn about its internal representations.

Of course, there is still not a systematic framework to connect the "slips of the tongue" made by a CNN with its internal representations, but we claim that this might be worth developing, as we look for ways to make these systems more transparent, in the face of their inherently subsymbolic nature.

## 2    Methodology

### 2.1    Taxonomic Identification of Animals

Animal species are organised according to a standardised taxonomic system that divides them into seven major taxa: species, genera, families, orders, classes, phylums and kingdoms, as shown in Fig. 1, along with further subdivision into minor taxa (i.e. superfamilies, suborders, superorders, subclasses and so on). Using this taxonomy, we can define a distance between any two animals based upon how far apart they are in the phylogenetic tree, where we represent each node as a minor taxon. For example, if we were to only consider the seven major taxa, the distance between two carnivores, such as a Red Fox (*Vulpes vulpes*)

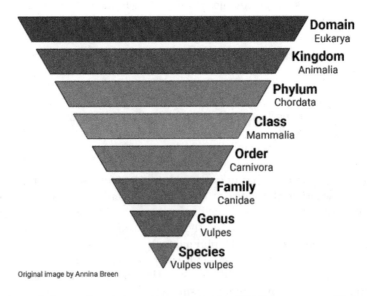

**Fig. 1.** Example hierarchy of the major taxa in the NCBI taxonomy of living species. Image by Annina Breen (Own work) [CC BY-SA 4.0], via Wikimedia Commons.

**Table 1.** List of the 54 Mammalian families, along with their common names used in the study.

| Family name | Common name example | Family name | Common name example |
| --- | --- | --- | --- |
| Aotidae | Night monkeys | Lemuridae | Ring-tailed lemurs |
| Atelidae | New world monkeys | Lepilemuridae | Sportive lemurs |
| Bathyergidae | Mole-rats | Leporidae | Rabbits and Hares |
| Canidae | Dogs | Macropodidae | Kangaroos |
| Caviidae | Guinea pigs | Macroscelididae | Elephant shrews |
| Cebidae | Capuchin monkeys | Manidae | Pangolins |
| Cercopithecidae | Baboons | Molossidae | Free-tailed bats |
| Cheirogaleidae | Dwarf lemurs | Muridae | Mice |
| Cricetidae | Hamsters | Mustelidae | Otters |
| Ctenomyidae | Tuco-tuco | Nesomyidae | Climbing mice |
| Dasyuridae | Quoll | Ochotonidae | Pika |
| Delphinidae | Dolphins | Octodontidae | Rock rats |
| Didelphidae | Opossums | Otariidae | Eared seals |
| Dipodidae | Jerboa | Phalangeridae | Cuscus |
| Echimyidae | Spiny rats | Phocidae | Earless seals |
| Emballonuridae | Sac-winged bats | Phyllostomidae | Leaf-nosed bats |
| Equidae | Horses | Pitheciidae | Titis |
| Erethizontidae | New world porcupines | Pteropodidae | Fruit bats |
| Erinaceidae | Hedgehogs | Rhinolophidae | Horseshoe bats |
| Felidae | Cats | Sciuridae | Squirrels |
| Geomyidae | Gophers | Soricidae | Shrews |
| Gliridae | Dormice | Spalacidae | Bamboo rats |
| Herpestidae | Mongeese | Talpidae | Moles |
| Heteromyidae | Kangaroo rats | Tupaiidae | Treeshrews |
| Hipposideridae | Old world bats | Vespertilionidae | Common bats |
| Hylobatidae | Gibbons | Viverridae | Civets |
| Indriidae | Lemurs | Ziphiidae | Beaked whales |

and a Red Panda (*Ailurus fulgens*), would be at the Order level, giving them a distance of 3 (Species → Genus → Family → Order).

This builds on recent work that has so far been done to assign the image of an animal to its correct species [2,9], while we focus here on the class of mammalians, and attempt to correctly assign each image of an animal to its appropriate family classification.

## 2.2 Data Curation

We aimed to select a number of mammalian families from the National Center for Biotechnology Information (NCBI) taxonomy [6] to obtain a representative, but broad coverage of all the species belonging to the mammalian class. Within the mammalian taxonomic tree, we found a total of 27 nodes at the Order level, and 140 nodes at the Family level. After filtering the species names to remove those containing special characters ('.' or '/') or with additional information, we compiled a list of every mammalian family containing at least 15 different species. Table 1 lists each of the 54 families resulting from this process, with an example of the common name for some of the animals that belong to the family.

Each family was additionally annotated with all of its child species from the NCBI taxonomy, along with the common names for the species, obtained by taking the title of the Wikipedia page after redirection from https://en.wikipedia.org/wiki/<species_name>. For example, the page for *Vulpes vulpes* (https://en.wikipedia.org/wiki/Vulpes_vulpes) redirects to the page entitled "Red fox".

Using the list of 3585 species and their common names for each of the 54 family categories, we queried for images from a popular search engine, retrieving an average of 98 images per query. We split the data for training and testing by assigning 100 randomly selected images per family to the hold-out test set. Following this, we performed data augmentation for any families which did not contain a minimum of 10,000 images in the training set. This included using a common technique in computer vision [15,17] of horizontally flipping and rotating images by [−15, −10, −5, 5, 10, 15] degrees in order to increase the number available images for training a classifier. In total, the "Family look" dataset contains 540,000 training and 5,400 test images, labelled by family. This dataset of labelled images, along with the NCBI taxonomy used, are available online at http://thinkbig.enm.bris.ac.uk/family-look

## 2.3 Learning a "Family Look" Classifier

Following the pre-processing steps in [13,17], each image in the "Family look" dataset was resized so the shortest length became 256 pixels. We took a center crop of size $256 \times 256$ from each image, ensuring the images were of a consistent size, before randomly cropping the images into $224 \times 224$ pixel patches for the training set, or taking the center $224 \times 224$ pixel patch for testing. Using the 18-layer ResNet CNN architecture proposed in [13] for object recognition and used in [9] for animal species identification[1], we trained the network on the 540,000 animal training images. The CNN model was trained from scratch using Stochastic Gradient Descent. We trained for 50 epochs using a mini-batch size of 64, an initial learning rate of 0.1 that decreased every 20 epochs by a factor

---

[1] The ResNet CNN used in their paper had the same architecture, but was much deeper. We trained a similar 152-layer network to [9] but found no clear difference with our 18-layer model, which aimed to strike a better balance between the depth of the network and the associated computational load.

of 0.1, along with a moment and weight decay being applied of 0.9 and 0.0001 respectively.

After training the CNN, we evaluated our model on the held-out test set of 5,400 test images, computing both an overall accuracy for the model and a classifier confusion matrix detailing specifically which family categories the model confused with one another for further analysis.

## 2.4    Constructing a Tree Representation

Once we had a trained classifier for discriminating between the different family categories in animal images, we wished to construct a tree representation of the mistakes made by the model. To construct a tree from the confusion matrix, we first needed to convert the matrix from a similarity matrix to a distance matrix, performed by subtracting each value in the confusion matrix from the maximal value. This distance matrix was then used to generate a tree of the mistakes by performing a furthest neighbour agglomerative hierarchical clustering [4], implemented using the `seqlinkage` command with the complete linkage option from the Bioinformatics toolbox [20]. The resulting tree represents each family category as a leaf node in the tree, with each internal node representing a cluster of animal families based upon how often they are mistaken for each other. In doing so, we can compare the categories which the model confused with each other with the taxonomic tree coming from the NCBI taxonomy [6] based upon taxonomic distance.

## 3    Results

In this experiment, we found that our deep CNN model could correctly classify the animals in our test set at the family level 53.22% of the time, well above the baseline of 1.85% one could trivially expect for a 54 category classification task. More interestingly, we find that there is a significant correlation ($\rho = 0.53$, $p < 0.0001$) between the family-similarities (as measured by error probability between them) and the taxonomic tree distance as indicated by the NCBI taxonomy [6], and shown in Fig. 2.

Considering this in terms of our Freudian slip concept, there was no reason for the network to learn any relation among these categories, and the expectation would be that there should be no correlation, unless the network learns weights corresponding with quantities that correlate at a higher taxon level than that of the family taxon, something that was not taught to it nor available from the data it has seen.

Probing deeper into the types of mistakes that the classifier made, we analysed how the distribution of errors made by the Neural Network over the 11 possible types of errors compared what would be expected if we were to assign the categories to the images uniformly at random. The ratio between the actual number of errors made by the network, and that expected under this null-model,

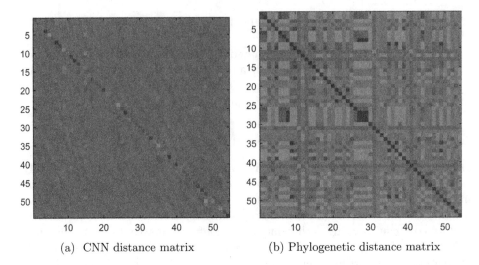

(a)  CNN distance matrix          (b) Phylogenetic distance matrix

**Fig. 2.** Comparison of the distance matrix calculated from the CNN confusion matrix and the phylogenetic distance matrix computed from the NCBI taxonomy.

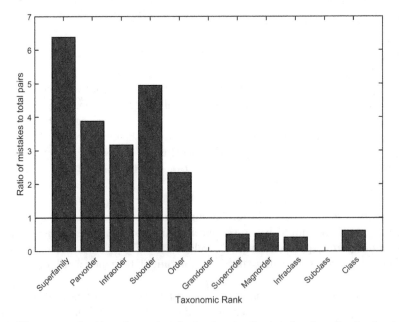

**Fig. 3.** Histogram showing the ratio of mistakes to the total pairs of animals at each phylogenetic distance. Values above 1 mean more mistakes were made at this taxonomic rank that we should expect, with values below 1 meaning less mistakes were made than expected.

is represented in Fig. 3, where values below 1 show that the network makes less mistakes than expected by chance, with those above 1 indicating the opposite.

We can immediately see that the classifier is not making mistakes uniformly across the 54 categories, but is making many more mistakes between animals sharing a taxonomic rank of Order or below, and making less mistakes than we should expect between animals sharing a taxonomic rank above that of the Order rank. With the notable exception of Suborder, we can also see that as the phylogenetic distance increase, we are less likely to confuse two animals with each other, suggesting that the network is indeed learning some internal representation of the phylogenetic taxonomy.

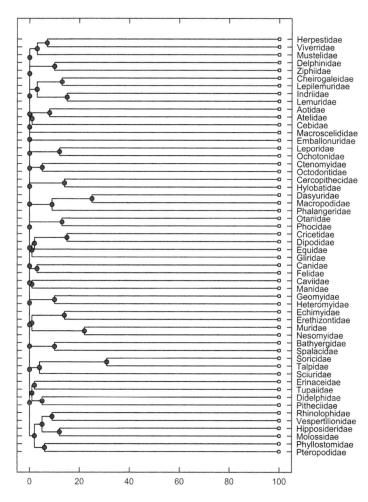

**Fig. 4.** Family tree built from the confusion matrix of the neural network showing the mammalian families which are most often mistaken for one another. We can see that this reflects the actual phylogenetic tree (Fig. 5) to some extend, mostly for relationships between animals below the Order rank.

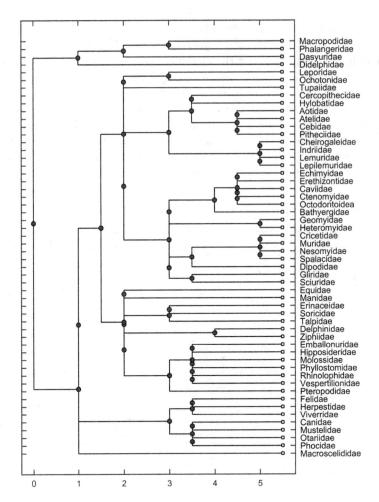

**Fig. 5.** Phylogenetic tree showing the taxonomic relationship between different mammalian families as recorded in the NCBI taxonomy [6]. Each level in the phylogenetic tree indicates a different major or minor taxon shared between animal families in the taxonomy.

Finally, examining the tree built from the confusion matrix (Fig. 4) and comparing this with the phylogenetic tree (Fig. 5), we can see that there is some taxonomic structure reflected in the mistakes of the network. For instance, we can see that the different families of microbats (Rhinolophidae, Vespertilionidae, Hipposideridae, Molossidae, Phyllostomidae) are grouped together with the megabat family (Pteropodidae). Similarly, the phylogenetic tree structure for the families of lemurs (Cheirogaleidae, Lepilemuridae, Indriidae, Lemuridae), seals (Phocidae, Otariidae), and marsupials native to Australia (Dasyuridae, Macropodidae, Phalangeridae) are also represented. While we can find many examples of this type, it is mostly only up to the Order rank, with no higher level taxonomic structure.

# 4   Discussion

While the question of understanding the internal representations in neural networks remains an important one, and it will probably be addressed by both surgery and theory, we think that our approach provides a simple way to examine the internal work representation of the network, just like the method of Freudian slips attempted to recognise internal structures based on the errors that a person made while speaking.

By no means do we think that this experiment settles the problem, but more that it points to a different way to organise a search for interpretable AI, one that can respect the inherently sub-symbolic and distributed representations that have made deep-networks so useful, while also respecting our need to understand how the network represents its knowledge.

Our representation of the world is also "deep", in the sense that it relies on a hierarchy of theoretical concepts, and communication between people requires that those concepts are shared by both. While this may be an elusive task in everyday experience, it is more simply demonstrated in certain scientific domains, such as taxonomy. But the general point about trying to match the abstractions used by humans with those used by deep networks might be more general than taxonomies.

Once we start developing methods to test certain properties of the internal knowledge representation, we are in a domain akin to 'psycho-metrics', and there is a lot of expertise that might be shared from that field. It would be useful to involve philosophers of science and psychologists in the discussions about readable AI.

One possible side-effect of this approach could be that – as we learn how to make sure that the network respects at least some of the internal constraints that we value, in our representation of the world – we might even be able to add this to the cost function used in training.

# References

1. Alain, G., Bengio, Y.: Understanding intermediate layers using linear classifier probes. arXiv preprint arXiv:1610.01644 (2016)
2. Chen, G., Han, T.X., He, Z., Kays, R., Forrester, T.: Deep convolutional neural network based species recognition for wild animal monitoring. In 2014 IEEE International Conference on Image Processing (ICIP), pp. 858–862. IEEE (2014)
3. Datta, A., Sen, S., Zick, Y.: Algorithmic transparency via quantitative input influence: theory and experiments with learning systems. In: 2016 IEEE Symposium on Security and Privacy (SP), pp. 598–617. IEEE (2016)
4. Day, W.H.E., Edelsbrunner, H.: Efficient algorithms for agglomerative hierarchical clustering methods. J. Classif. **1**(1), 7–24 (1984)
5. Denker, J., Schwartz, D., Wittner, B., Solla, S., Howard, R., Jackel, L., Hopfield, J.: Large automatic learning, rule extraction, and generalization. Complex Syst. **1**(5), 877–922 (1987)
6. Federhen, S.: The NCBI taxonomy database. Nucleic Acids Res. **40**(D1), D136–D143 (2012)

7. Felsenstein, J.: Inferring Phylogenies, vol. 2
8. Freitas, A.A.: Comprehensible classification models: a position paper. ACM SIGKDD Explor. Newsl. **15**(1), 1–10 (2014)
9. Gómez, A., Salazar, A., Vargas, F.: Towards automatic wild animal monitoring: identification of animal species in camera-trap images using very deep convolutional neural networks. arXiv preprint arXiv:1603.06169 (2016)
10. Goodman, B., Flaxman, S.: European union regulations on algorithmic decision-making and a "right to explanation". arXiv preprint arXiv:1606.08813 (2016)
11. Gunning, D.: Explainable artificial intelligence (XAI). Defense Advanced Research Projects Agency (DARPA) (2017)
12. Hassibi, B., Stork, D.G., Wolff, G.J.: Optimal brain surgeon and general network pruning. In: IEEE International Conference on Neural Networks, vol. 1, pp. 293–299 (1993)
13. He, K., Zhang, X., Ren, S., Sun, J.: Deep residual learning for image recognition. CoRR, abs/1512.03385 (2015)
14. Jia, S., Lansdall-Welfare, T., Cristianini, N.: Gender classification by deep learning on millions of weakly labelled images. In: 2016 IEEE 16th International Conference on Data Mining Workshops (ICDMW), pp. 462–467. IEEE (2016)
15. Jung, H., Lee, S., Park, S., Lee, I., Ahn, C., Kim, J.: Deep temporal appearance-geometry network for facial expression recognition. arXiv preprint arXiv:1503.01532 (2015)
16. Krause, J., Perer, A., Bertini, E.: Using visual analytics to interpret predictive machine learning models. arXiv preprint arXiv:1606.05685 (2016)
17. Krizhevsky, A., Sutskever, I., Hinton, G.E.: Imagenet classification with deep convolutional neural networks. In: Advances in Neural Information Processing Systems, pp. 1097–1105 (2012)
18. Le, Q.V.: Building high-level features using large scale unsupervised learning. In: 2013 IEEE International Conference on Acoustics, Speech and Signal Processing (ICASSP), pp. 8595–8598. IEEE (2013)
19. Lecun, Y., Bengio, Y., Hinton, G.: Deep learning. Nature **521**(7553), 436–444 (2015)
20. MathWorks. Bioinformatics toolbox: User's guide (r2017a) (2017). www.mathworks.com/help/pdf_doc/bioinfo/bioinfo_ug.pdf
21. Mnih, V., Kavukcuoglu, K., Silver, D., Rusu, A.A., Veness, J., Bellemare, M.G., Graves, A., Riedmiller, M., Fidjeland, A.K., Ostrovski, G., et al.: Human-level control through deep reinforcement learning. Nature **518**(7540), 529–533 (2015)
22. Rissanen, J.: Minimum description length principle. Wiley Online Library (1985)
23. Silver, D., Huang, A., Maddison, C.J., Guez, A., Sifre, L., Van Den Driessche, G., Schrittwieser, J., Antonoglou, I., Panneershelvam, V., Lanctot, M., et al.: Mastering the game of go with deep neural networks and tree search. Nature **529**(7587), 484–489 (2016)

# KAPMiner: Mining Ordered Association Rules with Constraints

Isak Karlsson, Panagiotis Papapetrou$^{(\boxtimes)}$, and Lars Asker

Department of Computer and Systems Sciences,
Stockholm University, Stockholm, Sweden
{isak-kar,panagiotis,asker}@dsv.su.se

**Abstract.** We study the problem of mining ordered association rules from event sequences. Ordered association rules differ from regular association rules in that the events occurring in the antecedent (left hand side) of the rule are temporally constrained to occur strictly before the events in the consequent (right hand side). We argue that such constraints can provide more meaningful rules in particular application domains, such as health care. The importance and interestingness of the extracted rules are quantified by adapting existing rule mining metrics. Our experimental evaluation on real data sets demonstrates the descriptive power of ordered association rules against ordinary association rules.

## 1 Introduction

Extracting rules from a set of transactions is a problem that has been studied extensively over the past two decades [1,2,16,28]. Transactional databases typically comprise sets of items or events grouped together in transactions. Each transaction can either be treated as a "bag" of items or as an ordered set of time-stamped events, hence, introducing a sequential order within each transaction. For example, in the former case, we can have transactions of customer activity in the form of basket data containing sets of items bought together from a store [1], while in the latter case, transactions appear in the form of sequential, time-stamped data, such as financial transactions or electronic health records [2].

Hence, sequential pattern and rule mining [2,18] extends traditional association rule mining to exploit the temporal information present in event datasets collected over time, by adding a time-stamp to each event. Such formulation might, however, fail to find interesting patterns, due to the fact that it is too restrictive by requiring a specific order of all items occurring in a sequential pattern or rule. For many real world applications, such as fault detection, or treatment recommendations in healthcare, it is not uncommon with procedures describing best practices or recommended actions that should be taken after certain preconditions are satisfied. Therefore, ordered association rules can be formed, where each rule is defined by a set of preconditions (i.e., antecedent) and a set of recommended actions (i.e., consequent). Such rules suggest that the set of recommended actions should be followed, irrespective of order, after the set

© Springer International Publishing AG 2017
N. Adams et al. (Eds.): IDA 2017, LNCS 10584, pp. 149–161, 2017.
DOI: 10.1007/978-3-319-68765-0_13

of preconditions is satisfied, irrespective of order. Depending on the application domain at hand, it would also be desirable to optionally include time constraints imposed on the temporal separation between each pair of items in the antecedent and consequent, respectively.

In this paper, we study the problem of mining ordered association rules with temporal constraints, and propose an efficient algorithm, called KAPMiner, to solve it. This problem has been studied in recent works in the form of temporal rule mining [5–7]. Nonetheless, KAPMiner can achieve competitive computational efficiency against state-of-the-art, while additionally supporting temporal constraints on the extracted rules. Next, we emphasize the importance of the problem at hand by providing an example from the healthcare domain.

**Example.** Consider a dataset containing administrative health records of patients having suffered from Heart Failure. Each patient record can be seen as a transaction, where time-stamped healthcare-related events can occur over time. Such events can be medical diagnoses, drug prescriptions, or treatment procedures. An example of an ordered rule of substantial interestingness and importance is the following:

$$\{Heart\ Failure\} \Rightarrow \{ACE\ inhibitors, \beta\text{-}blockers\}$$

This rule suggests that when a patient is diagnosed with Heart Failure (rule precondition or antecedent), a recommended action (rule consequence) is to follow a treatment including two drugs, ACE[1] inhibitors and $\beta$-blockers, which is in compliance with the guidelines issued by the Swedish National Board of Health and Welfare [17]. Since it is common that the two drugs included in the consequent part of the rule are not necessarily prescribed concurrently or in a particular order, a typical sequential rule would not be able to partially ignore their temporal order in the patient record. At the same time, a typical association rule would not be able to capture the particular temporal order between the antecedent precondition and the consequent treatment. Hence, ordered temporal association rules can be highly useful and meaningful in this domain.

**Related work.** Sequential pattern mining and association rule mining are very challenging tasks, since the search space is typically large [2]. Standard apriori-based algorithms employ a bottom-up search, enumerating every single frequent pattern, and then construct association rules based on the extracted patterns. The main characteristic of these approaches is that they apply the *Apriori principle* [1]. A more efficient approach, GSP [18], introduces time and window constraints into the pattern extraction process. At the same time, the notion of frequent episode mining [10,12] has been extensively studied in the literature. Frequent episode discovery has several applications in, among others, discovering local patterns in complex processes, conformance checking based on partial orders, medical process mining. Frequent episodes refer to event patterns occurring in a single sequence, that are partially ordered based on a predefined set of

---

[1] Angiotensin-converting-enzyme.

temporal relations. They are orthogonal to our formulation, since we are interested in patterns occurring frequently within a set of sequential transactions and not within a single one, while we have no predefined set of relations in our mining process.

An alternative candidate generation approach for frequent itemsets and sequential patterns employs tree-like data structures, such as an FP-tree [8] for frequent itemsets, or a set-enumeration tree [4] for sequences. The key idea is to traverse the candidate space in a depth-first search manner for enumerating all the candidate patterns. Examples of such algorithms include SPAM [3], SPADE [26], and GO-SPADE [11]. Finally, another class of sequential pattern mining algorithms includes prefix-based candidate generation approaches [15,21,25], Similar approaches, both tree-based and prefix-based, have been proposed for mining closed itemsets and closed sequential patterns [14,21,25,27].

In parallel, several studies have been focusing on alternative interestingness measures for association rules [19], except for support and confidence, aiming at removing redundancy and limiting the number of extracted rules to the most "interesting" ones. Alternative association rule measures and techniques have been proposed [9,13,20] for evaluating the importance of association rules in transactional databases, while generic and interactive techniques have been proposed [22,24] for effectively controlling the mining process and restricting the number of insignificant rules. Finally, there has been some work on constraint-based mining of frequent itemsets, where the goal is to mine the top $K$ patterns that maximize an interestingness measure (other than the typical support threshold) and satisfy a set of constraints [23]. More recently, a novel framework for mining the top-K sequential patterns under leverage has been proposed [16], where a novel definition of the expected support of a sequential pattern is presented along with an efficient branch-and-bound approach for mining sequential patterns under the new interestingness measure. Moreover, a statistical testing approach for association rules extracted from uncertain data combines an analytic with simulative processes for correcting the statistical test for distortions caused by data uncertainty [28].

In summary, the literature on mining rules of itemsets and sequential patterns is quite extensive, and a thorough review is far from the main objectives of this paper. Nonetheless, all existing approaches focus on more general types of rules, and are hence orthogonal to the formulation of this paper. The concept of *ordered rules* defined in this paper, suggests that the antecedent and consequent of a rule are seen as two bags of events separated by a temporal constraint regulating that the bag of antecedent events should be separated by at least $d$ time units from the bag of consequent events. To the best of our knowledge, we are the first to introduce this formulation. A similar formulation, with a looser temporal constraint (i.e., when $d = 0$, defined as Problem 1 in our paper), has been introduced along with three algorithmic solutions, i.e., RuleGrowth [7] and ERMiner [6], and CMDeo [5]. Still, the algorithm proposed in this paper is shown to be more efficient in terms of computational time than the competitors, while it can also solve the constrained version of the problem.

**Contributions.** Our main contributions in this paper include: (1) the formulation of the novel problem of mining ordered association rules with temporal constraints from transactions of time-stamped event sequences, (2) a novel and efficient algorithm to solve the problem by employing a tree-based rule enumeration process and by applying effective pruning techniques both on the antecedent and consequent sets of the rules, (3) an extensive experimental evaluation on several real datasets against state-of-the-art methods, where we demonstrate that our approach can achieve speedups of up to two orders of magnitude, when the dataset contains very dense or large sequences, and finally (4) a case-study in the area of healthcare, where our formulation can identify meaningful rules, when the temporal contraints are applied.

## 2   Problem Setting

Let $\Sigma$ be an alphabet of event labels. A *time-stamped event* is defined as a tuple $E = \langle l, t \rangle$, where $l \in \Sigma$ and $t$ is the time occurrence of the event. For notation purposes, the label and time stamp of an event $E$ are denoted as $E.l$ and $E.t$, respectively.

**Table 1.** Example of ordered rules extracted from the transaction database for $\mu = 0.2, \nu = 0.3$, and $\delta = 3$.

| Identifier | Transaction | | Identifier | Rule | Support |
|---|---|---|---|---|---|
| $T_1$ | $\{\langle 1,0 \rangle, \langle 2,1 \rangle, \langle 3,1 \rangle, \langle 1,1 \rangle, \langle 1,4 \rangle\}$ | | r1 | $\{1\} \Rightarrow \{3\}$ | 0.4 |
| $T_2$ | $\{\langle 2,0 \rangle, \langle 1,1 \rangle, \langle 3,3 \rangle, \langle 4,3 \rangle, \langle 3,4 \rangle\}$ | | r2 | $\{2\} \Rightarrow \{1\}$ | 0.2 |
| $T_3$ | $\{\langle 1,0 \rangle, \langle 2,1 \rangle, \langle 3,2 \rangle\}$ | | r3 | $\{2\} \Rightarrow \{3\}$ | 0.2 |
| $T_4$ | $\{\langle 1,0 \rangle, \langle 1,0 \rangle, \langle 3,4 \rangle\}$ | | r4 | $\{2,3\} \Rightarrow \{1\}$ | 0.2 |
| $T_5$ | $\{\langle 3,0 \rangle, \langle 1,1 \rangle, \langle 1,2 \rangle\}$ | | r5 | $\{1,2\} \Rightarrow \{3\}$ | 0.2 |

A *temporal transaction* $T$ is a set of events ordered by their respective time stamps, and the size of a transaction is the number of time-stamped events it contains. For example, temporal transaction $T = \{E_1, \ldots, E_N\}$ is of size $N$. A set of temporal transactions $\mathcal{D}$ constitutes a *temporal-transaction database*.

*Example 1.* An example of a temporal-transaction database (of 5 transactions) is shown in Table 1 (left). Each tuple represents an event consisting of a label (in the example, $\Sigma = \{1, 2, 3, 4\}$) and a time point. For instance, transaction $T_1$ shows that event label 1 occurred *one* time step before 2, 3, and 1 and *four* time steps before 1.

**Definition 1 (ordered rule).** *Given two subsets of event labels, i.e., $X \subseteq \Sigma$ and $Y \subseteq \Sigma$, with $X \cap Y = \emptyset$, we define an* ordered rule *as follows:*

$$r : X \Rightarrow Y ,$$

*such that $\forall E_i \in X$, $\nexists E_j \in Y$ with $E_i.t - E_j.t \geq \delta$, with $\delta \in \mathbb{R}$ and $\delta \geq 0$.*

One of the key tasks in our paper is to determine whether a rule occurs in a temporal transaction.

**Definition 2 (temporal occurrence of a rule).** *We say that an ordered rule* $r : \mathcal{X} \Rightarrow \mathcal{Y}$ *occurs in a temporal transaction* $\mathcal{T}$, *if all events in* $\mathcal{X} \cup \mathcal{Y}$ *occur in* $\mathcal{T}$ *at least once, and each and every event in* $\mathcal{X}$ *has at least one occurrence before each and every event in* $\mathcal{Y}$. *More formally,* $r \in \mathcal{T}$, *if*

- $\exists E \in \mathcal{T}, \forall E \in \mathcal{X} \cup \mathcal{Y}$
- $\forall (E, E')$ *with* $E \in \mathcal{X}, E' \in \mathcal{Y}, \nexists (E, E'), E.l, E'.l \in \mathcal{T}$, *s.t.* $E.t - E'.t \geq \delta$.

*Example 2.* For $\delta = 3$, the rule $\{2, 3\} \Rightarrow \{1\}$ is contained in transaction $\mathcal{T}_1$, whereas $\{1\} \Rightarrow \{3\}$ is not, because $\{3\}$ does not occur within at least 3 time-steps from $\{1\}$.

There are several ways of assessing the quality of an ordered rule. In this paper, our main objective is to emphasize the importance of ordered rules in terms of (1) the frequency of the rule in a given database, (2) the frequency of the particular temporal order for a given set of event labels compared to any other temporal order, (3) the conditional probability of the consequent set of labels given the precedent set, and (4) the degree of dependence between the occurrence probabilities of the precedent and consequent sets.

Next, we present four quality metrics for an ordered rule. Given an ordered rule $r : \mathcal{X} \Rightarrow \mathcal{Y}$ and a temporal transaction database $\mathcal{D}$, we have:

- the **support** of $r$ in $\mathcal{D}$ is the fraction of transactions of $\mathcal{D}$ containing at least one temporal occurrence of $r$, i.e.,

$$sup(r, \mathcal{D}) = \frac{|r \in \mathcal{D}|}{|\mathcal{D}|}$$

- the **support ratio** of $r$ in $\mathcal{D}$ is the fraction of transactions containing $r$ divided by the fraction of transactions containing itemset $\mathcal{X} \cup \mathcal{Y}$, i.e.,

$$r - sup(r, \mathcal{D}) = \frac{sup(r, \mathcal{D})}{sup(\mathcal{X} \cup \mathcal{Y}, \mathcal{D})} = \frac{|r \in \mathcal{D}|}{|\mathcal{X} \cup \mathcal{Y} \in \mathcal{D}|}$$

- the **confidence** of $r$ in $\mathcal{D}$ is the fraction of transactions containing $r$ divided by the fraction of transactions containing all items in the consequent set of the rule, i.e.,

$$conf(r, \mathcal{D}) = \frac{sup(r, \mathcal{D})}{sup(\mathcal{Y}, \mathcal{D})} = \frac{|r \in \mathcal{D}|}{|\mathcal{Y} \in \mathcal{D}|}$$

- the **lift** of $r$ in $\mathcal{D}$ is the frequency of $r$ in $\mathcal{D}$ divided by the product of the frequencies of the precedent and consequent sets, i.e.,

$$lift(r, \mathcal{D}) = \frac{sup(r, \mathcal{D})}{sup(\mathcal{X}, \mathcal{D}) \cdot sup(\mathcal{Y}, \mathcal{D})} = \frac{|r \in \mathcal{D}| \cdot |\mathcal{D}|}{|\mathcal{X} \in \mathcal{D}| \cdot |\mathcal{Y} \in \mathcal{D}|}$$

*Problem 1 (mining ordered association rules).* Given a temporal transaction database $\mathcal{D}$, a minimum support threshold $\mu$, and a minimum support ratio threshold $\nu$, we want to find the set of ordered rules $\mathcal{R}$, such that for each $r \in \mathcal{R}$, it holds that $sup(r, \mathcal{D}) \geq \mu$ and $r - sup(r, \mathcal{D}) \geq \nu$.

*Problem 2 (mining constrained association rules).* Given a temporal transaction database $\mathcal{D}$, a minimum support threshold $\mu$, a minimum support ratio threshold $\nu$, and a temporal constraint $\delta$, find the set of ordered rules $\mathcal{R}^\delta$, such that for each $r \in \mathcal{R}^\delta$, it holds that $sup(r, \mathcal{D}) \geq \mu$, $r - sup(r, \mathcal{D}) \geq \nu$ and $\forall E_i \in \mathcal{X}$, $\nexists E_j \in \mathcal{Y}$ with $E_i.t - E_j.t \geq \delta$. Note this problem reduces to Problem 1 for $\delta = 0$.

## 3  Mining Ordered Association Rules

We introduce KAPMiner, an algorithm for identifying ordered association rules with optional temporal constraints. The main operator of the algorithm (see Algorithm 1) is based on the concepts of antecedent and consequent matching and merging of rules, which are identified in the hierarchy of frequent itemsets.

**Definition 3 (Antecedent and consequent matching).** *A pair of rules $r$ : $\{X\} \Rightarrow \{Y\}$ and $r' : \{X'\} \Rightarrow \{Y'\}$ is said to be antecedent matching, if $\{X\} \setminus \{X'\} = \emptyset$, and said to be consequent matching, if $\{Y\} \setminus \{Y'\} = \emptyset$.*

The above definition can be extended to more than two rules, hence resulting in a set of antecedent or consequent matching rules, denoted as $r_a$ and $r_c$, respectively. The common antecedent in $r_a$ is denoted as $r_a.antecedent$, while the set of consequents is denoted as $r_a.consequents$. The notation is equivalent for $r_c$.

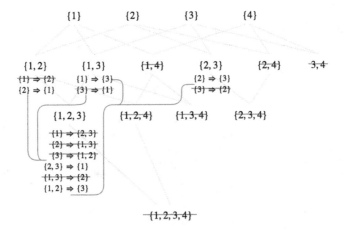

**Fig. 1.** Search space and pruning of the KAPMiner algorithm. For instance, the rule $\{2, 3\} \Rightarrow \{1\}$ is formed by merging the consequent matching rules $\{2\} \Rightarrow \{1\}$ and $\{3\} \Rightarrow \{1\}$, whereas rule $\{1, 2\} \Rightarrow \{3\}$ is formed by antecedent matching rules $\{1\} \Rightarrow \{3\}$ and $\{2\} \Rightarrow \{3\}$.

To explore the search-space of possible frequent constrained ordered rules our algorithm merges frequent antecedent or consequent matching rules (Fig. 1).

**Definition 4 (Antecedent and consequent merging).** *Given a set of antecedent or consequent matching rules $r_a$ or $r_c$, respectively, the intersection of transactions where rules with matching consequent or antecedent occur, is the support of the rule with antecedents or consequents merged, respectively.*

For instance, if $\{3\} \Rightarrow \{1\} \in \mathcal{R}_c$ occurs in transaction $\{\mathcal{T}_1\}$ and $\{2\} \Rightarrow \{1\} \in \mathcal{R}_c$ occurs in $\{\mathcal{T}_1\}$, then $\{2\} \cup \{3\} \Rightarrow \{1\}$ occurs in the intersecting transactions: $\{\mathcal{T}_1\} \cap \{\mathcal{T}_1\} = \{\mathcal{T}_1\}$.

---

**Algorithm 1.** The KAPMiner algorithm

---

**procedure** KAPMINER($\mathcal{D}$, minSup, minConf, $\delta$)
    Let $\mathcal{C} = [\mathcal{I}_1, \ldots, \mathcal{I}_m]$ be a vector of frequent event labels
    Let $\mathcal{I}_i.tid$ be a set that identifies which transactions an itemset occurs in
    $\mathcal{R} \leftarrow \emptyset, k \leftarrow 1$
    **do**
        $\mathcal{N} \leftarrow \emptyset$
        **for** $i \leftarrow 0$ until $|\mathcal{C}|$ **do**
            **for** $j \leftarrow 0$ until $|\mathcal{C}|$ such that PREFIXMATCH($\mathcal{I}_i, \mathcal{I}_j, k - 1$) **do**
                $\mathcal{I}' \leftarrow$ MERGEITEMSET($\mathcal{I}_i, \mathcal{I}_j$)
                $\mathcal{I}'.tid \leftarrow \mathcal{I}_i.tid \cap \mathcal{I}_j.tid$
                **if** $sup(\mathcal{I}') \geq minSup$ **then**
                    $\mathcal{N} \leftarrow \mathcal{N} \cup \{\mathcal{I}'\}$
                    **if** $k > 1$ **then**
                        Let $\mathcal{S}$ be the rules associated with each subset of $\mathcal{I}'$ in $\mathcal{C}$
                        $\mathcal{I}'.rules \leftarrow$ MERGERULES($\mathcal{S}, k, minSup$)
                    **else**
                        $\mathcal{I}'.rules \leftarrow$ ITEMRULES($\mathcal{I}_i, \mathcal{I}_j, \delta, minSup$)
                    **end if**
                    $\mathcal{R} \leftarrow \mathcal{R} \cup \{r \mid r \in \mathcal{I}'.rules \wedge conf(r) \geq minConf\}$
                **end if**
            **end for**
        **end for**
        $\mathcal{C} \leftarrow \mathcal{N}$
        $k \leftarrow k + 1$
    **while** $\mathcal{N} \neq \emptyset$
    **return** $\mathcal{R}$
**end procedure**

---

For a rule-merge to be valid for a rule consisting of elements from an itemset $\mathcal{I}$ on level $k$, the number of frequent rules in the antecedent or consequent matching rule set, identified from all subsets of itemsets on level $k - 1$, must satisfy the condition $|r_a.antecedents| + |X| = |\mathcal{I}|$ or $|r_c.consequents| + |Y| = |\mathcal{I}|$, for antecedent and consequent merges respectively. For an antecedent or consequent merge, the *antimonotonicity property* holds since the resulting rule contains exactly one item more in either the antecedent or the consequent, the rule can only appear in the same number of transactions or less. From the antimonicity property it follows that if a rule is infrequent, it should not be used in any antecedent or consequent merges.

The algorithm, which expects a temporal transaction database, a minimum support threshold, a minimum confidence threshold and temporal constraint ($\delta$), starts by scanning the temporal transaction database $\mathcal{D}$ once, to associate with each frequent item a set of transactions where the item occurs[2]. The algorithm then proceeds by identifies the frequent $k$-level itemsets using the $(k-1)$-level itemsets and prunes the infrequent ones. At level $k = 1$ rules are formed by the ITEMRULES-procedure which forms frequent ordered rules from two single item itemsets, by comparing the first and last position of each item in each transaction.

---

**Algorithm 2.** Constructing rules from 1-item itemsets

---

**procedure** ITEMRULES($\mathcal{I}_i, \mathcal{I}_j, \delta, minSup$)

    $r_1 \leftarrow \{I_i\} \Rightarrow \{I_j\}$ and $r_2 \leftarrow \{I_j\} \Rightarrow \{I_i\}$

    Let $r_1.tid$ be the set of transactions where LAST($\mathcal{I}_j, \mathcal{T}$) - FIRST($\mathcal{I}_i, \mathcal{T}$) $\geq \delta$ and let $r_2.tid$ be the set of transactions where LAST($\mathcal{I}_i, \mathcal{T}$) - FIRST($\mathcal{I}_j, \mathcal{T}$) $\geq \delta$

    **return** $\{r \mid r \in \{r_1, r_2\} \wedge sup(r) \geq minSup\}$

**end procedure**

---

To construct rules for itemsets of size larger than two, the MERGERULES-procedure is employed. This procedure expects a set of rules which are found among the $k-1$ level itemsets that are subsets of the newly formed itemset at level $k$. The procedure then iterates over the sets of antecedent matching rules and forms new rules by merging the consequents and taking the intersection of transaction identifiers as the support of the rule. The procedure proceeds similarly with the consequent matching rules and forms new rules by merging the antecedents and taking the intersection of transactions where the rules appears as the support. Finally, infrequent rules are pruned, and the size of search-space is reduced by pruning both infrequent itemsets (which are used to associate possible rule merges) and rules.

## 4    Evaluation

To evaluate the performance of the proposed ordered rule mining algorithm, we compare it against the current state-of-the-art, ERMiner [6], in terms of execution time for different values of minimum support threshold[3], using seven

---

[2] While not presented in the algorithm, the scan also records the first and last position of each item in each transaction, which is used to efficiently form rules with a single item as antecedent and consequent.

[3] The experiments was performed using a 2.5GHz Intel i7 processor and 16GB of RAM. Both algorithms are implemented in Java.

**Algorithm 3.** Merging the rules of $n$-item itemsets

---

**procedure** MERGERULES($\mathcal{S}, k, minSup$)

    Let $\mathcal{S}_a$ be a list of sets consisting of rules with matching antecedent and let $\mathcal{S}_c$ be a list of sets consisting of rules with matching consequent

    $\mathcal{R} \leftarrow \emptyset$

    **for** each set of rules with matching antecedent $r_a \in \mathcal{S}_a$ **do**

        **if** $|r_c.consequents| = k + 1 - |r_a.antecedent|$ **then**

            $r' \leftarrow \{r_a.antecedent\} \Rightarrow \{\text{MERGEITEMSET}(r_a.consequents)\}$

            $r'.tid \leftarrow \bigcap_{r \in r_a} r.tid$

            **if** $sup(r') \geq minSup$ **then**

                $\mathcal{R} \leftarrow \mathcal{R} \cup \{r'\}$

            **end if**

        **end if**

    **end for**

    **for** each set of rules with matching antecedent $r_c \in \mathcal{S}_c$ **do**

        **if** $|r_c.antecedents| = k + 1 - |r_c.consequent|$ **then**

            $r' \leftarrow \{\text{MERGEITEMSET}(r_c.antecedents)\} \Rightarrow \{r_c.consequent\}$

            $r'.tid \leftarrow \bigcap_{r \in r_a} r.tid$

            **if** $sup(r') \geq minSup$ **then**

                $\mathcal{R} \leftarrow \mathcal{R} \cup \{r'\}$         ▷ Set without duplicate rules.

            **end if**

        **end if**

    **end for**

    **return** $\mathcal{R}$

**end procedure**

---

datasets with varying properties. The datasets used for comparing the algorithms are: BIBLE (*text sequence*, 36k sequences/13905 distinct items/average sequence length of 17.84), BMS1 (*web logs*, 59k/497/2.42), BMS2 (*web logs*, 77k/3340/4.62), FIFA (*web logs*, 20k/2990/34.74), Kosarak25k (*web logs*, 25k/14804/8.04), MSNBC (*web logs*, 33k/17/13.3) and SIGN (*sign language*, 800/290/52).

To evaluate the importance of mining constraint ordered rules, we evaluate the proposed algorithm practically for identifying rules from administrative claims. In the experiment we use a real temporal transaction database extracted from the VAL database, and consists of 79028 patients diagnosed with heart failure. Each sequence represent the medical history of a patient and contains items from an alphabet of 13000 distinct items (diagnoses, drugs and actions), with an average sequence length of 300 items. We identify rules with $\delta = \{1, 7, 14\}$ days between the antecedent and consequent.

As seen in Fig. 2, the results indicate that the proposed algorithm is often significantly faster than ERMiner, but not consistently so. For instance, it can be observed that the performance gap generally expands when the minimum support threshold is lowered, which can be explained by the fact that the proposed algorithm is able to prune more rules than the competitor. More specifically, the algorithm proposed here will prune any rule for which union of the antecedent and consequent is infrequent in an hierarchical manner, which results in excellent

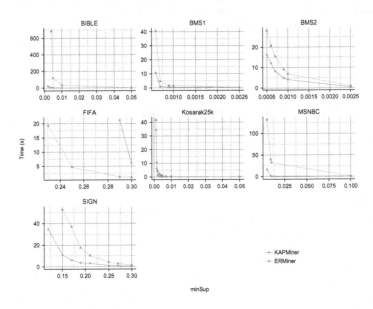

**Fig. 2.** Results for mining ordered rules from seven temporal transactions. Note that some points are missing, since the algorithms fail to complete within the memory constraint (e.g., BIBLE for ERMiner and FIFA for KAPMiner).

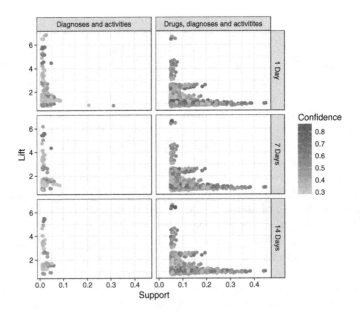

**Fig. 3.** Rule extraction for different values for $\delta$ with $\mu \geq 0.05$ and $\nu \geq 0.8$.

performance on datasets with many but infrequent items such as then `BIBLE` dataset. But results in low performance on the `FIFA` dataset, as the algorithm fails to prune infrequent itemsets early.

Figure 3, shows the support, lift and confidence for rules found during the case study, with a minimum support of 0.05 and a minimum support ratio of 0.8. The figure indicates that there are few high-lift rules, which has relatively low support, whereas there are a rather many high confidence rules. An example of an interesting rule that was found for $\delta = 7$, is:

$$\{hypothyroidism\} \Rightarrow \{levothyroxine, furosemide\}$$

with lift $= 6.6$, confidence $= 0.69$ and support ratio $= 0.88$, which indicates that a common treatment pattern for hypothyroidism is a manufactured form of thyroid hormone and a medication to treat fluid build up.

## 5   Conclusions

We proposed a novel algorithm for identifying ordered rules where the consequent and antecedent are temporally separated. The proposed algorithm uses an apriori style algorithm for identifying frequent itemsets and uses temporal orderings within those itemsets to construct ordered rules using a computationally efficient transaction intersection algorithm. The experimental evaluation shows that the algorithm is up an order of magnitude faster than current state of the art when the dataset contains dense or very long sequences, but slow in cases where the dataset contains many frequent itemsets. Furthermore, in a case study for identifying interesting rules in the treatment of heart failure patients, the temporal constraint is shown to help in identifying patient trajectories.

**Source code.** The source code for replicating the experiments is available at the supporting website (https://people.dsv.su.se/~isak-kar/orule/).

**Acknowledgments.** This work was partly supported by grants provided by the Stockholm County Council (SU-SLL). The work of Panagiotis Papapetrou was also partly supported by the VR-2016-03372 Swedish Research Council Starting Grant.

## References

1. Agrawal, R., Srikant, R.: Fast algorithms for mining association rules. In: Proceedings of VLDB, pp. 487–499 (1994)
2. Agrawal, R., Srikant, R.: Mining sequential patterns. In: Proceedings of IEEE ICDE, pp. 3–14 (1995)
3. Ayres, J., Gehrke, J., Yiu, T., Flannick, J.: Sequential pattern mining using a bitmap representation. In: Proceedings of ACM SIGKDD, pp. 429–435 (2002)
4. Bayardo, R.J.: Efficiently mining long patterns from databases. In: Proceedings of ACM SIGMOD, pp. 85–93 (1998)
5. Fournier-Viger, P., Faghihi, U., Nkambou, R., Nguifo, E.M.: Cmrules: mining sequential rules common to several sequences. Know.-Based Syst. **25**(1), 63–76 (2012). http://dx.doi.org/10.1016/j.knosys.2011.07.005

6. Fournier-Viger, P., Gueniche, T., Zida, S., Tseng, V.S.: ERMiner: sequential rule mining using equivalence classes. In: Blockeel, H., van Leeuwen, M., Vinciotti, V. (eds.) IDA 2014. LNCS, vol. 8819, pp. 108–119. Springer, Cham (2014). doi:10. 1007/978-3-319-12571-8_10

7. Fournier-Viger, P., Nkambou, R., Tseng, V.S.M.: Rulegrowth: mining sequential rules common to several sequences by pattern-growth. In: Proceedings of the 2011 ACM Symposium on Applied Computing, SAC 2011, NY, USA, pp. 956–961 (2011). http://doi.acm.org/10.1145/1982185.1982394

8. Han, J., Pei, J., Yin, Y., Mao, R.: Mining frequent patterns without candidate generation: a frequent-pattern tree approach. Data Min. Knowl. Discov. 8(1), 53–87 (2004)

9. Kamber, M., Shinghal, R.: Evaluating the interestingness of characteristic rules. In: Proceedings of ACM SIGKDD, pp. 263–266 (1996)

10. Laxman, S., Sastry, P.S., Unnikrishnan, K.P.: A fast algorithm for finding frequent episodes in event streams. In: Proceedings of the 13th ACM SIGKDD International Conference on Knowledge Discovery and Data Mining (KDD 2007), San Jose, USA, pp. 410–419. Association for Computing Machinery, Inc., August 2007. https://www.microsoft.com/en-us/research/publication/a-fast-algorithm-for-finding-frequent-episodes-in-event-streams/

11. Leleu, M., Rigotti, C., Boulicaut, J., Euvrard, G.: Go-spade: mining sequential patterns over databases with consecutive repetitions. In: Proceedings of MLDM, pp. 293–306 (2003)

12. Mannila, H., Toivonen, H., Verkamo, A.: Discovering frequent episodes in sequences. In: Proceedings of ACM SIGKDD, pp. 210–215 (1995)

13. Omiecinski, E.R.: Alternative interest measures for mining associations in databases. IEEE Trans. Knowl. Data Eng. 15(1), 39–79 (2003)

14. Pei, J., Han, J., Mao, R.: Closet: an efficient algorithm for mining frequent closed itemsets. In: Proceedings of DMKD, pp. 11–20 (2000)

15. Pei, J., Han, J., Mortazavi-Asl, B., Pinto, H., Chen, Q., Dayal, U., Hsu, M.C.: Prefixspan: mining sequential patterns efficiently by prefix-projected pattern growth. In: Proceedings of IEEE ICDE, pp. 215–224 (2001)

16. Petitjean, F., Li, T., Tatti, N., Webb, G.I.: Skopus: mining top-k sequential patterns under leverage. Data Min. Knowl. Discov. 30(5), 1086–1111 (2016). http://dx.doi.org/10.1007/s10618-016-0467-9

17. Socialstyrelsen: Nationella riktlinjer för hjärtsjukvård (2015). http://www.socialstyrelsen.se

18. Srikant, R., Agrawal, R.: Mining sequential patterns: generalizations and performance improvements. In: Apers, P., Bouzeghoub, M., Gardarin, G. (eds.) EDBT 1996. LNCS, vol. 1057, pp. 1–17. Springer, Heidelberg (1996). doi:10.1007/BFb0014140

19. Tan, P., Kumar, V.: Interestingness measures for association patterns: a perspective. Tech. Rep. TR00-036, Department of Computer Science, University of Minnesota (2000)

20. Tan, P., Kumar, V., Srivastava, J.: Proceedings of ACM SIGKDD, pp. 183–192, July 2002

21. Wang, J., Han, J.: Bide: efficient mining of frequent closed sequences. In: Proceedings of IEEE ICDE, pp. 79–90 (2004)

22. Webb, G.I.: Discovering significant rules. In: Proceedings of ACM SIGKDD (2006)

23. Webb, G.I., Zhang, S.: K-optimal rule discovery. Data Min. Knowl. Discov. 10(1), 39–79 (2005)

24. Xin, D., Shen, X., Mei, Q., Han, J.: Discovering interesting patterns through user's interactive feedback. In: Proceedings of ACM SIGKDD (2006)
25. Yan, X., Han, J., Afshar, R.: Clospan: mining closed sequential patterns in large databases. In: Proceedings of SDM (2003)
26. Zaki, M.: Spade: an efficient algorithm for mining frequent sequences. Mach. Learn. **40**, 31–60 (2001)
27. Zaki, M., Hsiao, C.: Charm: an efficient algorithm for closed itemset mining. In: Proceedings of SIAM, pp. 457–473 (2002)
28. Zhang, A., Shi, W., Webb, G.I.: Mining significant association rules from uncertain data. Data Min. Knowl. Discov. **30**(4), 928–963 (2016)

# A Dynamic Adaptive Questionnaire for Improved Disease Diagnostics

Xiaowei Kortum[1]([✉]), Lorenz Grigull[2], Werner Lechner[3], and Frank Klawonn[1,4]

[1] Department of Computer Science, Ostfalia University of Applied Sciences,
Salzdahlumer Str. 46/48, 38302 Wolfenbuettel, Germany
{x.kortum,f.klawonn}@ostfalia.de
[2] Department of Paediatric Haematology and Oncology,
Medical University Hanover, Carl-Neuberg Str. 1, 30625 Hannover, Germany
Grigull.Lorenz@mh-hannover.de
[3] Improved Medical Diagnostics IMD GmbH,
Ostfeldstr. 25, 30559 Hannover, Germany
werner.lechner@improvedmedicaldiagnostics.com
[4] Helmholtz Center for Infection Research,
Inhoffenstrasse 7, 38124 Braunschweig, Germany
Frank.Klawonn@helmholtz-hzi.de

**Abstract.** A diagnosis system assists medical doctors to figure out a disease that causes the symptoms of a sick patient. Standardized questionnaires are one way to gather patients' subjective perceptions about their present and also past healthy conditions. Such questionnaires often cover several question fields with the focus of significant symptoms as suitable disease indicators. Each question provides a different association strength with a disease group. And in most cases, a combination of question/answer pattern is needed to provide strong evidence for a particular disease. Nevertheless, people would like to keep the number of questions to be answered at a minimum for the patient. In this study, we address this aim and introduce an algorithm that reduces the size of a questionnaire by dynamically selecting patient-oriented questions to provide strong evidence for or against a suspected diagnosis tendency. It reduces the number of question to the most relevant ones. We evaluate our self-adapting questioning algorithm with 354 patients from a rare disease questionnaire survey.

**Keywords:** Machine learning · Dynamically adapted questionnaires · Fusion classifiers · Multiple classifier systems

## 1 Introduction

Finding a patient's disease through symptoms often relates to the medical doctor's knowledge and experience and can be overlooked by general practitioners, in particular for patients with rare diseases. A preliminary examination of a patient's physical condition and laboratory tests are normal proceedings

© Springer International Publishing AG 2017
N. Adams et al. (Eds.): IDA 2017, LNCS 10584, pp. 162–172, 2017.
DOI: 10.1007/978-3-319-68765-0_14

for delineating the range of possible conditions. Missing experience and data resources let a patient with a rare disease become an essential problem for general practitioners. Diagnosing rare diseases is a sophisticated process which can take months, years or decades starting with early symptoms to mature health issues. In the past years, adaptive solutions, like system-based diagnosis support have conquered the interests of medical institutions, doctors, and researcher.

Various studies in medical environments have been established, that approached these supportive diagnosis systems with beneficial results, like minimized risks for misdiagnoses or simplifying a symptoms interpretations through central experience bases [7,8]. Other studies gained new expertises in realizing or understanding of such systems in medical scenarios [4,11]. In one of these previous studies, we introduced an approach that supports the diagnosis process for particular types of rare diseases. The approach involved a computer-based diagnosis system that is trained with records from previously diagnosed patients. Based on questionnaires designed for particular groups of (rare and non-rare) diseases, a classifier is trained with questionnaires that filled in by patients with a corresponding diagnosis. Then the classifier can support the medical doctor by the prediction of possible diagnosis for a new patient who has filled in the questionnaire [2,5].

Our current approach presents an interdisciplinary collaboration between medical experts and data analysts, both with a shared aim of a reduced afford for patients with less time consumption when filling in questionnaires. Medical experts from Hannover Medical School (MHH) designed a locally used patient survey through interviews and clinical observations of patients who have been diagnosed in the past with a rare, a chronic disease or even a psychosomatic disorder. The questionnaire about rare diseases is based on 53 questions, most of them having answers on a four-level Likert scale. Since the questionnaires has to cover a wide variety of diseases, questions focus on very different aspects and for specific patients certain questions might not be useful for classifying whether the patient has a rare disease or not. For instance, questions that focus on neuromuscular rare diseases are not very helpful in identifying patients with immune deficiencies. Since the standardized patient questionnaire from the Hannover Medical School is not patient-centered, patients often need to fill in more survey questions than actually required for correctly classifying the patient. Furthermore, it can be very time consuming to write every issue of the questionnaire for the patient, which may lead to distraction or even falsely giving answers. Consequently, it would be highly desirable to improve the concept of the current patient questionnaires by a dynamic process with a self-adapting solution to select questions that focus on a patient's individual symptoms.

We introduce a method that improves our patient questionnaires with dynamically channelized questions based on patient's responses to previous questions. The iteratively chosen question set forms a patient substantiates questionnaire. This procedure minimizes the number of questions while gathering the maximal necessary symptom information to enable a system-based diagnosis for the patient. In the following chapters, we proved that a dynamic questionnaire has more benefits than a solely static survey. And evaluate our answer-oriented questioning algorithm with 354 patient questionnaires collected within the past years.

## 2    Related Work

Our research methodology about the question selection algorithm in combination with system-based diagnosis is based on the established founding from related work in fields of the system-based decision, fusion-classifier, and algorithm for questionnaire optimization.

Sboner et al. [10] describe early diagnoses approach through a decision support system that focuses on Melanoma (skin cancer) identifications. The author uses three different classifiers: linear discriminant analysis (LDA), k-nearest neighbor (k-NN) and decision tree. They use a computer-based diagnosis system to evaluate skin images, and the results revealed that the combination of different classifiers has better performance than any single classifier.

In our previous work [5], a fusion classifier diagnosis system has been established. The implemented system involves four classifiers: support vector machines (SVM) [1], linear discriminant analysis (LDA) [9], logistic regression (LR) [9] and random forests (RF) [6]. The system is based on a questionnaire that was designed to identify patients with a rare diseases. So far, it is based on a static questionnaire where a patient is expected to answer all questions. Which can be time-consuming and might also lead to distraction when patients are asked to answer questions that are not relevant for their specific disease.

By only presenting patient-oriented questions, the diagnosis will not be disturbed by physician's inherent medical bias about what constitutes a true disease, and will also decrease the time burden on the subject [3].

## 3    Methodology

Our research methodology about the question selection algorithm in combination with the classifier is based on the field of machine learning algorithms, fusion classifiers, and algorithm for questionnaire optimization. This section will introduce the algorithms for the self-adapting questionnaires and the computer-based diagnosis system.

### 3.1    The Diagnose Questionnaire Design and Data Collection

Over the last years, medical doctors from Hannover Medical School (MHH) have established various questionnaires with central focuses of appropriate indicators for rare diseases. The currently applied patient survey is a result from hundreds of interviews and clinical observations of patients who have been diagnosed with rare diseases, chronically disorders, or even psychosomatic illnesses. This questionnaire provides 53 questions about associable symptoms and the patient's

perceived intensity. In this approach, we consider a dataset with 354 fully itemized questionnaires, and each participant who completed the questionnaire was
diagnosed by a medical doctor. The survey structure requires users to complete
all questions, otherwise the submission would be counted as incomplete. Consequently, the data collection investigated in this approach excludes questionnaires
with missing values (answers).

## 3.2 Analysis of Commonly Used Static Questionnaire

There exist various ways to shrink the number of questions within a questionnaire. For instance, through statistical significance, impact or correlation measures that allow eliminating questions with weak or non-relevant influence for
a particular diagnosis. However, these strategies only provide a static reduction
for patient questionnaires and are therefore not appropriate, even if they seem to
fulfill their goals in shrinking the number of questions while increasing the system diagnosis overall accuracy. Figure 1 shows a comparison result that is based
on linear correlation measures between questionnaire answers and the patients'
final diagnosis.

**Fig. 1.** Question reduction through linear regression statistics and brute force

It was an early part of this approach and resolved question combinations
through a brute force method according to the correlations between questions
and final diagnosis. In fact, we could verify that a reduced questionnaire that contains 41 strongest questions has significance for deriving a system-based fusion
diagnosis, even results in a 1.4% higher overall accuracy compared with the
initial questionnaire that has 53 questions.

A diagnosis system assists medical doctors about the classification and interpretation of sick patient's indicating symptoms. Standardized questionnaires
become applied to gather a patients subjective perceptions about his or hers
present also past healthy conditions. These questionnaires often cover several
question fields with the focus on significant symptoms as appropriate disease
indicator. Every symptom question provides a different association strength,

sometimes only recognizable as a combined question set with relevance for a particular type of disease. Vice versa, patients are often held to fill in extensive and time-consuming surveys with symptom questions that have no linking for their proper disease. In this study, we address these issues and introduce an algorithm that improves diagnosis surveys with dynamically channelized question selections based on patient's responses to a previous question. The consequently resulting questionnaire is patient-centered and substantiates as to what constitutes his or her actual disease. Thus, it reduces the number of question only to the most relevant ones. We evaluate our self-adapting questioning algorithm with 354 patients and compare each treating doctor's diagnosis with the classified result from our medical diagnosis system. But the negative aspect of this strategy is the limitation to the same (reduced) questionnaire for all patients although certain questions might not be of relevance for some patients.

According to the limitation, a dynamic questionnaire with the capability of selecting questions based on the answers of previous questions is hereby highly desirable. This strategy of a dynamic adaptive survey closes the gap between maximal possible consideration of all questions and the respects of patients' given answers, which has the goal of asking as few as possible while gathering maximal possible information to derive a correct system-based diagnosis.

## 3.3  Dynamic Questionnaire with Self-Adapting Algorithm

Figure 2 illustrate our approach for the dynamic adaptive questionnaire. The system comprises three modules: a questionnaire initialization process, a question selection module, and a diagnose prediction module.

In case of two diagnoses, since the classifier is not designed to cope with a dataset that has missing values, while a patient's answer is a subset of the entire questionnaire (possibly empty), for different diagnoses, we derived the most frequent answer to each question from the raw data, as a universal answer. This will be later used to estimate the gap between the patient's answered questions subset and the initial questionnaire size.

In order to select the question that should be asked next, each diagnosis will obtain one probability, under the "optimistic" assumption that the patient answers the remaining question in accordance with the corresponding diagnosis. With the competing assumptions that the patient either answers consistently according to one diagnosis or the other diagnosis, when a patient start with an empty questionnaire, each of the two diagnoses will be assigned a probability of closely 100%, which would yield odds of two diagnoses 1:1. For questions that have not been asked so far, we assign the most frequent answer to each question in each diagnosis and compute the corresponding probability. So we obtain new odds for each diagnosis of a unanswered question. In the best case, all these odds sharply depart from 1:1, means that knowing the answer to this question would clarify the diagnosis significantly. From each of the odds we calculate the entropy and add to the corresponding entropies. The smaller the sum of these entropies the more the answer to this question leads to a unique decision for a diagnosis.

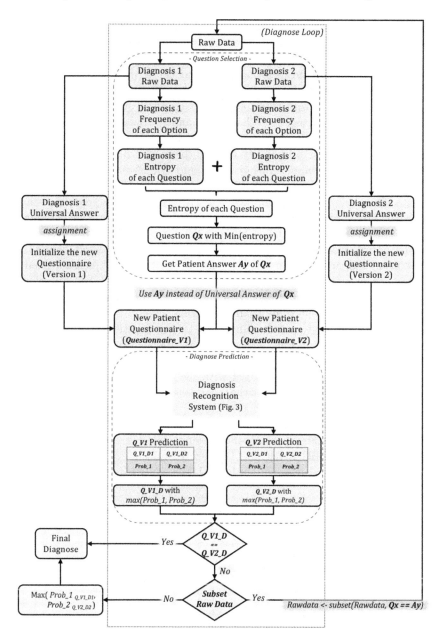

**Fig. 2.** Self-adapting question set selection with system-based diagnosis

Therefore, the dynamic questionnaire started with the question that has lowest entropy value.

When the question selection module is sorting out a question, the program will extract the content of the corresponding question and display to the new patient. The answer of extracted question, which given by the current patient, will be used to replace the corresponding response in the universal answer pattern of both diagnoses. We apply the diagnosis classification system to these artificially complete questionnaires and check with which probability the classifier votes for each diagnosis.

As illustrated above, from the diagnosis recognition system we will obtain probability $Prob\_1$ of Diagnose 1 ($Q\_V1\_D1$) and probability $Prob\_2$ of Diagnose 2 ($Q\_V1\_D2$) on $Questionnaire\_V1$. Regularly the system is taking the diagnosis with the highest probability as final diagnostic result. That is to say, $Questionnaire\_V1$ has diagnosis of $Q\_V1\_D$ and $Questionnaire\_V2$ has diagnosis of $Q\_V2\_D$. By comparing the value of two results, the system will decide whether to make the final decision. The best case is that both predictions have the same diagnostic results, which means one of the diagnosis result is deviates from its original one. Thereby the system can determine the final diagnosis. Otherwise, the information that system gained from the patient is not enough for making correct diagnose. Under this circumstances, the next question need to be selected for provide more evidence. Because the next appeared question is closely related to the patient's answer of the current question, by filtering out all data that has a different value than the patient choice, a subset of raw data is generated, and the system is looping back to the question selecting section for generating next question until getting enough information for confirming the diagnose. A special case is, when the diagnostic results from two questionnaires are not the same, and there is no subset with the current patient answer combination, the system can not derive further diagnose. In this situation, the system will choose the diagnosis with higher prediction probability as final result, based on the ratio of two different diagnose: $Max(Prob\_1_{Q\_V1\_D1}, Prob\_2_{Q\_V2\_D2})$.

As shown in Fig. 3, four classifiers have been applied in the diagnose prediction module, i.e. SVM, LDA, RF and LR. For diagnosing a new patient, each classifier has been trained independently with the same raw data and generate

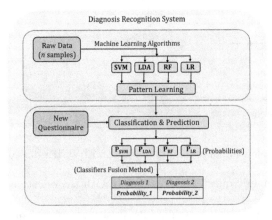

**Fig. 3.** Structure of diagnosis recognition system

an associated diagnose result, respectively. For merging the results of four classifiers, a fusion algorithm has been introduced. The fusion method derives final diagnosis by taking advantages of each classifier by evaluating their compatibility and accuracy. If a classifier has higher confidence about a diagnosis, it will gain higher weight in the fusion process. The fusion classification system takes an important role for the overall validation of disease diagnose when applying the adaptive questionnaire algorithm.

## 4   System Evaluation and Interpretation

*Self-Adapting Questionnaire Algorithm* - We evaluated our algorithm with 354 real patient questionnaires as referenced. Every questionnaire is anonymized and contains 53 questions with a 4-level Likert scale for the answers and the final diagnosis provided by a medical doctor. We use leave-one-out cross-validation (LOOCV) for evaluation. From this point on, the left out survey answers were used to respond the dynamically following and self-adapting question selections.

Based on the real testing answers, diagnosis recognition system analyzes whether the information value of this solely questions allow a correct system based diagnosis for the type of disease. If yes, the patient questioning ends and proposes a diagnosis. Otherwise, the algorithm will continue in selecting the question from a subset of 353 questionnaires with second lowest entropy within the diagnosis loop.

Figure 4 characterizes the textual result for the dynamically formed questionnaires based on patients real answers. In the case of the first patient, questions that need to be answered was reduced to 13 due to the patient's individual reply with the relevance for diagnosing the correct type of disease. The overview of the reduced question set reveals that some answers for a particular question, e.g. the *F32* provide substantial evidence for diagnosing the category of disease. Some other answers, like the ones given from the second patient, require even up to 39 questions for ensuring an adequate diagnosis.

| P Q | 1st | 2nd | 3rd | 4th | 5th | 6th | 7th | 8th | 9th | 10th | 11th | 12th | 13th | 14th | 15th | 16th | 17th | 18th | 19th | 20th | 21st | 22nd | 23rd | 24th | 25th | 26th | 27th | ... |
|---|---|---|---|---|---|---|---|---|---|---|---|---|---|---|---|---|---|---|---|---|---|---|---|---|---|---|---|---|
| 1 | F32 | F12 | F26 | F8 | F31 | F20 | F47 | F29 | F21 | F49 | F9 | F7 | F19 | | | | | | | | | | | | | | | |
| 2 | F32 | F12 | F26 | F8 | F31 | F47 | F20 | F29 | F9 | F7 | F19 | F14 | F53 | F10 | F52 | F50 | F41 | F22 | F15 | F36 | F1 | F23 | F48 | F11 | F27 | F2 | | |
| 3 | F32 | F12 | F26 | F8 | F31 | F47 | F20 | F29 | F21 | F49 | F9 | F7 | F19 | | | | | | | | | | | | | | | |
| 4 | F32 | F12 | F26 | F8 | F31 | F47 | F20 | F29 | F21 | | | | | | | | | | | | | | | | | | | |
| 5 | F32 | F12 | F26 | | | | | | | | | | | | | | | | | | | | | | | | | |
| 6 | F32 | F12 | F26 | F8 | F31 | F20 | F47 | F29 | F49 | F21 | F9 | F7 | F19 | F14 | F53 | F10 | F52 | F50 | F41 | F22 | F15 | F36 | F1 | F23 | F48 | F11 | F27 | F2 |
| 7 | F32 | F12 | F26 | F8 | F31 | F47 | F20 | F29 | F21 | F49 | F9 | F19 | F7 | F53 | F14 | F10 | F52 | F50 | F41 | F22 | F15 | | | | | | | |
| 8 | F32 | F12 | F26 | F8 | F31 | F20 | F47 | F29 | F21 | F49 | F9 | F19 | F7 | F14 | F53 | F10 | F52 | | | | | | | | | | | |
| 9 | F32 | F12 | | | | | | | | | | | | | | | | | | | | | | | | | | |
| 10 | F32 | F12 | F26 | F8 | F31 | F47 | F20 | F29 | F21 | F49 | F9 | F19 | F7 | F14 | F53 | F10 | F52 | F50 | F41 | F22 | F15 | F36 | F1 | F23 | F48 | F11 | F27 | F2 |
| 11 | F32 | F12 | F26 | F8 | F31 | F47 | F20 | F29 | F21 | F49 | F9 | F19 | F7 | F14 | F53 | F10 | F52 | F50 | F41 | F22 | F15 | F36 | F1 | F23 | F48 | F11 | F27 | F43 |

**Fig. 4.** Personalized questionnaire records with self-adapting algorithm

*Diagnosis Recognition System* - We evaluated the overall accuracy for correctly identified types of disease through our diagnosis recognition system. It has been

confirmed that the fusion method produces fewer errors than any single classifier
[5]. However, to derive the system-based diagnosis overall accuracy, we examined
354 dynamically formed questionnaires with the implementation of LOOCV.
Within the dataset of 354 rows, LOOCV sequentially divides the original questionnaire raw data into a training set and testing part to avoid biasing effects.
In each loop, a single questionnaire has been left out as new coming patient's
questionnaire. The remained 353 samples were applied for training the diagnosis
recognition system, as shown in Fig. 3. The fusion algorithm is evaluating the
classification of *new patient* associated with training data.

Figure 5 shows the comparison between obtained results from the diagnostic
system and confirmed doctor's diagnosis. The third column records the amount
of questions that a patient needs to fill in to get a final diagnosis, and the right
most column recorded whether the system-based diagnosis is correct. Our study
reveals that the average of patients' answered questions could be reduced to
21 out of initially applied 53 questions to predict the diagnosis. Furthermore,
the diagnosis system capability in deriving correct patient diagnosis due to a
dynamic channelized questionnaire increased the system's overall accuracy to
83.33%, while the previous system with the standard questionnaire that involves
53 questions achieves 82.70% correct diagnosis rate.

| | Doctor_Diagnose | System_Result | Questions_Amount | Answered Question / Total | Correct_Diagnose |
|---|---|---|---|---|---|
| 1 | 1 | 1 | 13 | 0.245283019 | T |
| 2 | 1 | 1 | 39 | 0.735849057 | T |
| 3 | 1 | 1 | 13 | 0.245283019 | T |
| 4 | 1 | 1 | 9 | 0.169811321 | T |
| 5 | 1 | 1 | 3 | 0.056603774 | T |
| 6 | 1 | 2 | 33 | 0.622641509 | F |
| 7 | 1 | 1 | 21 | 0.396226415 | T |
| 8 | 1 | 1 | 17 | 0.320754717 | T |
| 9 | 1 | 1 | 2 | 0.037735849 | T |
| 10 | 1 | 1 | 39 | 0.735849057 | T |
| 11 | 1 | 1 | 42 | 0.79245283 | T |

**Fig. 5.** Diagnosis results for self-adapting questionnaires

As shown in Fig. 6, the changing of prediction probability under the circum-
stance that the amount of answered questions continuously increases. As the
course of probabilities reveals, when the accuracy of one diagnose tends to be
stable, and the accuracy of another diagnose declines sharply, the diagnostic
system can make the final decision, and the current question is the last question
that dynamic questionnaire asked.

*Threats to Validity* - Patient's answers depend on their interests. Potential biases
while filling the patient questionnaires can be argued with understanding problems and decreasing focuses that most likely could appear due to the size of
the survey. Whether a symptoms question and its answer options really can
be associated with a particular type of disease remains on the skill of medical experts who designed the questionnaire. The diagnosis systems capability is

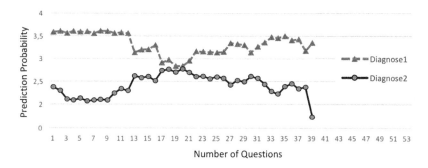

**Fig. 6.** The changes of prediction probability

limited to the patient's questionnaires collection, its associations between questions, answers and the medical doctors independently given diagnosis about each patient's type of disease. The diagnosis system also the self-adapting questionnaire algorithm are not limited to only medical diagnosis and surveys. Most importantly, the quality of diagnosis will always remain on data quality and its mutual information given through the significant existing data associations.

# 5   Conclusions and Future Perspectives

This study covers the concept and realization for a self-adapting questionnaire with computer-based diagnosis support for medical doctors. In contrast to conventionally used questionnaires with preselect question sets in static order, the patient's dynamic questioning algorithm introduced in this approach enables a time and afford saving processing, also to increase the level of accuracy for our diagnosis system with the medical purpose. The questionnaire size dynamically increase through channelizing next questions, that reflect symptoms with categorical relevance for a medical diagnosis that consequently depends on the patient's current question responses. Our diagnosis recognition system evaluates both continuously enriched questionnaire sequences parallel during each iteration, for collecting sufficient evidence to diagnose a patient with minimizing issues. The evaluation of patients condition depends on the given answers that linked to the typical response pattern of a particular disease group.

In particular, we provide 354 itemized patient questionnaires from clinical studies that are used to train a fusion classifier diagnosis system. The diagnose is based on patients individually responses about healthy conditions with particular relevance to some diseases. We applied all given answers to the self-adapting questionnaire algorithm, to verify whether a patient correctly given diagnosis could be reached by a reduced but channelized amount of question due to the improvement of the dynamic unit. Indeed, we could register an overall increase of accuracy up to 83.33% when applying the patient's answer-oriented questionnaire to our diagnosis recognition system. As a comparison, the applied static minimization of the questionnaire through linear regression and brute forced question combination only enhanced the diagnosis systems accuracy to 82.7% with a fixed question set of 41.

We think that the dynamic adaptive questionnaire could help scientists, researchers and medical doctors to reveal their patients' information, enhance questionnaires structure and improve their prediction models by using channelized information with direct relevance.

# References

1. Auria, L., Moro, R.A.: Support vector machines (SVM) as a technique for solvency analysis (2008)
2. Grigull, L., Lechner, W., Petri, S., Kollewe, K., Dengler, R., Mehmecke, S., Schumacher, U., Lücke, T., Schneider-Gold, C., Köhler, C., et al.: Diagnostic support for selected neuromuscular diseases using answer-pattern recognition and data mining techniques: a proof of concept multicenter prospective trial. BMC Med. Inform. Decis. Mak. **16**(1), 31 (2016)
3. Huyn, N., Melmon, K., Perrone, A.: Computerized clinical questionnaire with dynamically presented questions. US Patent Ap. 09/910,463, 20 July 2001
4. Jansen, A.C., van Aalst-Cohen, E.S., Hutten, B.A., Büller, H.R., Kastelein, J.J., Prins, M.H.: Guidelines were developed for data collection from medical records for use in retrospective analyses. J. Clin. Epidemiol. **58**(3), 269–274 (2005)
5. Kortum, X., Grigull, L., Muecke, U., Lechner, W., Klawonn, F.: Diagnosis support for orphan diseases: a case study using a classifier fusion method. In: Yin, H., Gao, Y., Li, B., Zhang, D., Yang, M., Li, Y., Klawonn, F., Tallón-Ballesteros, A.J. (eds.) IDEAL 2016. LNCS, vol. 9937, pp. 379–385. Springer, Cham (2016). doi:10.1007/978-3-319-46257-8_41
6. Liaw, A., Wiener, M.: Classification and regression by random forest. R News **2**(3), 18–22 (2002)
7. Ma, L., Liu, X., Song, L., Zhou, C., Zhao, X., Zhao, Y.: A new classifier fusion method based on historical and on-line classification reliability for recognizing common CT imaging signs of lung diseases. Comput. Med. Imaging Graph. **40**, 39–48 (2015)
8. Madabhushi, A., Agner, S., Basavanhally, A., Doyle, S., Lee, G.: Computer-aided prognosis: predicting patient and disease outcome via quantitative fusion of multi-scale, multi-modal data. Comput. Med. Imaging Graph. **35**(7), 506–514 (2011)
9. Pohar, M., Blas, M., Turk, S.: Comparison of logistic regression and linear discriminant analysis: a simulation study. Metodoloski zvezki **1**(1), 143 (2004)
10. Sboner, A., Eccher, C., Blanzieri, E., Bauer, P., Cristofolini, M., Zumiani, G., Forti, S.: A multiple classifier system for early melanoma diagnosis. Artif. Intell. Med. **27**(1), 29–44 (2003)
11. Zhang, X., Klawonn, F., Grigull, L., Lechner, W.: VoQs: a web application for visualization of questionnaire surveys. In: Fromont, E., De Bie, T., van Leeuwen, M. (eds.) IDA 2015. LNCS, vol. 9385, pp. 334–343. Springer, Cham (2015). doi:10.1007/978-3-319-24465-5_29

# ABIDE: Querying Time-Evolving Sequences of Temporal Intervals

Orestis Kostakis[1]([⊠]) and Panagiotis Papapetrou[2]

[1] Microsoft, Redmond, USA
orkostak@microsoft.com
[2] Department of Computer and Systems Sciences,
Stockholm University, Stockholm, Sweden
panagiotis@dsv.su.se

**Abstract.** We study the problem of online similarity search in sequences of temporal intervals; given a standing query and a time-evolving sequence of event-intervals, we want to assess the existence of the query in the sequence over time. Since indexing is inapplicable to our problem, the goal is to reduce runtime without sacrificing retrieval accuracy. We present three lower-bounding and two early-abandon methods for speeding up search, while guaranteeing no false dismissals. We present a framework for combining lower bounds with early abandoning, called ABIDE. Empirical evaluation on eight real datasets and two synthetic datasets suggests that ABIDE provides speedups of at least an order of magnitude and up to 6977 times on average, compared to existing approaches and a baseline. We conclude that ABIDE is more powerful than existing methods, while we can attain the same pruning power with less CPU computations.

## 1 Introduction

Sequences of temporal intervals, also known as *e-sequences* [18], are constructed by labeled events that have a time duration. Multiple events may be concurrent, and, in general, any event may start or end independently of another. Sequences of temporal intervals provide advantages over traditional types of sequences. For example, employing an interval sequence-based temporal abstraction on multivariate time series representing the medical history of patients [17] has shown substantial improvement in terms of predictive performance against traditional classifiers on real datasets from the domains of diabetes, intensive care, and infectious hepatitis. Similar temporal abstractions have been employed for the exploration and classification of renal-damage risk factors in patients with diabetes type II [9]. Moreover, such sequences appear in several other application areas, such as computational linguistics, geo-informatics, cognitive science, and music informatics (see [10] for a detailed set of references).

In this paper, we study the following problem of similarity search in time-evolving sequences of temporal intervals: given a standing query $Q$ and a time-evolving sequence $S$, we want to monitor the sequence and identify whether a

© Springer International Publishing AG 2017
N. Adams et al. (Eds.): IDA 2017, LNCS 10584, pp. 173–185, 2017.
DOI: 10.1007/978-3-319-68765-0_15

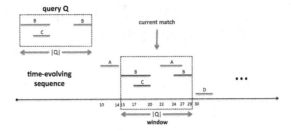

**Fig. 1.** Example of a query and a time-evolving e-sequence. The query is found in the indicated section (in red).

subsequence of $S$ matches $Q$, also allowing for a matching error. An example of a time-evolving sequence and a query is depicted in Fig. 1. The time-evolving e-sequence is an unbounded sequence of labeled events that have duration; while the query is bounded. The query is identified in the area indicated in red colour. Our main objective in this work is to reduce the CPU resources required for this task compared to a brute-force approach.

The problem of similarity search in time-evolving sequences of temporal intervals is crucial for several application areas. For example, in healthcare, when monitoring heart failure patients, we are interested in identifying particular diagnoses/treatment patterns in their medical history, such as whether the standard treatment procedures for heart failure have been applied in the proper order [3]. Solving this problem requires two key steps: the definition of an appropriate distance function (that takes into account the temporal relations between the events) and the development of fast and effective techniques for speeding up search under that distance function.

**Related work.** Most work on temporal interval sequences has so far been focusing on frequent pattern and association rule mining [15,16,18,19,22,23]. Moreover, three distance measures have been proposed for event-interval sequences. They all require knowledge of the whole e-sequence in advance, and none of them solves the problem of sub-sequence search. In other words, they are incapable of determining to which extent a query e-sequence exists within another e-sequence. Two of these distance measures, Relation Matrix [12] and Artemis [11], focus only on the temporal relations among intervals, and disregard completely the absolute values of the interval durations or the time between intervals. The third distance measure, IBSM [14], follows the same model as this work. However, it is not applicable to time-evolving sequences, as it requires knowledge of the length of both e-sequences, as well as the length of the longest e-sequence in the dataset. To an extent this can be retrofitted for the purpose of subsequence matching by employing a sliding window equal to the query length. Even then, such approach reduces to computing the Euclidean distance (ED) between the vector representations of the involved e-sequences. More importantly, speedup techniques proposed for IBSM may result in false dismissals. On the contrary, all our techniques provide guarantees for no false dismissals. Finally, related to subsequence matching, a method has been recently proposed [10] for finding the longest common sub-pattern in sequences of temporal intervals. It can be used for determining the existence of a query e-sequence in a target e-sequence. However, this method, too, focuses only on Allen's relations [2] among intervals, disregarding the intervals' duration, and is NP-hard.

The problem of subsequence matching has been studied in other sequential data domains, such as time series [20, 21] or evolving relational databases [4, 5]. However, there is yet no principled framework or distance function for querying time-evolving sequences of labeled temporal intervals. More importantly, indexing techniques for e-sequences, such as the one proposed by Kostakis et al. [13] for exact subsequence matching, are not directly applicable to our problem for three reasons: first, we are interested in handling very fast the future values of the time-evolving e-sequence, and not the past values (we cannot index values we have not yet seen). Second, we want to support arbitrary query lengths, hence indexing becomes intractable [20]. And finally, suppose that we could tolerate a delay so that we input all new observed values into an index database. Most indexing techniques (such as kd-trees) would require to map our comparison problem to a geometric (e.g., Euclidean) space. In our case, however, such mapping will suffer from the same limitations of lower bounds used in this work.

**Contributions.** We formulate the problem of similarity search in sequences of temporal intervals, and propose three lower-bounds and two early-abandon techniques that achieve substantial speedups. An extensive experimental evaluation on a collection of eight real and two synthetic datasets demonstrates that our overall approach provides greater speed-ups (of at least one order of magnitude and up to a factor of 6977, on average) than competitor methods [6], and can achieve similar pruning power with fewer CPU resources. Unlike recent work on event-interval sequences [14], our framework guarantees 100% recall.

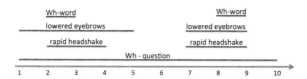

**Fig. 2.** Example of a query taken from the American Sign Language (ASL) domain, encoding the phrase "who drove the car, who?".

## 2   Preliminaries

Let $\Sigma = \{e_1, \ldots, e_{|\Sigma|}\}$ denote an alphabet of event labels. An event-interval is a triplet $S = (e, t_{start}, t_{end})$, where $e \in \Sigma$ is the event-interval label, and $t_{start}, t_{end}$ are the start and end time-points of $S$. We use $S.t_{start}$ and $S.t_{end}$ to denote that start and end time-points of $S$, respectively. Moreover, an *event-interval sequence*, or *e-sequence*, $\mathcal{S} = \{S_1, \ldots, S_n\}$ is a set of $n$ event intervals. Events of different label may be concurrent and overlapping in any manner.

A time-evolving e-sequence, or *e-stream*, is an unbounded e-sequence evolving over time; a stream of event-intervals. An example of an e-stream is shown in Fig. 1; the set of visible intervals define the following e-sequence:

$$S = \{(A, 10, 14), (B, 15, 20), (C, 17, 20), (A, 22, 27), (D, 30, 33)\}.$$

For a given e-stream $S$, we denote as $S_{ij}$ the e-sequence starting at time-point $i$ and ending at time-point $j$ of $S$. We assume that temporal data are produced by sampling at regular intervals that are defined by a fixed resolution. Monitoring the activity of multiple binary sensors provides us with a vector-based representation of an e-stream, that carries information about the status of the event labels; this approach is generally employed for e-sequences [14,15].

**Definition 1 (active event).** *Event $e_i \in \Sigma$ is active at time-point $t$, if there exists an event-interval $S = (e, t_{start}, t_{end})$, such that $S.t_{start} \leq t$, $S.t_{end} \geq t$, and $S.e = e_i$. Trivially, an event is inactive at time-point $t$, if it is not active.*

Given the set of all event labels $\Sigma$, we can define for each of the e-stream's time-points the vector of active events as a $|\Sigma|$-dimensional binary vector.

**Definition 2 (active event vector).** *Given e-stream $S$ and time-point $t$, the active event vector of $S$ at $t$, is a $|\Sigma|$-dimensional binary vector $V_S^t$, such that $V_S^t(i) = 1$, if $e_i$ is active in $S$ and 0, otherwise.*

Hence, the history of an e-stream $S$ can be represented as an ordered set of active event vectors $\mathcal{V}_S = \{V_S^1, \ldots, V_S^T\}$, where $T$ is the current time point of the e-stream. Moreover, the set of distinct event labels that are contained in $S$, or in other words the set of event labels that are active for at least one time point, is denoted as $\Sigma_S$, where $\Sigma_S \subseteq \Sigma$. We call $\Sigma_S$ the *projection* of $S$ over $\Sigma$.

**Problem (nearest neighbor query in e-streams).** Given an e-stream $S$, a query e-sequence $Q$, and a distance function $d$ we want to find the subsequence $S_{ij}$ of length $|Q|$ that minimizes $d(Q, S_{ij})$, or more formally, we want to identify and maintain $S_{xj}$ such that:

$$x = \arg\min_i d(Q, S_{ij}), i = 1, 2, \ldots \text{ and } j = i + |Q| - 1.$$

This problem generalizes to the following two problems of similarity search in e-streams: **k-NN query**, finding the $k$ substreams most similar to $Q$; and **range query**, given $\epsilon$, find all substreams $S_{ij}$ such that $d(Q, S_{ij}) \leq \epsilon$.

The choice of distance function $d$ determines the computational complexity of the task. If we choose as $d$ the sub-sequence function employed in [10,13], then the problem is NP-hard. For other distance functions [11,12,14] the task can be solved in polynomial time. In this work, we choose the Euclidean distance, since it is the basis of IBSM [14] that has demonstrated superior performance in tasks such as $k$-NN classification.

**A Naive Approach.** A straightforward solution is to employ a sliding window $W$ of length $|Q|$ over $S$. The goal is to find the substream defined by $W$, that minimizes the distance between the vector-based representation of $Q$, $\mathcal{V}_Q$, and the vector-based representation of the corresponding substream, $\mathcal{V}_W$. Due to the binary nature of the vectors, all $L_p$ norms are equivalent. For reasons of simplicity, in this work we employ $L_1$ as the distance metric between the vector-representations. We will refer to it as the Euclidean distance (ED). Hence, the distance between $Q$ and $W$ is

$$ED(Q, W) = \sum_{t=1}^{|Q|} \sum_{i=1}^{|\Sigma_Q|} |V_Q^t(i) - V_W^t(i)| \ . \tag{1}$$

The computation of ED is restricted to $\Sigma_Q$, the alphabet of $Q$. This does not oblige the user to define the values for intervals of labels that are not of interest. It does penalize intervals in $W$ that appear out of place in comparison to $Q$. Furthermore, our problem setting can be extended to handle negative queries (penalizing the existence of intervals with specific label in the streams). The time complexity of the naive algorithm, when comparing $Q$ to a window $W$ of length $|Q|$, is $O(|Q| \cdot |\Sigma_Q|)$.

Our main objective is to reduce the total CPU time and resources required to search for $Q$ in the stream. Since consecutive windows contain a significant overlap ($|W| - 1$ time-points), we would like to re-use as much of the computations as possible from previous instances. However, with ED this is not possible, since with every window slide, each vector of $Q$ gets compared to a new (the next) vector of $W$. As mentioned already, it is neither a meaningful option to apply indexing techniques in a stream setting.

Instead of computing the full ED between each possible pair of sliding window and query, we may use a cheaper lower-bounding technique to prune as many such comparisons as possible [7,8]. Given $best$, the minimum score over the computed instances so far, we can avoid computing the expensive $ED(Q, W)$ if the value of the lower bound is greater or equal to $best$, since the value of ED is guaranteed to be greater or equal to $best$.

## 3   Methods

We first present three lower-bounding techniques, which are then applied in a *filter-and-refine* manner; thus, achieving zero false dismissals; 100% recall rate. In addition, we propose two early-abandon approaches that may be applied on top of the lower bounds for achieving further speedups. Finally, we present ABIDE, a framework for combining early abandon techniques with lower bounds for querying time-evolving sequences of temporal intervals.

**TB_LB: Time-Based Lower Bound.** The first lower bound, TB_LB, summarizes the time dimension of the e-sequences. It takes into consideration only the number of active intervals in an e-sequence $\mathcal{S}$ for each time point $t_i$, denoted as $\alpha^{\mathcal{S}}(t_i)$. If the query were the example in Fig. 2, by counting the active interval for each time-point we acquire: $\alpha^Q = [2, 4, 4, 3, 2, 1, 3, 4, 4, 1]$. Then, TB_LB is computed as follows:

$$TB\_LB(Q, W) = \sum_{t=1}^{|Q|} |\alpha^Q(t) - \alpha^W(t)|. \tag{2}$$

**Theorem 1.** *TB_LB is a lower bound of the Euclidean distance; TB_LB* $(Q, W) \leq ED(Q, W) \ \forall Q, W$.

Given the counts $\alpha^W(t_i)$ for all time-points of an existing window $W$, updating the values for the next window can be performed in $\Theta(|\Sigma_Q|)$ time. Computing the value of TB_LB requires $\Theta(|Q|)$ time, using Eq. 2, yielding a total time complexity of $\Theta(|Q| + |\Sigma_Q|)$ for each time window.

**AB_LB: Alphabet-Based Lower Bound.** As opposed to TB_LB, the AB_LB lower bound summarizes the involved e-sequences with respect to each alphabet label. AB_LB counts the number of time-points during which intervals of each event label are active in $Q$ and in $W$, and then compares those counts without considering the actual positions of the corresponding time-points. Let $c_i^\Sigma(S)$ be the number of time-points in an e-sequence $S$ for which the $i$-th symbol of alphabet $\Sigma$ is active. For the example in Fig. 2, $c^\Sigma(Q) = [WhW : 4, LE : 8, RHS : 6, WhQ : 10]$. The lower bound is then computed as follows:

$$AB\_LB(Q, W) = \sum_{i}^{\Sigma_Q} |c_i^{\Sigma_Q}(Q) - c_i^{\Sigma_Q}(W)|. \tag{3}$$

**Theorem 2.** *AB_LB is a lower bound of the Euclidean distance; AB_LB* $(Q, W) \leq ED(Q, W) \ \forall Q, W$.

Instead of recomputing the values of $c_i^{\Sigma_Q}(W)$ for each window, which would require $\Theta(|\Sigma_Q||Q|)$ time, we can maintain their correct values in $\Theta(|\Sigma_Q|)$ time for each window slide. Given the existing values, we simply subtract the values of the expiring vector and add the values of the new vector. Computing the value of the lower bound (Eq. 3) requires $\Theta(|\Sigma_Q|)$ time per window.

**An $O(1)$ lower bound.** The two lower bounds, AB_LB and TB_LB, require $\Theta(|\Sigma_Q|)$ and $\Theta(|Q|)$ time to compute, respectively. A cheaper, yet less tight, lower bound, that requires $O(1)$ time can be derived by computing the difference between the sums of active labels per time point in the query and the corresponding e-stream window. More formally, the lower bound is computed as follows:

$$\sum_{i}^{|\Sigma_Q|} |c_i^{\Sigma_Q}(Q)| - \sum_{i}^{|\Sigma_Q|} |c_i^{\Sigma_Q}(W)| \text{ or } \sum_{t=1}^{|Q|} |\alpha^Q(t)| - \sum_{t=1}^{|Q|} |\alpha^W(t)|.$$

Given a comparison between $Q$ and a window $W$ not pruned by any of the two lower bounds, the computation of $ED(Q, W)$ is abandoned as soon as the score becomes higher than the current best. This method is known as *early abandoning*, and has been extensively used for speeding up NN search for time series [20]. The benefit of such methods compared to lower bounding is the lack of additional computational overhead for each window slide.

**TiDE: Time-point Density Early Abandon.** The first early-abandoning technique we propose is TiDE. It does not compute $ED(Q, W)$ from left to right, but attempts to identify those time-points that can lead to the highest contribution towards the total ED score. The intuition is that the time-points participating the most in the ED score are those time-points in $Q$ that contain the most active intervals. Hence, we may start the computation of ED from those points; the method may be applied equally by starting from the time points that contain the least active intervals. It suffices to compute the values $\alpha^Q(t_i)$ and to sort them once offline. This approach is inspired by [20], which is inapplicable to e-sequences, since z-normalization becomes an identity function. For example, if the query were the e-sequence in Fig. 2, with $\alpha^Q = [2, 4, 4, 3, 2, 1, 3, 4, 4, 1]$, the Euclidean distance would be computed in the following order: $[t_2, t_3, t_8, t_9, t_4, t_7, t_1, t_5, t_6, t_{10}]$. Assume we have computed the partial value of ED up to $t_4$ and it is greater than the best so far, then the computation can be terminated without computing the remaining part of ED (corresponding to time-points $t_7, t_1, \ldots, t_{10}$), since the remaining time-points can only add to the score.

**AiDE: Alphabet-Distribution Early Abandon.** AiDE is based on the density of each alphabet symbol. AiDE adapts to the evolving alphabet distribution of the e-stream, resulting in a better performance compared to learning the distribution from the beginning or knowing the overall stream distribution in advance. As in AB_LB, instead of computing the distance given the vectors at each time point, we compute the ED on the values of each interval-label channel, i.e., $ED(W, Q) = \sum_{i=1}^{|\Sigma_Q|} \sum_{t=1}^{|Q|} |Q_i^t - W_i^t|$.

ED is computed by selecting the alphabet labels in descending order starting from the one expected to provide the maximum contribution to the distance score. Let $r^S(\Sigma_i)$ denote the ratio of the total duration in e-sequence $S$ of all intervals labeled as $\Sigma_i$ divided by the length of $S$. Given $Q$ and a window $W$, we compute $r^W(\Sigma_i)$ and $r^Q(\Sigma_i), \forall i$, and sort the labels in descending order with respect to $|r^W(\Sigma_i) - r^Q(\Sigma_i)|$. Then, ED is computed based on this ordering. The procedure is described in Algorithm 1. For example, if $W$ were the sequence in Fig. 2 and query Q were the sub-sequence of the first two time-points $[t_1, t_2]$, $|Q| = 2$, it would be $r^W = [Whw : 0.4, LE : 0.8, RHS : 0.6, WhW : 1.0]$ and $r^Q = [Whw : 0.5, LE : 1.0, RHS : 0.5, WhW : 1.0]$. Then, the label-order for computing the ED would be: $LE, Whw, RHS, WhQ$, since the $|r^W(\Sigma_i) - r^Q(\Sigma_i)|$ values are $0.2, 0.1, 0.1, 0.0$ respectively. Assume that the partial cost of ED for labels $LE$ and $Whw$ is greater than the best so far, AiDE would break early without computing the cost contributed by the remaining labels. We may, in fact, immediately derive an indicator for the label ordering that should be followed, given our current knowledge of the alphabet distribution in the window. This information is maintained in the data structure of AB_LB.

---

**Algorithm 1.** Applying AiDE

---

   **function** APPLY_AIDE
      **for** $t \in 1, 2, ..$ **do**
         $W = shiftWindow(W)$;
         $update\_AB\_LB\_struct(W)$;
         **if** $t$ mod $sortPeriod == 0$ **then**
            Sort($AB\_LB\_struct$);
         $score = $ AiDE$(Q, W, AB\_LB\_struct, best)$;
         **if** $score < best$ **then**
            $best = score$;

---

This data structure already provides the numerators of the ratios $r^W(\Sigma_i)$. For $r^Q(\Sigma_i)$ we may compute it once offline. The denominators are known in advance; equal to $|Q|$. Thus, it is necessary to only sort the values of that data structure.

Sorting the data structure of AB_LB requires $\Omega(|\Sigma| \cdot log|\Sigma|)$ steps, which is a high cost to pay at every window slide. However, from one window slide to the other, the relative order between most pairs of interval-labels should remain the same. We may use other more expensive (worst-case) sorting algorithms, such as Bubblesort, that for certain instances require less than the fixed $|\Sigma| \cdot log|\Sigma|$ cost. Hence, we apply an optimized version of Bubblesort for computing the order of interval-labels based on their $|r^W(\Sigma_i) - r^Q(\Sigma_i)|$-values. Since the order is only an indicator (a conjecture) and not the ground truth for optimally computing any early-abandon method, we do not need to re-compute the order (sort) at every window slide. Instead, we perform the sorting periodically after a predefined number of window slides The benefit of AiDE is that it adapts to the evolution of the e-stream; it handles the *concept drift*. It turns out, that even if we knew in advance the alphabet distribution of the whole e-stream, or monitored the distribution from the beginning of the stream and sorted the $r$-values accordingly, the pruning power would still be lower. We demonstrate this in Sect. 4, and also show that AiDE is still faster due to the higher pruning power, despite the extra cost of sorting the $r$-values.

**The ABIDE framework.** In general, we may combine one or more lower bounds with any of the early-abandon methods. As also discussed in [20], the application of consecutive lower bounds should follow the *skyline* of the complexity vs. tightness trade-off. In other words, we should start by applying the computationally cheapest, but at the same time tightest for that time-complexity, lower bound. Following this guideline, first we apply the $O(1)$ lower bound, followed by AB_LB, and finally by AiDE. We will refer to this combination as ABIDE. In other words, AB_LB should be applied if the $O(1)$ lower bound did not prune the comparison, and AiDE only if neither $O(1)$ nor AB_LB pruned the comparison.

An additional advantage of the three methods in ABIDE over TB_LB and TiDE is their support for *negative queries*; penalizing the existence of particular events in $W$ that do not appear in $Q$. This is done simply by augmenting $\Sigma_Q$.

Finally, ABIDE works in a *filter-and-refine* manner and guarantees no false dismissals (100% recall) due to the fact that both its filtering step (the $O(1)$-LB and AB_LB) and its refine step (AiDE) guarantee no false dismissals. By trivially modifying how the LBs are used, we can switch between k-NN and range queries; replacing *best* with the range value, or the $k$-th best value seen so far.

## 4   Experiments

In this section, we benchmark the performance of the proposed methods.

**Datasets.** We used 8 real datasets from various application fields including American Sign Language (ASL), Health Informatics, and sensor networks. These datasets are used extensively in works related to e-sequences [10,15,16,18,19]. For each dataset, we created e-streams by concatenating all its e-sequences into one single e-sequence. Moreover, using our Dataset Replicator (DR)[1], we selected our largest real dataset, ASL2, and created two synthetic datasets of 100 million ($10^8$) time-points and $|\Sigma| = 254$ interval labels; so 25.4 billion data points. The 'density multiplier' was set to 1 and 3, respectively.

**Setup.** We benchmark our methods by creating an e-sequence query $Q$ and searching over the whole stream. If $Q$ is found within the stream, we set the value of *best* (so far) variable to $\infty$ and resume the process for the remainder of the stream. We do this in order to avoid favoring non-tight lower-bounds when the value of *best* would be 0; they would prune the comparison regardless of their tightness. We employ two ways of selecting $Q$: (1) by randomly selecting an existing (contiguous) segment of the stream, and (2) by randomly selecting and concatenating $|Q|$ individual time points (vectors) from the stream. We refer to the former case as "segment" queries and to the latter as "random" queries. We created query e-sequences of sizes 10, 100, 1000, and repeated the random experiment 20 times for each window size. We monitored the run-time and pruning power of our methods. As run-time, we excluded the time required to fetch the stream data from disk or memory, since disk I/O might distort the values. The source code as well as the complete results on all datasets as provided online [1]. In the remainder of this section, we will provide a sample of the results and describe the overall outcomes.

***Comparison vs competitors.*** We benchmarked ABIDE against MS-distance [6], a competitor method for lower bounding ED. Figure 4 depicts the results over ASL2. MS-distance utilizes the mean values and variance of vectors and iteratively converges to the full ED value, in $|\Sigma_Q|$ steps. When applied to our problem, it becomes similar to ABIDE but without any early-abandon priority optimization, and with significantly more overhead computations than $O(1)$ and AB_LB (or TB_LB) combined.

---

[1] DR generates synthetic datasets, given the statistical properties of an input dataset, i.e., the e-sequence length, count of intervals per event label, and the total duration of the e-stream. We can also create denser streams by multiplying the number of intervals and their total duration by a given scalar value, the 'density multiplier'.

Moreover, Fig. 3a depicts the pruning power of each method applied independently, on the ASL2 dataset. Figure 3b depicts the running times, normalized by the running time of the naive method (17.8, 126.3, and 2030.9 seconds for $|Q| = 10, 100, 1000$ respectively). We present our results in this way to facilitate examination of how the methods behave w.r.t. the window length.

We observe that ABIDE provides at least an order of magnitude speedup over the naive approach for all the datasets of significant size, i.e., ASL, ASL2, Hepatitis, and Skating. The best speedup was achieved for ASL2 and $|Q| = 100$, which was $1/0.0001433$ (this is a speedup of 6977.35 times; for $|Q| = 1000$ the speedup was $1/0.00182$, it took 3.696 seconds versus 33min50sec of the naive). For the remaining datasets, ABIDE provides at least one order of magnitude speedup for small query sizes; for larger query sizes the e-stream becomes very short and the result is non-informative. AB_LB, compared to TB_LB, provides higher pruning power, for 23 out of 24 combinations of datasets and query size (all except for Blocks with $|Q| = 1000$), and thus it is able to achieve higher speedups. Our experiments (omitted) also showed that performing TB_LB in conjunction with AB_LB is not sensible; TB_LB fails to prune any significant amount of comparisons when applied after AB_LB.

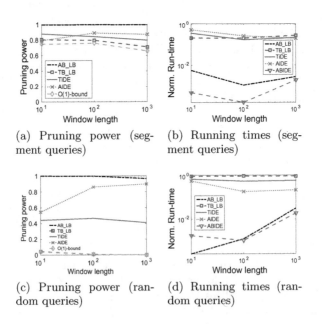

(a) Pruning power (segment queries)

(b) Running times (segment queries)

(c) Pruning power (random queries)

(d) Running times (random queries)

**Fig. 3.** Benchmark on ASL2 for 'segment' (a-b) and 'random' queries (c-d).

Regarding the early-abandon methods, we witness that AiDE beats TiDE in 5 out of 8 datasets (Auslan2, Context, Blocks, Hepatits, Pioneer), while for the remaining 3 datasets their performance is similar. Figure 3c depicts results of benchmarking the methods over the ASL2 dataset with queries formed by

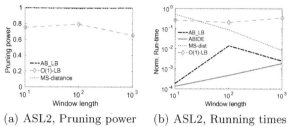

(a) ASL2, Pruning power     (b) ASL2, Running times

**Fig. 4.** ABIDE versus MS-distance method [6]. Both AB_LB and ABIDE outperform MS-distance. Even using $O(1)$-LB is faster for small query sizes.

randomly choosing $|Q|$ vectors from the stream. We witness that the pruning power decreases. While AB_LB suffers from pruning power reduction, surprisingly, TB_LB's pruning power falls close to zero for all datasets. Similarly, the pruning power of early abandon methods decreases, but AiDE now performs better than TiDE in more datasets.

The main benefit of AiDE is that it adapts to the stream's distribution, but has to apply a sorting algorithm. We benchmarked AiDE against possible variants where: (1) instead of following the window distribution, it simply learns the distribution from the stream's history (monitoring from $t_0$), and (2) it knows the distribution of the whole e-stream in advance. The results over the ASL2 dataset are illustrated in Fig. 5. We observe that even if AiDE sorted its data structure for every time-point, it would achieve similar pruning power but would be slower. AiDE provides higher pruning power, which in turn pays off the cost of sorting. In both cases it is faster than the learning variant, and for $|Q| > 10$ it even outperforms knowing the whole distribution in advance.

***Large Synthetic Dataset.*** The results of ABIDE's benchmark on large synthetic datasets are depicted in Fig. 6. In particular, Figs. 6a and 6b depict the results for Dataset Replicator's "density multiplier" equal to 1 and 3, respectively. As before, the time is normalized be the absolute running times of the naive ED. Computing that, however, would be impractical. For that reason we

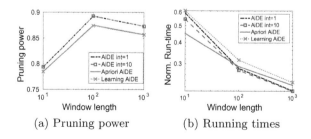

(a) Pruning power     (b) Running times

**Fig. 5.** AiDE variations on ASL2. We outperform other variations; even knowing the distribution of the whole e-stream in advance.

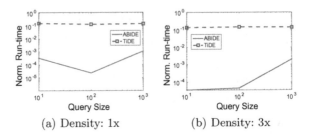

(a) Density: 1x        (b) Density: 3x

**Fig. 6.** Normalized running times of TiDE and ABIDE on two synthetic datasets of $10^8$ time-points and alphabet size $|\Sigma| = 254$, for varying stream densities.

used TiDE and extrapolated the naive running time values; this is easy since for TiDE pruning power translates to equal run-time improvement. For ABIDE, we witness that as the dataset density increases, the run-time increases for small query sizes ($|Q| = 10$), but decreases for larger query sizes. For all cases however, ABIDE achieves at least three orders of magnitude speedup over the naive ED. In the best cases, the speedup is over four orders of magnitude ($1/2.352 \cdot 10^{-5}$).

## 5    Conclusions

We studied the problem of online similarity search in time-evolving e-sequences. Since applying indexing methods is not an option, the goal was to reduce the CPU time required to search for the existence of a given query in future inputs. We presented three lower bounding techniques, AB_LB, TB_LB, and $O(1)$, as well as two early-abandon methods, AiDE and TiDE. We demonstrated a framework for combining lower bounds with early abandoning, called ABIDE, that is able to provide speedups of at least one order and up to 6977 times, on average, against the naive ED computation, on real datasets. The lower bound AB_LB is simple to implement and is able to outperform more complicated state of the art methods such as the $MS$-distance for ED.

**Acknowledgments.** This work was partly supported by the VR-2016-03372 Swedish Research Council Starting Grant.

## References

1. Source Code and Datasets (2017). http://goo.gl/fFnTEw
2. Allen, J.F.: Maintaining knowledge about temporal intervals. Commun. ACM **26**(11), 832–843 (1983)
3. Asker, L., Boström, H., Papapetrou, P., Persson, H.: Identifying factors for the effectiveness of treatment of heart failure: a registry study. In: CBMS 2016, pp. 205–206 (2016)
4. Babu, S., Widom, J.: Continuous queries over data streams. SIGMOD Rec. **30**(3), 109–120 (2001)

5. Hartzman, C., Watters, C.: A relational approach to querying data streams. IEEE TKDE **2**(4), 401–409 (1990)
6. Hwang, Y., Ahn, H.-K.: Convergent bounds on the euclidean distance. In: NIPS, pp. 388–396 (2011)
7. Keogh, E., Ratanamahatana, C.A.: Exact indexing of dynamic time warping. KAIS **7**(3), 358–386 (2005)
8. Kim, S., Park, S., Chu, W.: An index-based approach for similarity search supporting time warping in large sequence databases. In: Proceedings of ICDE, pp. 607–614 (2001)
9. Klimov, D., Shknevsky, A., Shahar, Y.: Exploration of patterns predicting renal damage in patients with diabetes type ii using a visual temporal analysis laboratory. J. Am. Med. Inform. Assoc. **22**(2), 275–289 (2015)
10. Kostakis, O., Papapetrou, P.: Finding the longest common sub-pattern in sequences of temporal intervals. Data Min. Knowl. Disc. **29**(5), 1178–1210 (2015)
11. Kostakis, O., Papapetrou, P., Hollmén, J.: Artemis: assessing the similarity of event-interval sequences. In: Proceedings of ECML/PKDD, pp. 229–244 (2011)
12. Kostakis, O., Papapetrou, P., Hollmén, J.: Distance measure for querying sequences of temporal intervals. In: Proceedings of PETRAE, pp. 40:1–40:8. ACM (2011)
13. Kostakis, O.K., Gionis, A.G.: Subsequence search in event-interval sequences. In: Proceedings of ACM SIGIR, pp. 851–854. ACM (2015)
14. Kotsifakos, A., Papapetrou, P., Athitsos, V.: Ibsm: Interval-based sequence matching. In: Proceedings of SDM, pp. 596–604 (2013)
15. Moerchen, F., Fradkin, D.: Robust mining of time intervals with semi-interval partial order patterns. In: Proceedings of SDM, pp. 315–326 (2010)
16. Moskovitch, R., Shahar, Y.: Medical temporal-knowledge discovery via temporal abstraction. In: Proceedings of the AMIA Annual Symposium, pp. 452–456 (2009)
17. Moskovitch, R., Shahar, Y.: Classification-driven temporal discretization of multivariate time series. Data Min. Knowl. Disc. **29**(4), 871–913 (2014)
18. Papapetrou, P., Kollios, G., Sclaroff, S., Gunopulos, D.: Mining frequent arrangements of temporal intervals. KAIS **21**, 133–171 (2009)
19. Patel, D., Hsu, W., Lee, M.: Mining relationships among interval-based events for classification. In: Proceedings of SIGMOD, pp. 393–404. ACM (2008)
20. Rakthanmanon, T., Campana, B., Mueen, A., Batista, G., Westover, B., Zhu, Q., Zakaria, J., Keogh, E.: Searching and mining trillions of time series subsequences under dynamic time warping. In: Proceedings of KDD, pp. 262–270 (2012)
21. Sakurai, Y., Faloutsos, C., Yamamuro, M.: Stream monitoring under the time warping distance. In: Proceedings of ICDE, pp. 1046–1055 (2007)
22. Winarko, E., Roddick, J.F.: Armada - an algorithm for discovering richer relative temporal association rules from interval-based data. DKE **63**(1), 76–90 (2007)
23. Wu, S.-Y., Chen, Y.-L.: Mining nonambiguous temporal patterns for interval-based events. IEEE TKDE **19**(6), 742–758 (2007)

# The Actors of History: Narrative Network Analysis Reveals the Institutions of Power in British Society Between 1800-1950

Thomas Lansdall-Welfare[1], Saatviga Sudhahar[1], James Thompson[2], and Nello Cristianini[1(✉)]

[1] Intelligent Systems Laboratory, University of Bristol, Bristol, UK
nello.cristianini@bristol.ac.uk
[2] Department of History, University of Bristol, Bristol, UK

**Abstract.** In this study we analyze a corpus of 35.9 million articles from local British newspapers published between 1800 and 1950, investigating the changing role played by key actors in public life. This involves the role of institutions (such as the Church or Parliament) and individual actors (such as the Monarch). The analysis is performed by transforming the corpus into a narrative network, whose nodes are actors, whose links are actions, and whose communities represent tightly interacting parts of society. We observe how the relative importance of these communities evolves over time, as well as the centrality of various actors. All this provides an automated way to analyze how different actors and institutions shaped public discourse over a time span of 150 years. We discover the role of the Church, Monarchy, Local Government, and the peculiarities of the separation of powers in the United Kingdom. The combination of AI algorithms with tools from the computational social sciences and data-science, is a promising way to address the many open questions of Digital Humanities.

**Keywords:** Big data · Network analysis · Digital humanities · Narrative analysis · Natural language processing

## 1 Introduction

Previous successes in the application of Artificial Intelligence (AI) techniques to data analysis date back many decades [15], and have enabled massive progress in fields such as Bioinformatics, Physics and Social Sciences. This intelligent data analysis (IDA) has only recently started to be advocated as part of historical research in the burgeoning fields of the Digital Humanities, mostly fueled by the ongoing digitization efforts that involve many libraries around the world. Early attempts [19] have been met with skepticism by part of the historical community, whose main criticism could be summarized as the need for digital humanities to go beyond counting words [14].

© Springer International Publishing AG 2017
N. Adams et al. (Eds.): IDA 2017, LNCS 10584, pp. 186–197, 2017.
DOI: 10.1007/978-3-319-68765-0_16

Here, we look at the different roles played in British society by religion, monarchy and the various components of governance (judiciary, legislature, central and local government) and how they have changed over time, a change that has been extensively studied using traditional methodologies (see, for example [16]). The boundaries between the various sources of legal authority (legislature, executive and judiciary) have not always been clear, and have been the object of discussion among scholars, particularly the separation between executive and legislative power, in the 19th century [2,6]. Furthermore, if we focus on "soft power" (influence), then the boundaries between the sphere of action of legal and other moral authorities are even more indistinct, with the role of the Church and the Monarchy being crucial, yet ever shifting over the centuries.

The relation between these political power structures and their representation in language is studied in Critical Discourse Analysis [10], a discipline interested in what our use of language can reveal about political power relations in society. One way in which this can be done is to determine who controls and shapes the narrative - or discourse - of the news.

We are interested in adding a data-driven element to this discussion about the evolution of the spheres of power and influence in British society, by analyzing the discourse found in historical local newspapers published over 150 years. In particular, we are interested in charting the various spheres of action of different narrative communities in the overall narrative network of local-news content, and following the evolution of their boundaries and key actors.

As such, we present here the first application of large-scale Narrative Network Analysis [27] to a corpus of 35.9 million articles, made possible by the deployment of a dependency parser in a map-reduce framework, and the study of topological properties of the resulting narrative networks, a method built on the pioneering work of [12] and automated by the use of Natural Language Processing by [26]. The resulting analysis involves 29 networks made up of a total of 156,738 nodes and 230,879 edges connecting them extracted from 150 years of newspapers. Different spheres of inter-actions (entire sectors of society and governance), as well as the role of individual actors, are mapped by analyzing these narrative networks extracted from vast amounts of text. We are interested in the changes and continuities, in the role of individual actors as well as in that of entire communities.

Note that, in accordance with Critical Discourse Analysis, this method goes beyond the mere political and legal balance of powers, and includes the moral authority and influence on the collective imaginary exercised by authorities such as the Church and the Monarchy. Furthermore, it can include intellectuals, opinion leaders or commercial players (though this might be the subject of analysis for later corpora). In other words, we want to see how public discourse – and the narration of reality – are organized and segmented in the period 1800 to 1950.

## 2   Data Description

The newspaper collection used in this study is composed of 35.9 million newspaper articles from a combination of the British Newspaper Archive

(FindMyPast, 2017) and digitized newspaper records provided by the Joint Information Systems Committee (JISC) from the same geographic regions and time period.

The collection was curated from the full archive so that the selected subset would allow for a comprehensive data-driven study of Britain in the 19th century on a representative sample of newspapers. The selection was performed by committee, and the criteria for selection included: the completeness of the digitization of a given newspaper title, the number of years that a newspaper title covers, the geographical region that the newspaper is from, the quality of the OCR output for the newspaper title and the political bias of the newspaper.

Our aim was to represent all geographical regions and time intervals as fairly as allowed by the available data. Newspaper titles were first split into their different geographical regions, and then within each region newspaper titles were ranked by a combination of the years covered (favoring titles with many years of continuous coverage), their average OCR quality, and the total size of data available for the title. Newspaper titles were then selected from this ranking until each geographical region had good coverage. We further used domain knowledge to take into consideration the balance of political opinion in the regional press at the time.

The newspaper collection we assembled, first reported on in [18], includes 28.6 billion words from 120 titles selected to best cover the United Kingdom, covering approximately 14% of the total output of U.K. regional newspapers during the period.

## 3  Methodology

Our methodology begins with the extraction of semantic triplets (subject-verb-object triplets where subject and object are noun phrases, referred to as 'actors', and the verb is transitive) from the corpus. The semantic triplets are used to generate a network of actors, linked by the transitive verbs connecting them in narrative. Analysis of the topology of these networks is used to discover important narrative information about the corpus [12, 26].

### 3.1  Extracting Semantic Triplets and Networks

Events are actions performed by actors that can be summed up by a verb or a name of an action [12]. An event can therefore be thought of as a narrative or story, where someone does something (the action) on or to someone or something else. Linguistically, these events can be represented as sequential sets of semantic triplets. Here, we consider semantic triplets of the form Subject-Verb-Object (SVO) which consists of a subject S as an actor, the verb V as the action performed by S, and O as the target of the action [11] (e.g., Police (S) Arrest (V) Thief (O)).

Before extracting semantic triplets, we first pre-process the textual data with a co-reference and anaphora resolution procedure. Co-reference resolution identifies whether two actors in the text refer to the same entity in the world [25].

These actors are then resolved to the most common short representation for the actor, using the Orthomatcher module in the ANNIE plugin of GATE, which reports an average precision of 96% and recall of 93% [5]. Anaphora resolution is used to resolve pronouns in the text to the specific actor that is being referred to, and is performed using the Pronominal resolution module in ANNIE, which reports an average precision of 66% and recall of 46% [5].

After pre-processing, the resulting text is split into sentences and passed into a dependency parser [22], outputting the dependency tree of each sentence. From the dependency tree, we extract the sentence subject, the sentence verb, and the object of the verb into the appropriate form for the semantic triplets. In total, 140,225,349 semantic triplets were extracted from the newspaper content.

From these semantic triplets, we generate semantic networks where the nodes of the network represent subjects and objects and the edges represent the verbs connecting them. We prune the triplet networks to reduce noise, keeping only nodes (actors) which occur a minimum of three times in the extracted triplets.

Networks were generated for each decade covered by the corpus, with an overlap of five years, giving us a total of 29 networks. Overlapping of networks was chosen to allow us to extract the communities which persist throughout the time period under investigation.

### 3.2  Centrality of Nodes

The centrality of a node in a network can be measured in many ways, and can be used as one measure of the importance of a node to the connected structure of the network. Here we consider the betweenness centrality of nodes, which represents the number of shortest paths between nodes in the network that pass through a given node [13].

### 3.3  Community Detection

Community detection is the task of partitioning a network into different communities, where densely connected nodes are placed within the same community, with nodes belonging to different communities being sparsely connected. Within the semantic networks presented here, community detection therefore corresponds with finding the communities of actors within the text that often perform actions on or to each other, with sparser interaction between different communities of actors. Communities in our networks were discovered by computing the modularity class of each node within the Gephi software [3], which attempts to find communities based upon finding high modularity partitions of the network using Blondel's fast unfolding algorithm, which has been shown to outperform other community detection algorithms [4].

Once the communities for each node within each of the 29 networks had been computed, we linked the communities across time, allowing us to form "macro-communities" which represent persistent community structures that are exhibited in the majority of the different networks across the 150 year period.

For each actor, we built a vector of their network communities, where each element in the vector refers to the actors community in each year, resulting in a 29-length vector encapsulating all the communities in which that actor participates. The similarity between each node and every other node was then computed using the cosine similarity, generating a similarity matrix of the communities for each actor. Using this similarity matrix, a 'macro-network' of actors is generated, with an edge joining every two actors that have a cosine similarity greater than 0.5, corresponding to the two actors participating in the same community more often than not.

On the 'macro-network', we once again perform community detection using the same algorithm as before [4], partitioning the network into smaller partitions with high modularity. The resulting communities from this process are referred to as 'macro-communities' and represent the persistent community structures over time.

### 3.4 Sentimental Classification of Actions

Actions in the narrative can fall into one of three sentimental categories: positive, negative or neutral. We can classify each of the actions being performed in the narrative into one of these categories to determine how sentimental the portrayal of the actions performed by or to the actors of a given community are reported on average in the news. We use the linguistic resource SentiWordNet [1], containing the positive and negative polarity for a set of 13,767 verbs. For each verb extracted as part of a semantic triplet, we score the sentimental polarity of the action as the average difference in positive and negative polarity for the verb across the possible verb synsets. Sentimental polarity for a given community or macro-community is then calculated as the average polarity of all actions performed by or to that community respectively.

## 4    Results

In this study, we investigate the narrative structure of the news over 150 years by network analysis of actors and the actions that connect them. Each network we generated contains all narrative triplets in a 10 year period that occur at least three times, with a five year overlap, giving us a total of 29 networks in all. An example[1] of one of these networks for the triplets extracted between 1905 and 1915 can be seen in Fig. 1. These narrative networks are formed by actors (entities who act or are acted upon) linked by their actions. In these networks, there are two possible notions of salience for an actor: its raw frequency within the corpus or its centrality within the network. The second notion, that of centrality, reflects how closely embedded this actor is into the overall structure, or how well linked it is to all other actors. It is a measure of how close all actors are to this actor.

---

[1] The full list of networks, along with their properties can be seen at http://thinkbig. enm.bris.ac.uk/supp-info-actors-of-history/.

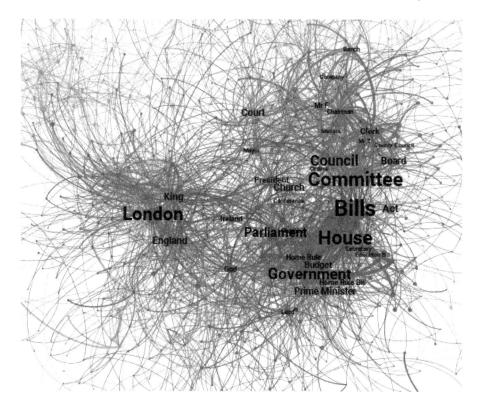

**Fig. 1.** Narrative network of the actors (nodes) and actions (edges) performed by them between 1905 and 1915 in the British newspaper corpus. Nodes are coloured based upon the community to which they belong.

Using these two measures of salience for an actor, Table 1 shows an overview of the 20 most frequent actors, along with the 20 most central actors, extracted from the news corpus.

We found that simply using the frequency of an actor does not generate a very meaningful list of actors, with "this", "one", "the", "defendant", "men" being those most frequent in all the triplets. However, we discover that the most-central actor in the narrative is "Bills", speaking to the centrality of politics in the coverage of local and national events, and in shaping public discourse in Britain. Following this, we discovered the next most central actors are "King", "House", "Government", "God", and "Queen", additionally showing the influence of both the monarchy and religion in the image of society communicated to the public via the mass media. This difference in meaning between notions of salience for an actor shows the importance of the linkage of an actor with the overall narrative.

Building up from this, communities in a narrative network are subsets of actors that are interacting tightly with each other in the narrative, and interacting much less with others, so that they can be separated from the rest of the

**Table 1.** List of the most salient actors in the narrative networks found in the 150 years of historical British newspaper corpus using the two different measures of salience: frequency of an actor and the centrality of an actor.

| | |
|---|---|
| Most frequent actors | One, This, Place, The, Mr, Defendant, Men, Prisoner, Members, Man, People, Nothing, Government, Committee, Bill, House, Some, More, Council, Attention |
| Most central actors | Bills, King, House, Government, God, Queen, Chairman, Committee, Duke, Court, England, London, President, Council, Bench, Prince, Jury, Clerk, Port, Men |

network with a relatively small number of cuts. By identifying the key communities, we partition the set of all actors into groups of highly interacting players, that signal power (hard and soft) structures within public life. For example, one community found in the discourse contains the actors "Court", "Jury", "Bench", "Judge", "Magistrates", etc. which we shall label as the Judiciary community.

Through the analysis of the central actors, and of the communities found in the narrative network, we discover three clear findings, relating to the division of powers of the State, the significance of local government intervention, and role of influence and moral authority in society.

## 4.1  Division of Powers

We discovered that the macro-communities within the narrative networks clearly identify the main spheres of action within society: Judiciary, Church, Monarchy, Local Government - but that they do not separate between the legislature and executive (Central Government). This can be seen in Fig. 2 where we show the macro-communities discovered for the 1000 most central actors over the 150 year period, finding that the orange community (labeled as 'Central Government') of nodes contain actors such as "Government", "Prime Minister", "Home Secretary" and "Lord Chancellor", but also "Bills", "House", "Commons" and "Parliament". This is in line with a long-standing position among scholars [2] according to which the U.K. has in practice been lacking the separation between those two powers. The key actors for the other structures of society include "Council", "Chairman", "Board" and "Clerk" (labeled as 'Local Government'), "Judge", "Jury", "Court" and "Counsel" (labeled as 'Judiciary'), "King", "Queen", "Prince" and "Princess" (labeled as 'Royalty') and finally "Bishop", "Rev", "Lord Bishop" and "Archbishop" (labeled as 'Church').

## 4.2  Local Government

As one might expect, we find that a prominent role is played by the 'central' (e.g. 'London'-centered) actors, such as "House", "Commons" and 'Prime Minister", but we also see, particularly from the 1870s onwards, the increasing accessibility and activism of Local Government over time. We can see that the actor

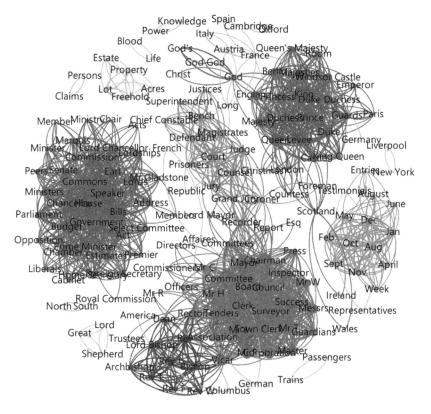

**Fig. 2.** Macro-communities discovered for the 1000 actors with the highest centrality over the 150 year period. It can be seen that the largest communities correspond with the broad topics of executive/legislative (orange, left), local authority (green, bottom right), royalty (purple, top right), judiciary (yellow, centre), months (pink, right) and the church (blue, bottom). (Color figure online)

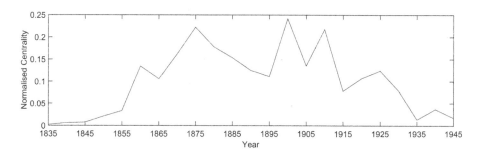

**Fig. 3.** Normalised centrality of the actor "Board" between 1830 and 1950 showing the prominence of the actor "Board" after the Public Health legislation of 1848.

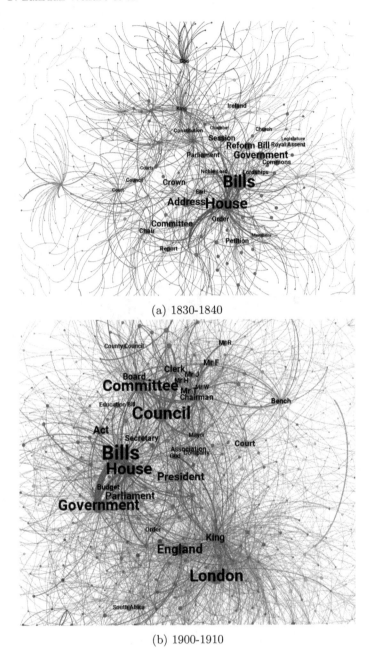

(a) 1830-1840

(b) 1900-1910

**Fig. 4.** The growth in the narrative role played by Local Government can be seen in (a) and (b), showing in (a) that the community as a whole, centred around the "Committee" and "Chair" nodes (shown in green, bottom left) did not play a major role in the 1830s, but, as shown in (b), by the start of the 20th century was an integral part of society (centred around the nodes "Committee" and "Council", top). (Color figure online)

"Board" gains in prominence after the Public Health legislation of 1848, but with some added salience from roughly the 1870s, that reflects the creation of School Boards, elected bodies tasked with creating and providing elementary education for children in areas that were under-served by the existing schools (see Fig. 3). By the last two decades of the 19th century, a nexus appears to emerge around "Committee", "Council", "Chairman", "Clerk", sometimes "Mayor", persisting into the mid-20th century. This growth in the relevance of Local Government can be seen in the change between the 1830s (Fig. 4a) when the main communities in the narrative are Central Government and Royalty and the 1900s (Fig. 4b) where we can clearly see that Local Government has become a major player in the narrative of the media.

### 4.3 The Role of Influence and Moral Authority

While the actual power of the Monarchy is known to have been declining during the period under investigation [8,17], we see that its role, along with that of the other members of royalty, in public discourse is very prominent throughout the period, manifesting as a distinct macro-community in the narrative network, populated by actors such as "King", "Queen", "Majesty", "Prince", and "Princess". From the late 19th century the Monarchy is known to have made a deliberate effort to maintain a presence in the media [17]. The centrality of the actors in the community – especially that of the monarchs – speaks to the cultural prestige and popularity that the Crown retained even as its actual power was limited. This is further supported by an analysis of the sentimental classification of the actions (Sect. 3.4), where we found that, on average, the Royalty community is portrayed as having the most positive actions performed on them. This coverage of royalty does surely bring out the differences between narrative appeal and political power, as the latter for the monarchy is known to be much less by 1950 than it had been in 1800 [9].

A similar case can be seen in the role of the Anglican community, formed by actors such as "Lord Bishop", "Archbishop", "Bishop", and "Rev". The Church of England as the established church held long-term significance to local life of the people, but its role in society was declining over the years, slowly becoming less central in the narrative of our society as presented in the news, supporting the view presented in [7], where it is argued that the 1960s are the decade of real change for the Church in which it loses considerable cultural salience. Notice that we do see, as expected in the case of the United Kingdom, a separation between Church and Government.

We also found that another centre of power in a State, the military, appears to form a large community during times of war but is not a constant presence throughout the period, and so does not appear in the macro-communities.

## 5 Conclusion

This study is aimed at addressing the criticism leveled at digital humanities approaches to historical corpora, which had called for efforts to go "beyond counting words". We show that the deployment of AI techniques to this data-analysis task allows us to access valuable semantic information that would not be accessible without Intelligent Data Analysis.

We also observe that this approach represents a step towards the full automation of the literary approach that Franco Moretti called distant reading [20]: the combination of AI, big data and data-visualization methods to large corpora has the potential to turn massive textual resources into semantically meaningful representations. This has an enormous potential in a community where most standard methods still involve various forms of statistical counting on words [19, 21, 23, 24].

**Acknowledgments.** Thomas Lansdall-Welfare, Saatviga Sudhahar and Nello Cristianini are supported by the ERC Advanced Grant "ThinkBig" awarded to NC. The authors would like to thank FindMyPast for making the original corpus available for study, as well as Dr. Gaetano Dato for his helpful comments.

# References

1. Baccianella, S., Esuli, A., Sebastiani, F.: SentiWordNet 3.0: an enhanced lexical resource for sentiment analysis and opinion mining. In: LREC, vol. 10, pp. 2200–2204 (2010)
2. Bagehot, W.: The English Constitution, vol. 3. Kegan Paul, Trench, Trübner, London (1900)
3. Bastian, M., Heymann, S., Jacomy, M.: Gephi: An open source software for exploring and manipulating networks (2009)
4. Blondel, V.D., Guillaume, J.-L., Lambiotte, R., Lefebvre, E.: Fast unfolding of communities in large networks. J. Stat. Mech. Theor. Exp. **2008**(10), P10008 (2008)
5. Bontcheva, K., imitrov, M., Maynard, D., Tablan, V., Cunningham, H.: Shallow methods for named entity coreference resolution. In: Chaines de références et résolveurs danaphores, Workshop TALN (2002)
6. Bradley, A.W., Ewing, K.D.: Constitutional and Administrative Law, vol. 1. Pearson Education, London (2007)
7. Brown, C.G.: The Death of Christian Britain: Understanding Secularisation, 1800-2000. Routledge, New York (2009)
8. Cannadine, D.: The context, performance and meaning of ritual: the british monarchy and the invention of tradition, c. 1820-1977. In: The Invention of Tradition, pp. 101–164. Cambridge University Press, Cambridge (1983)
9. Craig, D.M.: The crowned republic? monarchy and anti-monarchy in britain, 1760-1901. Hist. J. **46**(01), 167–185 (2003)
10. Fairclough, N.: Critical Discourse Analysis: The Critical Study of Language. Routledge, London (2013)
11. Franzosi, R.: Narrative as data: linguistic and statistical tools for the quantitative study of historical events. Int. Rev. Soc. Hist. **43**(S6), 81–104 (1998)
12. Franzosi, R.: Quantitative Narrative Analysis, vol. 162. Sage, Thousand Oaks (2010)
13. Freeman, L.C.: A set of measures of centrality based on betweenness. Sociometry **40**, 35–41 (1977)

14. Gooding, P.: Mass digitization and the garbage dump: the conflicting needs of quantitative and qualitative methods. Literary Linguist. Comput. **28**(3), 425–431 (2013)

15. Hand, D.J.: Intelligent data analysis: issues and opportunities. In: Liu, X., Cohen, P., Berthold, M. (eds.) IDA 1997. LNCS, vol. 1280, pp. 1–14. Springer, Heidelberg (1997). doi:10.1007/BFb0052825

16. Harris, J.: The Penguin Social History of Britain: Private Lives, Public Spirit: Britain 1870-1914. Penguin, UK (1994)

17. Hobsbawm, E.: The invention of tradition edited by eric hobsbawm and terence ranger (1983)

18. Lansdall-Welfare, T., Sudhahar, S., Thompson, J., Lewis, J., FindMyPast Newspaper Team, Cristianini, N.: Content analysis of 150 years of british periodicals. In: Proceedings of the National Academy of Sciences, p. 201606380 (2017)

19. Michel, J.-B., Shen, Y.K., Aiden, A.P., Veres, A., Gray, M.K., Pickett, J.P., Hoiberg, D., Clancy, D., Norvig, P., Orwant, J., et al.: Quantitative analysis of culture using millions of digitized books. Science **331**(6014), 176–182 (2011)

20. Moretti, F.: Distant Reading. Verso Books, London (2013)

21. Nicholson, B.: Counting culture; or, how to read victorian newspapers from a distance. J. Victorian Cult. **17**(2), 238–246 (2012)

22. Nivre, J., Hall, J., Nilsson, J.: MaltParser: a data-driven parser-generator for dependency parsing. In: Proceedings of LREC, vol. 6, pp. 2216–2219 (2006)

23. Pechenick, E.A., Danforth, C.M., Dodds, P.S.: Characterizing the google books corpus: strong limits to inferences of socio-cultural and linguistic evolution. PLoS One **10**(10), e0137041 (2015)

24. Roth, S., Clark, C., Berkel, J.: The fashionable functions reloaded: an updated google Ngram view of trends in functional differentiation (1800-2000) (2016)

25. Soon, W.M., Ng, H.T., Lim, D.C.Y.: A machine learning approach to coreference resolution of noun phrases. Comput. Linguist. **27**(4), 521–544 (2001)

26. Sudhahar, S.: Automated analysis of narrative text using network analysis in large corpora. Ph.D. thesis, University of Bristol (2015)

27. Sudhahar, S., De Fazio, G., Franzosi, R., Cristianini, N.: Network analysis of narrative content in large corpora. Nat. Lang. Eng. **21**(01), 81–112 (2015)

# Learning DTW-Preserving Shapelets

Arnaud Lods[1], Simon Malinowski[2(✉)], Romain Tavenard[3],
and Laurent Amsaleg[4]

[1] IRISA, Rennes, France
[2] Univ. Rennes 1, IRISA, Rennes, France
smalinow@irisa.fr
[3] Univ. Rennes 2, CNRS, UMR LETG, IRISA, Rennes, France
[4] CNRS-IRISA, Rennes, France

**Abstract.** Dynamic Time Warping (DTW) is one of the best similarity measures for time series, and it has extensively been used in retrieval, classification or mining applications. It is a costly measure, and applying it to numerous and/or very long times series is difficult in practice. Recently, Shapelet Transform (ST) proved to enable accurate supervised classification of time series. ST learns small subsequences that well discriminate classes, and transforms the time series into vectors lying in a metric space. In this paper, we adopt the ST framework in a novel way: we focus on learning, without class label information, shapelets such that Euclidean distances in the ST-space approximate well the true DTW. Our approach leads to an ubiquitous representation of time series in a metric space, where any machine learning method (supervised or unsupervised) and indexing system can operate efficiently.

## 1 Introduction

Time series analysis and mining is a wide research domain becoming increasingly popular over the last decades for tasks as diverse as classification, clustering, indexing or retrieval (see [4] for a survey). One popular similarity measure to compare time series is the Dynamic Time Warping (DTW), due to its capacity to cope with time shifts and warpings. Its complexity being quadratic with the length of time series, it is difficult to use DTW against very long time series and/or very large sets of time series. In turn, many research works have attempted to reduce that complexity and/or have tried to run DTW onto a very limited subset of candidate sequences [7,9,11,12]. Note furthermore that DTW is not a distance as the triangular inequality does not hold, which induces sub-optimality when used with traditional optimizations for indexing, or with kernel-based classifiers for instance.

Recently, a new family of approaches, based on the concept of *shapelets* [5,6,13], has been proposed for time series classification. Shapelets are time series subsequences selected (or learned) so as to discriminate classes. Amongst these approaches, the Shapelet Transform (ST) [6] uses shapelets as surrogates for representing time series: each time series is projected against the set of shapelets,

© Springer International Publishing AG 2017
N. Adams et al. (Eds.): IDA 2017, LNCS 10584, pp. 198–209, 2017.
DOI: 10.1007/978-3-319-68765-0_17

resulting in a vector in which components represent the distances between the time series and the shapelets.

This paper proposes a different approach at shapelet learning time: we do not try to best discriminate classes, but instead we aim at learning shapelets that best preserve the DTW. In other words, shapelets are selected such that the Euclidean distance between transformed time series best approximates the DTW between raw time series. The objective function we use to learn the shapelets is hence different from the ones of traditional works based on shapelets.

Learning shapelets to best approximate the true DTW between pairs of time series from a training set has several nice properties. First, it becomes possible to get a good estimation of the true DTW between any two time series by simply computing the Euclidean distance between the resulting shapelet transform vectors. Second, shapelet transformed vectors being a good proxy for the DTW, it becomes possible to use them, not only for supervised classification of time series, but also for many other tasks such as time series clustering, retrieval or indexing. This novel shapelet representation becomes quite ubiquitous as it can feed a wide range of machine learning or indexing methods for time series.

This paper first presents how to determine shapelets such that the DTW between all pairs of a training set is well captured in the high-dimensional space by the resulting vectors. We then demonstrate the validity of our approach by measuring how good surrogates are such vectors for the DTW. We also highlight the performance of this novel representation for time series clustering.

## 2    Related Work

A very large number of contributions aim at reducing the cost of the DTW, either by relying on lower bounds or by applying sophisticated pruning strategies. All this helps, only to some extent [4,11]. This paper, however, follows an entirely different direction as it builds on *vectorial representations* of time series. This related work section hence mainly focuses on such techniques and essentially discusses shapelet transforms for time series analysis.

Shapelets were introduced by Ye and Keogh in [13] for time series classification where a shapelet is an existing subsequence of a time series that best discriminate classes. Hills *et al.* proposed the *shapelet transform* in [6]. It consists in transforming a time series into a vector, its components representing the distances between the time series and shapelets determined beforehand. This vectorial representation of time series then feeds a classifier.Instead of using *existing* subsequences as shapelets, Grabocka *et al.* in [5] propose to rather forge the shapelets by learning the subsequences that minimize a classification loss. The learning step relies on a gradient descent. Then, the learnt shapelets are used to transform the time series into vectors, as proposed by Hills *et al.* Unsupervised extraction of shapelets has also been proposed in the literature for clustering purposes. In [14], Zakaria *et al.* extract the shapelets so that they divide the set of time series into well separated groups. Zhang *et al.* [15] propose to combine the learning of shapelets with pseudo-class labels.

To the best of our knowledge, only a very recent work tackles the same objective as ours which is to learn a mapping such that the Euclidean distance between the transformed time series preserves at best the DTW between the original time series. Compared to the approach we describe in this paper, Lei et al. in [8] reach this goal with very different means, as no shapelets are involved in their work while they are at the core of the technique we describe here. Lei et al. learn a vectorial representation of time series such that the dot product between these representations well preserves the similarity between the original time series. The representations are obtained by matrix factorization. They also propose another learning strategy based on a gradient descent that is faster, but less accurate in preserving the original similarities. An extremely severe drawback of their contribution is that it learns the transformed time series, and not the transformation itself. It is therefore impossible to transform a new and unknown time series once a database of time series has been fully transformed. Their method can therefore not be used in the many applications where queries are not known in advance or where the database of time series has to be updated.

In contrast, with the approach detailed below, not only each time series in the database is transformed into a vector, but any unknown new time series that probes or that is to add to the dataset can undergo such a transform. It is the process of transforming the time series into a DTW-preserving high-dimensional vector that we overall learn.

## 3   Learning DTW-Preserving Shapelets (LDPS)

In this section, we detail LDPS an algorithm that embeds time series into a metric space such that the Euclidean distance in the transformed space approximates the DTW in the original time series space. An illustration of the LDPS algorithm is schematically given in Fig. 1.

### 3.1   Definitions and Notations

We define here notions and quantities that are used in most of the papers that deal with time series shapelets [5,6]. Let $\mathcal{T} = \{T_1, \ldots, T_N\}$ be a set of $N$ time series. We assume here that all time series in $\mathcal{T}$ have the same length $Q$, but the method presented in this paper is also valid for time series of different lengths. A time series $T_i$ in $\mathcal{T}$ is hence composed of $Q$ elements: $T_i = T_{i,1}, \ldots, T_{i,Q}$. In the following, $T_{i,m:L}$ will denote the $m^{\text{th}}$ segment of length $L$ of $T_i$: $T_{i,m:L} = T_{i,m}, \ldots, T_{i,m+L-1}$.

**Definition 1.** A shapelet $S$ of length $L$ is an ordered sequence of $L$ values. In the following, $\mathcal{S}$ will denote a set of $K$ such shapelets: $\mathcal{S} = \{S_1, \ldots, S_K\}$, where $S_k = S_{k,1:L}$ for all $k \in \{1, \ldots, K\}$.

**Definition 2.** The Euclidean score between $S_k$ and $T_{i,j:L}$ is defined as

$$D_{i,k,j} = \frac{1}{L} \sum_{l=1}^{L} (T_{i,j+l-1} - S_{k,l})^2. \tag{1}$$

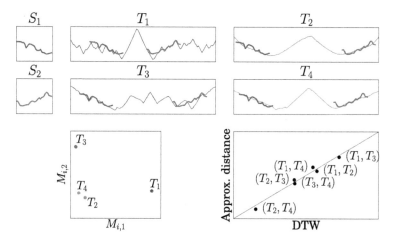

**Fig. 1.** Illustration of LDPS. Learned shapelets $S_1$ and $S_2$ are used to embed time series in a 2-dimensional space (*cf.* bottom-left figure), in which the Euclidean distance is a good approximation of the DTW between original time series. In real settings, more than 2 shapelets are learned to obtain higher-dimensional embeddings and hence richer representations.

*The Euclidean shapelet match between $S_k$ and $T_i$ is defined as*

$$M_{i,k} = \min_{j \in \{1,\dots,Q-L+1\}} D_{i,k,j} \,. \tag{2}$$

The Shapelet Transform of time series has been defined in [6]. It consists in, given a set $\mathcal{S}$ of $K$ shapelets, transforming $T_i$ into a $K$-dimensional vector $M_i$ whose components are $\{M_{i,k}\}_{1 \leq k \leq K}$ (Eq. (2)).

### 3.2   Loss Function to be Minimized

Most works dealing with shapelets first select the best set of shapelets before doing the shapelet transformation. Best shapelets are selected to discriminate classes. In this paper, we adopt a different approach. We aim at learning a set of shapelets such that the Shapelet Transform preserves as well as possible the Dynamic Time Warping measure. In other words, we would like that the Euclidean distance in the transformed space approximates the DTW. We turn this problem into the minimization of a loss function, as explained in the following. Let $\mathcal{S}$ be a set of $K$ shapelets. Let $T_{i_1}$ and $T_{i_2}$ be two time series in $\mathcal{T}$. The Shapelet Transform of $T_{i_1}$ and $T_{i_2}$ is denoted $M_{i_1}$ and $M_{i_2}$ respectively. The Dynamic Time Warping between $T_{i_1}$ and $T_{i_2}$ is denoted $DTW(T_{i_1}, T_{i_2})$. The loss $\mathcal{L}(T_{i_1}, T_{i_2})$ induced by the approximation of $DTW(T_{i_1}, T_{i_2})$ by the Euclidean distance between $M_{i_1}$ and $M_{i_2}$ is defined as:

$$\mathcal{L}(T_{i_1}, T_{i_2}) = \frac{1}{2}\left(DTW(T_{i_1}, T_{i_2}) - \beta \|M_{i_1} - M_{i_2}\|_2\right)^2, \tag{3}$$

where $\beta$ is a scale parameter that is learned by the proposed method, as explained below. The overall loss for a dataset $\mathcal{T}$ of $N$ time series is given by:

$$\mathcal{L}(\mathcal{T}) = \frac{2}{N(N-1)} \sum_{i_1=1}^{N-1} \sum_{i_2=i_1+1}^{N} \mathcal{L}(T_{i_1}, T_{i_2}). \qquad (4)$$

### 3.3   Stochastic Gradient Descent

The LDPS method aims at learning, for a training set $\mathcal{T}$ of time series, a set $\mathcal{S}$ of $K$ shapelets and a scale parameter $\beta$ that minimize the overall loss defined in Eq. (4). For the sake of clarity, we assume here that the shapelets of $\mathcal{S}$ have the same length $L$, but the method can be straight-forwardly extended to shapelets of different lengths. We adopt the stochastic gradient descent framework to learn the $K \cdot L + 1$ coefficients that lead to minimize $\mathcal{L}(\mathcal{T})$. In this framework, the gradients of $\mathcal{L}(T_{i_1}, T_{i_2})$ with respect to these coefficients need to be computed. If we denote $\hat{Y}_{i_1,i_2} = \|M_{i_1} - M_{i_2}\|_2$ and $\Delta_{i_1,i_2,k} = M_{i_1,k} - M_{i_2,k}$, we get:

$$\frac{\partial \mathcal{L}_{i_1,i_2}}{\partial \beta} = \hat{Y}_{i_1,i_2} \left( \beta \hat{Y}_{i_1,i_2} - DTW(T_{i_1}, T_{i_2}) \right) \qquad (5)$$

$$\frac{\partial \mathcal{L}_{i_1,i_2}}{\partial S_{k,l}} = \frac{\partial \mathcal{L}_{i_1,i_2}}{\partial \hat{Y}_{i_1,i_2}} \frac{\partial \hat{Y}_{i_1,i_2}}{\partial \Delta_{i_1,i_2,k}} \left( \frac{\partial M_{i_1,k}}{\partial S_{k,l}} - \frac{\partial M_{i_2,k}}{\partial S_{k,l}} \right) \ \forall k, l. \qquad (6)$$

Straight-forward derivations give:

$$\frac{\partial \mathcal{L}_{i_1,i_2}}{\partial \hat{Y}_{i_1,i_2}} = \beta \left( \beta \hat{Y}_{i_1,i_2} - DTW(T_{i_1}, T_{i_2}) \right) \qquad (7)$$

$$\frac{\partial \hat{Y}_{i_1,i_2}}{\partial \Delta_{i_1,i_2,k}} = \frac{\Delta_{i_1,i_2,k}}{\|M_{i_1} - M_{i_2}\|_2} \quad \forall M_{i_1} \neq M_{i_2} \qquad (8)$$

$$\frac{\partial M_{i,k}}{\partial S_{k,l}} = \sum_j \frac{\partial M_{i,k}}{\partial D_{i,k,j}} \frac{\partial D_{i,k,j}}{\partial S_{k,l}}. \qquad (9)$$

In practice, we extend the formula provided in Eq. (8) in the case where $M_{i_1} = M_{i_2}$ by: $\frac{\partial \hat{Y}_{i_1,i_2}}{\partial \Delta_{i_1,i_2,k}} = 0$. We observe experimentally that this case is sufficiently rare not to impair the convergence process.

In our implementation, we do not use soft-minimum approximation as done in [5] for the computation of $M_{i,k}$. Indeed, we observe that authors of [5] tend to use an $\alpha$ parameter so large (in absolute value) that they almost end up with a hard minimum computation. We then consider the limit case when $\alpha \to -\infty$ of the soft-minimum formula to get back to a hard-minimum setup, which gives:

$$\frac{\partial M_{i,k}}{\partial D_{i,k,j}} = \delta_{j,j*},$$

where $j*$ is the argmin of Eq. (2). Finally, for the computation of $\frac{\partial D_{i,k,j}}{\partial S_{k,l}}$, derivations from [5] can be used:

$$\frac{\partial D_{i,k,j}}{\partial S_{k,l}} = \frac{2}{L}(S_{k,l} - T_{i,j+l-1}).$$

These gradients are used to update the coefficients at each iteration of the algorithm with a learning rate of $\alpha$, like for any gradient descent algorithm.

### 3.4 Model Initialization

The loss function presented in Eq. (4) is not convex, as illustrated in Fig. 2. It is therefore of prime importance to ensure proper initialization of the model parameters for the optimization process not to get stuck in highly suboptimal local minima. In our setting, $k$-means clustering is used to generate the set of initial shapelets. Once the initial shapelets fixed, an initial value $\beta_{\text{init}}$ is selected for $\beta$ by randomly sampling a set $\mathcal{P}$ of 100 time series pairs and computing the corresponding optimal least square solution to the monodimensional regression problem that relates distances between Shapelet Transforms to DTW between original time series. We could, theoretically, update $\beta$ the same way at each iteration, but this update rule has $O(|\mathcal{P}|)$ complexity, which contradicts our will to use stochastic gradient descent to ensure a fast update of the model.

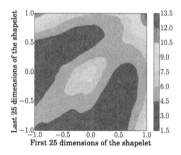

**Fig. 2.** Heat map of our loss function $\mathcal{L}$ with respect to shapelet coefficients for Swedish-Leaf dataset. For visualization purposes, we consider a single shapelet with a single value assigned to the first 25 coefficients ($x$-axis) and another value for the remaining 25 coefficients ($y$-axis). Best viewed in color. (Color figure online)

### 3.5 Convolutional Variant of LDPS

The Shapelet Transform on which we rely in this paper is very similar in spirit to what is learned by Convolutional Neural Networks. More precisely, the shapelet match presented in Definition 2 is very similar to a convolutional layer in a Neural Network. The computed Shapelet Transform corresponds to the output of a single-layer convolutional neural network with infinite max pooling in which the convolution operation would be replaced by a sliding window distance computation (and hence, the max pooling would be replaced by a min pooling). In this comparison, convolution filters are the equivalent of shapelets. We can then consider a unified framework in which both approaches can be used and compared experimentally. To do so, we introduce the convolutional shapelet match between a shapelet and a time series, which consists in using the following definition in place of Definition 2 (and its related Eqs. (1) and (2)):

**Definition 3.** *The convolutional score between $S_k$ and $T_{i,j:L}$ is defined as*

$$D_{i,k,j} = \frac{1}{L} \langle S_k, T_{i,j:L} \rangle = \frac{1}{L} \sum_{l=1}^{L} S_{k,l} \cdot T_{i,j+l-1}. \tag{10}$$

*The convolutional shapelet match between $S_k$ and $T_i$ is defined as*

$$M_{i,k} = \max_{j \in \{1,\dots,Q-L+1\}} D_{i,k,j}. \tag{11}$$

As a consequence, the computation of $\frac{\partial D_{i,k,j}}{\partial S_{k,l}}$ for this convolutional variant of our model differs from the one presented above, and we get:

$$\frac{\partial D_{i,k,j}}{\partial S_{k,l}} = \frac{1}{L} T_{i,j+l-1}.$$

In the following, we will refer to this convolutional variant of our model as LDPS-C, while the Euclidean one will be denoted LDPS-E.

Extending this analogy between neural networks and Shapelet models, our proposition can be seen as a *siamese* architecture [2] for Shapelets, *i.e.* two time series are provided as inputs to the same Shapelet model and distance between the corresponding outputs is used as a proxy for time series similarity. However, LDPS models are learned to minimize discrepancy between a target metric and the obtained distance whereas, in [2], the idea is to threshold the obtained distance for classification purposes.

### 3.6 Summary of the LDPS Algorithm

A summary of the learning phase of the LDPS algorithm, i.e. the learning of $KL + 1$ coefficients (the set $S$ of shapelets and the parameter $\beta$) is given in Algorithm 1. After this phase, the shapelet set $S$ can be used to transform any time series into a vector of dimension $K$. The complexity of the shapelet transform (once the shapelets learned) is $\mathcal{O}(NLK)$, where $N$ is the number of time series to transform.

## 4   Experimental Results

In this section, we present experiments to evaluate the performance of our method. We first study the quality of DTW reconstruction reached by LDPS and then compare it to state-of-the-art competitors for a clustering task.

| | |
|---|---|
| **Input** | : A set $\mathcal{T}$ of time series |
| | The number $K$ and length $L$ of shapelets |
| | The learning rate $\alpha$ for the gradient descent algorithm |
| **Output:** | A set $\mathcal{S}$ of $K$ shapelets of length $L$ |
| | The scale coefficient $\beta$ |

1  Initialize $\mathcal{S}$ and $\beta$ according to Sect. 3.4
2  **for** $i \leftarrow 1$ **to** $n_{iter}$ **do**
3     |  Randomly pick two time series $T_1$ and $T_2$ from $\mathcal{T}$
4     |  Compute the DTW between $T_1$ and $T_2$
5     |  Compute the gradients of $\mathcal{S}$ and $\beta$ from Eqs. (5) to (9)
6     |  Update $\mathcal{S}$ and $\beta$ (using their gradients and the learning rate $\alpha$)
7  **end**

**Algorithm 1.** Learning phase of LDPS

**Experimental Setup.** Following the principles of reproducible research, the Python code used in these experiments (including both variants of our model) is made publicly available for download[1].

Unless otherwise stated, each of our models makes use of shapelets of different lengths to better learn scale-specific patterns. Shapelet lengths $L$ are set to 15%, 30% and 45% of time series lengths. Inspired by [5], we use a number $K$ of shapelets for each length equal to $K = 10 \cdot \log(Q - L)$. Finally, as our method is stochastic, for each experiment, 5 different models are fitted. All models are fitted for 500,000 stochastic gradient descent steps, and we use the AdaGrad [3] algorithm to adapt the learning rate during the convergence process. Datasets used for the experiments are publicly available [1].

**Comparison Between LDPS-E and LDPS-C.** We analyze in this section the difference between LDPS-E and LDPS-C in terms of performance.

In practice, we observe that there does not seem to exist a consistently better variant on all datasets. Figure 3 presents model losses (*i.e.* mean squared DTW reconstruction error) as a function of the number of iterations. It shows that, for this criterion, LDPS-C outperform LDPS-E for Synthetic Control data set, while the opposite observation holds true when considering SwedishLeaf data set. Similar conclusions can be drawn when considering clustering performance as presented in Table 1.

**Quality of DTW Approximation.** Figure 4a presents the fit between DTW values and their approximations through the LDPS-E algorithm. Each dot in this figure corresponds to a pair of training time series. We can see that fully fitted model drastically improves the quality of DTW approximation over partially fitted ones. Another important point is to observe that all distance magnitudes are reproduced with similar accuracy, meaning that our method is able to reproduce

---

[1] https://github.com/rtavenar/LDPS/.

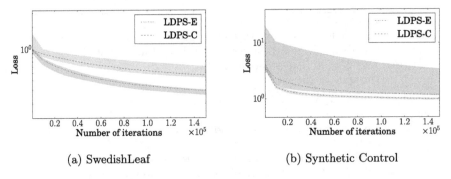

(a) SwedishLeaf  (b) Synthetic Control

**Fig. 3.** Compared convergence of LDPS-E and LDPS-E on two different datasets. Shaded areas illustrate the loss span between best and worse models and dashed lines correspond to the median loss model for each variant. Best viewed in color. (Color figure online)

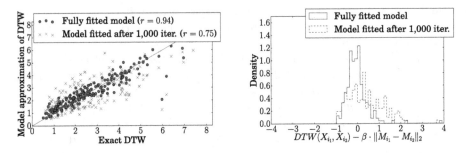

**Fig. 4.** Left: LDPS-E against exact DTW on dataset *SwedishLeaf*. $r$ is the Pearson correlation coefficient. Right: Histogram of the difference between the exact DTW and the corresponding LDPS-E values.

both similarities and dissimilarities between time series. Finally, we observe in Fig. 4b that there is no strong bias towards overestimation (resp. underestimation) observed for the fully fitted model, which indicates that the learned scale parameter $\beta$ is reasonable.

**Time Series Clustering with LDPS.** As LDPS embeds time series in a Euclidean space, it can be used for various machine learning tasks, including unsupervised ones, since class labels are not required to fit our models. We evaluate in this section the quality of the clustering induced by LDPS. LDPS can be used for clustering by feeding a standard Euclidean $k$-means algorithm with the transformed time series.

Before going into clustering results, we address the issue of unsupervised model selection. The question we are asking here is the following: Is there a way to select a model that is likely to lead to a good clustering without using any *a priori* ground truth information? Figure 5 depicts the relationship between clustering quality, evaluated in terms of Normalized Mutual Information (NMI)

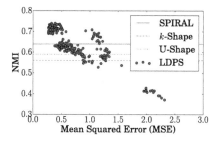

**Fig. 5.** Clustering quality as a function of model loss for dataset *SwedishLeaf*.

score and model loss. The NMI score measures the coherence between the true labels of time series and the estimated cluster indices. In this figure, each dot corresponds to a partially fitted model. Dots that have high losses correspond to small numbers of iterations and the loss decreases when the number of iterations increases, as observed previously in Fig. 3. An important point here is that we can observe a strong negative correlation between the loss associated to a model and clustering quality obtained with this model. This seems to indicate that the value of the loss can be used as a model selection criterion without using any ground truth information. For a given data set, several models can be learned (different initialization and variants of LDPS). The one leading to the smallest loss will be selected. This model selection criterion is applied in the clustering results presented below.

We conducted experiments on 15 datasets from the UEA & UCR repository [1]. Selected datasets cover a wide range of time series lengths, with varied numbers of classes and dataset sizes. For all these datasets, training and test sets are gathered, as we do not tackle the usual classification task in this piece of work. Performance of LDPS are compared with the following competitive methods. SPIRAL is the method proposed in [8] that has the same objective as LDPS but using a different approach to learn the transformation. It is combined with a $k$-means algorithm for clustering purposes. Reported results for SPIRAL have been obtained using the code made available by the authors[2]. $k$-Shape is a time series clustering algorithm proposed in [10] based on the cross-correlation measure. Reported results for $k$-Shape have been obtained using the dtwclust library of the R software, in which $k$-Shape is implemented. U-Shape corresponds to the clustering method using unsupervised shapelets presented in [14], and for which the code is available on a dedicated webpage[3]. For the sake of fair comparisons, we use the same shapelet lengths for this competitor and LDPS. All these competitor methods have been shown to be very efficient for time series clustering. Table 1 presents NMI scores for LDPS and competitive methods. Presented scores are medians obtained over 20 clustering runs for each method. Per-dataset performance as well as average ranks reported in this Table show the benefit of

---

[2] https://github.com/cecilialeiqi/SPIRAL.
[3] https://sites.google.com/site/ushapelet/.

**Table 1.** Comparison of Normalized Mutual Information (NMI) scores. Best performance is marked as bold. When the difference cannot be considered significant using a Mann-Whitney rank test with $p = 5\%$, several models are bolded.

| Datasets | LDPS-E | LDPS-C | SPIRAL | U-Shape | $k$-Shape |
|---|---|---|---|---|---|
| CBF | **0.83** | 0.71 | 0.39 | 0.61 | 0.76 |
| CricketX | 0.34 | 0.35 | 0.30 | 0.37 | **0.38** |
| ElectricDevices | 0.34 | 0.35 | **0.35** | 0.31 | 0.25 |
| FaceAll | **0.63** | 0.60 | **0.63** | 0.53 | 0.60 |
| FaceFour | 0.63 | 0.63 | 0.60 | **1.00** | 0.48 |
| FiftyWords | 0.68 | 0.64 | **0.68** | 0.56 | 0.66 |
| Lightning2 | **0.13** | 0.12 | 0.08 | 0.05 | 0.11 |
| Lightning7 | **0.53** | **0.55** | 0.48 | 0.50 | **0.54** |
| OSULeaf | **0.42** | 0.34 | 0.26 | 0.33 | **0.42** |
| StarLightCurves | 0.68 | **0.68** | 0.61 | 0.51 | 0.60 |
| SwedishLeaf | **0.70** | 0.63 | 0.64 | 0.59 | 0.56 |
| SyntheticControl | 0.97 | **0.98** | 0.81 | 0.83 | 0.72 |
| Trace | **0.75** | **0.75** | 0.50 | 0.73 | **0.75** |
| TwoPatterns | 0.69 | **0.86** | 0.11 | 0.32 | 0.30 |
| UWaveGestureLibraryX | 0.44 | 0.43 | **0.47** | 0.31 | 0.45 |
| Average rank | 2.13 | 2.33 | 3.40 | 3.87 | 3.20 |

using LDPS models for this task, as they tend to get higher clustering performance. Moreover, one should note that contrary to $k$-Shape, our method is not specifically designed for clustering and could be used for many other machine learning tasks. Also, when compared to SPIRAL, LDPS has the key property that it learns a transformation for time series that can later be applied to new data, which SPIRAL cannot do, hence strongly limiting its application scope.

## 5     Conclusion

In this paper, we present LDPS, an algorithm that aims at embedding time series into an Euclidean space, in which distances approximate the Dynamic Time Warping measure between raw time series. The embedding we design is based on the Shapelet Transform, that maps time series into high-dimensional vectors. The originality of our approach is that we learn shapelets using a stochastic gradient descent so that they best preserve the DTW between time series pairs. We show that the original DTW can be accurately captured by Euclidean distances in the transformed space. Clustering performance using this novel time series representation outperforms competitive methods designed specifically for this task. An interesting property of LDPS is that it leads to an ubiquitous time series representation that can feed a wide range of machine learning or indexing methods. As a future work, we will in particular aim at designing time series indexing schemes based on LDPS. As time series are embedded in a metric space, we can benefit from efficient indexing systems designed specifically in such spaces.

# References

1. Bagnall, A., Lines, J., Vickers, W., Keogh, E.: The UEA and UCR time series classification repository. www.timeseriesclassification.com
2. Bromley, J., Guyon, I., LeCun, Y., Säckinger, E., Shah, R.: Signature verification using a "siamese" time delay neural network. In: Advances in Neural Information Processing Systems, pp. 737–744 (1994)
3. Duchi, J., Hazan, E., Singer, Y.: Adaptive subgradient methods for online learning and stochastic optimization. JMLR **12**, 2121–2159 (2011)
4. Esling, P., Agon, C.: Time-series data mining. ACM Comput. Surv. **45**(1) (2012)
5. Grabocka, J., Schilling, N., Wistuba, M., Schmidt-Thieme, L.: Learning time-series shapelets. In: Proceedings of KDD (2014)
6. Hills, J., Lines, J., Baranauskas, E., Mapp, J., Bagnall, A.: Classification of time series by shapelet transformation. DMKD **28**(4), 851–881 (2014)
7. Keogh, E., Ratanamahatana, C.A.: Exact indexing of dynamic time warping. KAIS **7**, 358–386 (2005)
8. Lei, Q., Yi, J., Vaculín, R., Wu, L., Dhillon, I.S.: Similarity preserving representation learning for time series analysis (2017). http://arxiv.org/abs/1702.03584
9. Lemire, D.: Faster retrieval with a two-pass dynamic-time-warping lower bound. Pattern Recogn. **42**(9), 2169–2180 (2009)
10. Paparrizos, J., Gravano, L.: k-Shape: efficient and accurate clustering of time series. In: Proceedings of SIGMOD (2015)
11. Rakthanmanon, T., Campana, B., Mueen, A., Batista, G., Westover, B., Zhu, Q., Zakaria, J., Keogh, E.: Searching and mining trillions of time series subsequences under dynamic time warping. In: Proceedings of KDD (2012)
12. Tan, C.W., Webb, G.I., Petitjean, F.: Indexing and classifying gigabytes of time series under time warping. In: Proceedings of SIAM ICDM (2017)
13. Ye, L., Keogh, E.: Time series shapelets: a new primitive for data mining. In: Proceedings of KDD (2009)
14. Zakaria, J., Mueen, A., Keogh, E.: Clustering time series using unsupervised-shapelets. In: Proceedings of ICDM (2012)
15. Zhang, Q., Wu, J., Yang, H., Tian, Y., Zhang, C.: Unsupervised feature learning from time series. In: Proceedings of IJCAI (2016)

# Droplet Ensemble Learning on Drifting Data Streams

Pierre-Xavier Loeffel[1,2(✉)], Albert Bifet[4], Christophe Marsala[1,2],
and Marcin Detyniecki[1,2,3]

[1] Sorbonne Universités, UPMC Univ Paris 06, UMR 7606, LIP6, 75005 Paris, France
{pierre-xavier.loeffel,christophe.marsala,marcin.detyniecki}@lip6.fr
[2] CNRS, UMR 7606, LIP6, 75005 Paris, France
[3] Polish Academy of Sciences, IBS PAN, Warsaw, Poland
[4] LTCI, Télécom ParisTech, Université Paris-Saclay, 75013 Paris, France
albert.bifet@telecom-paristech.fr

**Abstract.** Ensemble learning methods for evolving data streams are extremely powerful learning methods since they combine the predictions of a set of classifiers, to improve the performance of the best single classifier inside the ensemble. In this paper we introduce the Droplet Ensemble Algorithm (DEA), a new method for learning on data streams subject to concept drifts which combines ensemble and instance based learning. Contrarily to state of the art ensemble methods which select the base learners according to their performances on recent observations, DEA dynamically selects the subset of base learners which is the best suited for the region of the feature space where the latest observation was received. Experiments on 25 datasets (most of which being commonly used as benchmark in the literature) reproducing different type of drifts show that this new method achieves excellent results on accuracy and ranking against SAM KNN [1], all of its base learners and a majority vote algorithm using the same base learners.

**Keywords:** Concept drift · Ensemble learning · Online-learning · Supervised learning · Data streams

## 1 Introduction

The explosion of data generated in real-time from streams has brought to the limelight the learning algorithms able to handle them. Sensors, stock prices on the financial markets or health monitoring are a few example of the numerous cases in real life where data streams are generated. It is therefore important to devise learning algorithms that can handle this type of data.

Unfortunately, these data streams are often non-stationary and their characteristics can change over time. For instance, the trend and volatility of the stock prices can suddenly change as a consequence of an unexpected economic event. This phenomenon, referred as *concept drift* (when the underlying distribution which generates the observations on which the algorithm is trying to learn

© Springer International Publishing AG 2017
N. Adams et al. (Eds.): IDA 2017, LNCS 10584, pp. 210–222, 2017.
DOI: 10.1007/978-3-319-68765-0_18

changes over time), raises the need to use adaptive algorithms to handle data streams.

In this paper we propose a novel ensemble method which aims at obtaining good performances regardless of the dataset and type of drift encountered. One of the main characteristic of this method is that, it determines the regions of expertise of its base learners (BL) in the feature space and selects the subset of BL which is the best suited to predict on the latest observation. This new method outperforms SAM-KNN [1], a new classifier algorithm for data streams that won the Best Paper award at ICDM 2016.

The main contributions of the paper are the following:

- a new streaming classifier for evolving data streams, which weights its base learners according to their local expertise in the feature space.
- an extensive evaluation over a wide range of datasets and type of drifts.
- a discussion on how the new method, DEA over-performs the best state of the art algorithms.

The paper is organized as follows: Sect. 2 lays down the framework of our problem and goes through the related works. Section 3 details the proposed algorithm while Sect. 4 presents the datasets used as well as the experimental protocol. Section 5 presents and discuss the results of the experiments and finally Sect. 6 concludes.

## 2   Framework and Related Work

In this section we present the framework of our problem and we discuss related works on learning algorithms handling concept drift.

### 2.1   Framework

The problem being addressed here is supervised classification on a stream of data subject to concept drifts. Formally, a stream endlessly emits observations $\{x_1, x_2, ...\}$ (where $x_i = \{x_i^1, ..., x_i^k\} \in X = \mathbb{R}^k$, $k$ designates the dimension of the feature space and $i$ designates the time step at which the observation was received) which are unlabeled at first but for which a label $y_i \in Y = \{1, ..., c\}$ is being received a constant amount of time $u \in \mathbb{R}^{+*}$ after $x_i$. We will work in the framework where the label of $x_i$ is always received before reception of $x_{i+1}$. The goal is to create an on-line classifier $f : X \rightarrow Y$ which can predict, as accurately as possible, the class $y_i$ associated to $x_i$.

An on-line classifier is a classifier which can operate when data are received in sequence (as opposed to a batch classifier which needs a full dataset from scratch to operate) and can evolve over time (i.e. its learned model is constantly updated with the latest observations). Formally, the operating process of an on-line classifier is described thereafter:

– When an observation $x_t$ is received at time $t$, it outputs a prediction $\hat{y}_t$. The true class $y_t$ is then released and, after computation of the prediction error according to the 0-1 loss function: $\mathscr{L}(y, \hat{y}) = \mathbb{I}_{\{y \neq \hat{y}\}}$ (where $\mathbb{I}$ is the indicator function), the classifier is updated with the latest observation: $f_t = Update(f_{t-1}, \{x_t, y_y\})$. Our goal then, is to minimize the average error $\left(\frac{1}{n} \sum_{i=1}^{n} \mathscr{L}(y_i, \hat{y}_i)\right)$ over the $n$ observations received so far.

In the considered framework, the hidden joint distribution $P(X, Y)$ (called *Concept*) which generates the couples $(x_i, y_i)$ at each time step, is also allowed to unexpectedly change over time: a phenomenon referred as *Concept Drift*. Formally [1], concept drift occurs at time $t$ if $P_{t-1}(X, Y) \neq P_t(X, Y)$. According to Bayes rule: $P(X, Y) = P(Y/X) P(X)$. Thus, a drift of concept can result either in a change of the posterior probability of the classes $P(Y/X)$ (called *real drift*) either in a change of the distribution of the features $P(X)$ (called *virtual drift*) either in both.

The types of drifts can be further categorized according to the speed at which they occur. We say that a drift is *abrupt* when the drift last for one observation ($P_{t-1}(X, Y) \neq P_t(X, Y)$ and the concept is stable before $t - 1$ and after $t$) or conversely that it is *incremental* when the drift last more than one observation ($P_{t-k}(X, Y) \neq \ldots \neq P_{t-1}(X, Y) \neq P_t(X, Y)$ and the concept is stable before $t - k$ and after $t$). *Reoccurring drifts*, happen when a previously learned concept reappears after some time ($\exists k \in \mathbb{N} / P_{t-k}(X, Y) = P_t(X, Y)$).

## 2.2   Related Work

Several methods have been proposed in order to deal with the issue of drifting concepts on data streams, the majority of which being ensemble methods.

**ADACC** was introduced in [11]. It maintains a set of BL which are weighted every $\tau$ time steps according to their number of wrong predictions. It then randomly selects one BL from the worst half of the ensemble and replaces it by a new one which is protected from deletion for a few time steps. The final prediction is given by the current best performer. The algorithm also includes a mechanism to remember past concepts.

**Dynamic Weighted Majority** (DWM) is an ensemble method introduced in [6]. Each of its BL has a weight which is reduced in case of a wrong prediction. When a BL's weight drops bellow a given threshold, it is deleted from the ensemble. If all the BL output a wrong prediction on an instance, a new classifier is added to the ensemble.

**ADWIN Bagging** (Bag Ad) was introduced in [9] and improves the On-line Bagging algorithm proposed by Oza and Rusell [10] by adding the ADWIN algorithm as a change detector. When a change is detected, the worst performing BL is replaced by a new one.

Similarly, **ADWIN Boosting** (Boost Ad) improves the on-line Boosting algorithm of Oza and Russell [10] by adding ADWIN to detect changes.

**Leveraging Bagging** (Lev Bag) was introduced in [7] and further improves the ADWIN Bagging algorithm by increasing re-sampling (using a value $\lambda$ larger than 1 to compute the Poisson distribution) and by adding randomization at the output of the ensemble by using output codes.

**Hoeffding Adaptive Tree** (Hoeff Tree) was introduced in [12] and uses ADWIN to monitor the performance of the branches on the tree. When the accuracy of a branch decreases, it is replaced with a more accurate one.

**AccuracyUpdatedEnsemble** (AUE) described in [4] maintains a weighted ensemble of BL and uses a weighted voting rule for its final prediction. It creates a new BL after each chunk of data which replaces the weakest performing one. The weights of each BL are computed according to their individual performances on the latest data chunk.

Finally, **SAM KNN** [1], best paper award at ICDM 2016, is a new improvement method of the KNN algorithm. It maintains the past observations into 2 types of memories (short and long term memory). The task of the short term memory is to remain up to date according to the current concept whereas the long term memory is in charge of remembering the past concepts. When a concept change is detected, the observations from the short term memory are transfered to the long term memory.

## 3   The Droplets Ensemble Algorithm

Our main goal in designing our new ensemble algorithm dealing with a data stream subject to concepts drift, is to take into account the local expertise of each of its BL on the region of the feature space where the latest observation was received. This means that it gives more weight to the predictions of the BL which demonstrated an ability to predict accurately in this region.

We propose DEA (Droplets Ensemble Algorithm), an ensemble learning algorithm which dynamically maintains an ensemble of $n$ BL $\left(F = \left\{f^1, ..., f^n\right\}\right)$ along with an ensemble of $p$ Droplets $\left(Map = \left\{D^1, ..., D^p\right\}\right)$ up to date with respect to the current concept.

The BL can be any learning algorithms, as long as they are able to classify on a data stream subject to concept drifts.

A Droplet is an object which can be represented as a $k$-dimensional hypersphere (with $k$ the dimension of the feature space). Each Droplet $D^t$ is associated with an observation $x_t$ and holds a pointer to a BL: $f^i$ ($i \in \{1, ..., n\}$). The values taken by $x_t$ correspond to the coordinates of the center of the Droplet in the feature space whereas $f^i$ corresponds to the BL which managed to achieve the lowest prediction error on a region of the feature space defined around $x_t$.

Figure 1 shows an example of Map learned where the numbers represent the time step at which each Droplet has been received.

We now go through the algorithm in details.

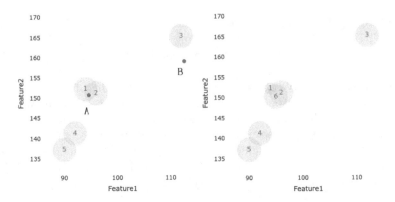

**Fig. 1.** Example of map learned in 2 dimensions. Left: before update of the model with the $6^{th}$ observation (received at point A). Right: after update of the model with the $6^{th}$ observation.

### 3.1   Model Prediction

At time $t$, upon reception of a new unlabeled observation $x_t$ the first step is to normalize the values of $x_t$ according to a vector of normalization constants $x_{const}$ found on the initialization step[1]. Then $OD_t$, the set of Droplets which contains the normalized coordinates of the latest observation is computed. If $OD_t \neq \emptyset$, the predicted value for this observation is given by a simple majority vote of the BL associated with the overlapped Droplets in $OD_t$. On the other hand, if $OD_t = \emptyset$, the learner associated with the nearest Droplet $D^{nn}$ is used for prediction. For instance, in the left plot of Fig. 1, if an observation is received at the position of point A, the BL associated with $D^1$ and $D^2$ will be used for prediction whereas if an observation is received at the position of point B, only the BL associated with $D^3$ will be used for prediction.

The prediction process is summarized in Algorithm 1.

### 3.2   Model Update

Once the true label $y_t$ associated with the latest observation $x_t$ is released, each BL $f^i$ (with $i \in \{1, ..., n\}$) predicts on the latest observation and the vector of the prediction errors $e_{t+1} = \{e^1_{t+1}, ..., e^n_{t+1}\}$ (with $e^i_{t+1} \in \{0, 1\}$) is set aside. The BL are then updated with $\{x_t, y_t\}$.

The next step is to search for the BL which will be associated to the new Droplet $D^t$. This is done by summing the prediction errors achieved by each BL on the $N$ nearest Droplets, where $N$ is a parameter defined by the user. If an unique BL minimizes this sum, it is associated to the new Droplet, otherwise (if at least 2 BL minimizes the sum of prediction error) the search space is

---

[1] This is simply done by computing the average $\mu^i$ as well as the standard deviation $\sigma^i$ of each feature on the initialization set and by transforming the $i^{th}$ feature of $x_t$ into $\frac{x^i_t - \mu^i}{\sigma^i}$.

**Algorithm 1.** Model Prediction

---

**Inputs:** $F = \{f^1, ..., f^n\}$: Ensemble of base learners,
$Map = \{D^1, ..., D^p\}$: Ensemble of existing Droplets,
$x_t$: Latest unlabeled observation,
$x_{const}$: Normalization constants
**Output:** $\hat{y}_t$: Estimated class for $x_t$

$x_t^{norm} \leftarrow Normalize\,(x_t, x_{const})$
$OD_t \leftarrow Get\ overlapped\ Droplets\,(Map, x_t^{norm})$
**If** $(OD_t \neq \emptyset)$
    **Foreach** $D^h \in OD_t\ (h \in \{a, ..., u\})$
        $\hat{y}_t^h \leftarrow Predict\,(D^h, x_t)$
    **End Foreach**
    $\hat{y}_t \leftarrow Majority\ Vote\,(\hat{y}_t^a, ..., \hat{y}_t^u)$
**Else**
    $D^{nn} \leftarrow Get\ Nearest\ Droplet\,(Map, x_t^{norm})$
    $\hat{y}_t \leftarrow Predict\,(D^{nn}, x_t)$
**End If**

---

expanded in turns to the $N + 1, N + 2, N + 3, ...$ nearest Droplets until a single best performer is found.

The new Droplet $D^t$ is then added to the feature space at the coordinates of $x_t^{norm}$. This Droplet is given a default radius $R_{default}$ (where $R_{default}$ is a parameter defined by the user), stores the vector of prediction errors $e_{t+1}$ and creates a pointer to the best BL $f^k$ found on the previous step.

The algorithm then goes through the set of overlapped Droplets $OD_t$ and if it is not empty, it decreases the influence of the Droplets in $OD_t$ which have outputted a wrong prediction on $x_t$. This is done by shrinking their radius which will make them less likely to predict on a future observation received in this region of the feature space. Formally, for each Droplet $u$ in $OD_t$:

1. Compute the overlap between $D^u$ and the latest Droplet:
   $Overlapp_u = R_{default} + R_u - \|x_u^{norm} - x_t^{norm}\|$ (where $\|.\|$ denotes the Euclidean distance)
2. Update the radius of $D^u$: $R_{u,t+1} = R_{u,t} - \dfrac{Overlapp_u}{2}$.
3. Delete $D^u$ if $R_{u,t+1} \leq 0$.

For instance the right plot of Fig. 1. shows the updated model after reception of an observation at the position of point A and where the BL associated with $D^1$ outputted a wrong prediction on the $6^{th}$ observation whereas the BL associated with $D^2$ predicted correctly.

Finally, a memory management module is ran at each time step to ensure that $p$, the user defined parameter for the maximum number of Droplets allowed in memory is not exceeded. If the memory is full, the algorithm uses 3 different criteria to select the Droplet which will be removed:

1. Remove the Droplet with the smallest radius.
2. If all the Droplets have the same radius, remove the Droplet which has outputted the highest number of wrong prediction.
3. If criteria 1. and 2. failed, remove the oldest Droplet.

Algorithm 2 summarizes the model update process.

---

**Algorithm 2.** Model Update

---

**Inputs:** $R_{default}$: Default radius of a Droplet,
$F = \{f^1, ..., f^n\}$: Ensemble of base learners,
$Map = \{D^1, ..., D^p\}$: Ensemble of existing Droplets,
$x_t$: Latest unlabeled observation,
$y_t$: True label latest observation,
$p$: Maximum number of Droplets allowed in memory,
$OD_t$: Set of overlapped Droplets at time $t$
**Output:** Updated DEA

**Foreach** $f^i$ in $F$
 $e_{t+1}^i \leftarrow Get\ Prediction\ Error\left(f_t^i, \{x_t, y_t\}\right)$
 $f_{t+1}^i \leftarrow Update\ Base\ Learner\left(f_t^i, \{x_t, y_t\}\right)$
**End foreach**
$f^k \leftarrow Search\ best\ base\ learner\left(Map, \{x_t, y_t\}\right), k \in \{1, ..., n\}$
$D^t \leftarrow Create\ Droplet\left(R_{default}, x_t^{norm}, f^k, e_{t+1}, sum\ errors = 0\right)$
$Map \leftarrow Add\ Droplet\left(Map, D^t\right)$
**Foreach** $D^u \in OD_t$
 $R_{u,t+1} \leftarrow Update\ Radius\left(R_{u,t}\right)$
 **If** $R_{u,t+1} \leq 0$
  $Map \leftarrow Remove\ Droplet\left(Map, D^u\right)$
 **End if**
**End foreach**
**If** $(Card\left(Map\right) \geq p)$
 $Map \leftarrow Memory\ Management\left(Map\right)$
**End if**

---

### 3.3 Running Time and Space Requirements

*Running time:* Provided that each of the base learner runs in constant time at each time step, the temporal complexity of both the prediction and update steps of DEA is $\mathcal{O}\left(i.p\right)$ with $i$ is the number of observations generated by the stream so far and $p$ the maximum number of Droplets allowed in memory.

*Space requirements:* As previously explained, the maximum number $p$ of Droplets saved into computer memory is constrained and so is the number $n$ of base learners. This means that, as long as each of the $n$ base learner constrains its memory consumption at each time step, the space complexity of DEA will be $\mathcal{O}\left(n + p\right)$ which is independant of the number of observations generated by the stream so far.

# 4   Experimental Framework

In this section, we describe the datasets on which the experiments have been conducted, their characteristics as well as the experimental protocol used.

## 4.1   Datasets

A total of 25 artificial and real world datasets have been used. These datasets have been chosen for the diversity of their characteristics, which are summarized thereafter (Table 1):

**Table 1.** Datasets used and their characteristics.

| Dataset | Features | Classes | Observations | # Drifts | Type of drift |
|---|---|---|---|---|---|
| Agrawal | 9 | 2 | 100 000 | 0 | N/A |
| Random Tree | 10 | 6 | 100 000 | 15 | Abrupt, Reoc. |
| Waveform | 21 | 3 | 100 000 | Continuous | Real, Local |
| LED | 7 | 10 | 100 000 | Continuous | Real, Local |
| KDD Cup | 41 | 2 | 494 000 | N/A | N/A |
| Rotating Check | 2 | 2 | 419 600 | Continuous | Real, Gradual |
| SPAM | 500 | 2 | 9324 | N/A | N/A |
| Usenet | 658 | 2 | 5 931 | N/A | N/A |
| Airlines | 7 | 2 | 153 200 | N/A | N/A |
| Multi Dataset | 3 | 2 | 200 000 | 3 | Abrupt, Real |
| Multi Dataset NO | 3 | 2 | 200 000 | Continuous | Abrupt, Real, Virt. |
| Weather | 8 | 2 | 18 100 | N/A | N/A |
| SEA | 3 | 2 | 50 000 | 3 | Real, Abrupt |
| Chess | 2 | 8 | 200 000 | 0 | N/A |
| Transient Chess | 2 | 8 | 200 000 | Continuous | Virt., Reoc. |
| Mixed Drift | 2 | 15 | 600 000 | Continuous | Incr., Abrupt, Virt. |
| Moving Square | 2 | 4 | 200 000 | Continuous | Real, Gradual |
| Interchanging RBF | 2 | 15 | 200 000 | 9 | Real, Abrupt |
| Moving RBF | 10 | 5 | 200 000 | Continuous | Real, Gradual |
| Cover Type | 54 | 7 | 581 000 | N/A | N/A |
| Electricity | 8 | 2 | 45 300 | N/A | N/A |
| Outdoor Stream | 21 | 40 | 4 000 | N/A | N/A |
| Poker Hand | 10 | 10 | 829 200 | N/A | N/A |
| Rialto | 27 | 10 | 82 200 | N/A | N/A |
| Rotating Hyp. | 10 | 2 | 200 000 | Continuous | Real, Gradual |

Most of these datasets have frequently been used in the literature dedicated to streams subjects to concept drifts. Also, please note that in this table, an

"N/A" value doesn't mean that there is no concept drift. It means that, because the dataset comes from the real world, it is impossible to know for sure the number of drifts it includes as well as their type.

The first 4 datasets from Agrawal to LED have been generated using the built-in generators of MOA[2]. A precise description of theses datasets can be found in the following papers [13–15]. The KDD Cup 10 percent dataset was introduced and described in [2]. Rotating Check board was created in [5] and the version CB (Constant) dataset was used (constant drift rate). SPAM was introduced in [3] and Usenet was inspired by [8]. Airlines was introduced in the 2009 Data Expo competition. The dataset has been shrinked to the first 153 000 observations.

Multidataset is a new synthetic dataset created for this paper. Every 50 000 observations, the concept drifts to a completely new dataset, starting with Rotating checkboard, then Random RBF, then Rotating Hyperplane and finally SEA. In the basic version, the successive concepts overlap each other whereas in the No Overlap (NO) version the datasets are shifted and the data are randomly generated on each dataset.

Finally, all the datasets listed after Weather have been retrieved from the repository[3] given in the paper of Losing et al. [1].

All the datasets used as well as the code of the DEA algorithm and the results of the experiments are available at the following link[4].

## 4.2 Experimental Setting

MOA have been used to conduct the experiments and provide the implementation of the classifiers. DEA was also implemented in MOA. The code for SAM KNN was directly retrieved from the link provided in their paper [1] (See Footnote 3).

All the parameters of all the classifiers were set to default values (except for the training period which was set to 100 observations for all the learners and for the number of observations allowed in memory which was set to 400 for DEA and SAM KNN) and for all the datasets. In the case of the Droplets algorithm, the default radius was set to 0.1 and the minimum number of neighbors considered was set to 20 for all the experiments. We used all the algorithms described in this paper as BL for DEA (they were chosen because of their availability on MOA) with the exception of SAM KNN and Majority Vote. The simple majority vote algorithm (which uses the same BL as DEA) was used as a base-line for performance comparison.

Leaving all the parameters to default values for all the datasets is required because there is no assumptions regarding the structure of the data or the type of drifts the classifiers will have to deal with. Therefore, it wouldn't be relevant to optimize parameters that would be suitable for a particular concept, at a particular time and for a particular dataset.

---

[2] http://moa.cms.waikato.ac.nz/.

[3] https://github.com/vlosing/driftDatasets.

[4] https://mab.to/o5iNvZdhH.

The goals of the experiments were to compare the performance of DEA against one of the currently best adaptive algorithm (SAM KNN), assess how DEA was faring against another ensemble algorithm which is given the same BL (Majority Vote) and assess whether DEA was able to outperform each of its BL.

**Table 2.** Percentage of correct classification achieved on each dataset.

| Dataset | DEA | SAMKNN | ADACC | DWM | Bag Ad | Lev Bag | Hoeff Tree | Boost Ad | AUE | Maj Vote |
|---|---|---|---|---|---|---|---|---|---|---|
| Agrawal | 94.20 | 92.53 | 84.83 | 81.46 | **94.87** | 93.97 | 94.24 | 90.13 | 90.80 | 93.62 |
| Random Tree | 51.75 | 32.27 | 29.08 | 30.25 | 46.93 | **52.36** | 35.76 | 19.80 | 48.35 | 43.89 |
| Waveform | 83.26 | 82.15 | 76.35 | 77.37 | 82.95 | **84.84** | 81.55 | 81.35 | 80.49 | 82.87 |
| LED | 73.51 | 71.12 | 64.67 | 63.46 | 73.94 | 73.88 | **73.95** | 73.69 | 73.93 | 73.86 |
| KDD Cup | **99.91** | 99.90 | 99.84 | 99.62 | 99.87 | 99.91 | 99.78 | 98.91 | 82.44 | 82.71 |
| Rotating Check | 92.94 | 91.39 | 79.65 | 76.70 | 84.68 | 93.03 | 84.39 | **93.95** | 82.68 | 87.41 |
| SPAM | **96.60** | 94.95 | 94.91 | 92.30 | 90.95 | 96.04 | 90.60 | 95.54 | 65.36 | 69.45 |
| Usenet | 60.30 | 57.24 | 61.00 | 60.85 | 56.43 | 61.95 | 56.91 | 59.17 | 61.07 | **62.28** |
| Airlines | 66.40 | 65.16 | 62.44 | 60.25 | 68.14 | 65.41 | 66.26 | 62.72 | 67.36 | **68.54** |
| Multi Dataset | 95.11 | 92.56 | 90.99 | 89.53 | 93.80 | **95.31** | 92.52 | 93.71 | 92.90 | 93.96 |
| Multi Dataset NO | 90.25 | 85.09 | 55.00 | 54.39 | 88.28 | **90.86** | 88.73 | 87.93 | 89.19 | 88.02 |
| Weather | 76.52 | 76.02 | 73.56 | 72.22 | 75.00 | **78.14** | 73.53 | 74.40 | 74.55 | 76.74 |
| SEA | 87.77 | 85.45 | 85.12 | 84.72 | 86.77 | **88.33** | 86.76 | 82.33 | 86.85 | 87.64 |
| Chess | **94.92** | 78.22 | 14.45 | 13.84 | 55.35 | 94.84 | 58.49 | 12.64 | 88.02 | 76.67 |
| Transient Chess | **94.98** | 85.37 | 58.03 | 57.07 | 56.11 | 89.52 | 37.27 | 56.12 | 26.03 | 39.68 |
| Mixed Drift | 76.64 | **91.57** | 37.56 | 35.63 | 59.54 | 75.17 | 55.70 | 42.76 | 68.23 | 64.91 |
| Moving Square | 99.08 | 97.36 | **99.14** | 83.31 | 88.02 | 87.85 | 74.89 | 79.66 | 67.18 | 88.67 |
| Inter RBF | 98.45 | 98.00 | **98.59** | 97.14 | 88.92 | 94.10 | 56.81 | 94.14 | 91.61 | 96.17 |
| Moving RBF | 59.18 | **86.98** | 44.65 | 45.80 | 52.31 | 55.02 | 38.72 | 45.43 | 54.70 | 53.09 |
| Cover Type | 92.69 | **93.58** | 90.38 | 84.29 | 84.02 | 90.40 | 80.54 | 92.55 | 82.69 | 86.65 |
| Electricity | **90.66** | 82.54 | 89.55 | 82.34 | 83.63 | 89.49 | 82.83 | 87.34 | 78.98 | 87.32 |
| Outdoor Stream | 69.43 | **88.25** | 66.88 | 58.32 | 58.91 | 60.21 | 57.20 | 57.70 | 40.08 | 38.63 |
| Poker Hand | 88.09 | 79.77 | 79.36 | 75.64 | 73.46 | 85.68 | 65.54 | **91.74** | 68.65 | 80.47 |
| Rialto | 70.92 | **81.90** | 71.21 | 45.27 | 49.46 | 60.43 | 30.65 | 18.34 | 47.35 | 40.76 |
| Rotating Hyp. | 86.89 | 81.42 | 82.78 | 83.46 | 88.02 | 86.87 | 86.45 | 76.58 | 87.20 | **88.44** |
| Average Accuracy | **83.62** | 82.83 | 71.60 | 68.21 | 75.21 | 81.75 | 70.00 | 70.74 | 71.87 | 74.10 |
| Average Rank | **2.48** | 4.72 | 6.44 | 7.96 | 5.28 | 2.96 | 7.12 | 6.72 | 6.36 | 4.96 |

For each dataset, the performance of the algorithms was computed using the prequential method (interleaved test-then-train): when an unlabelled observation is received, the algorithm is first asked to predict its label and the prediction error is recorded (test). Once the true labelled is released, the classifier is trained with this labelled observation (train). This method has the advantage of making use of the whole dataset.

## 5     Results and Discussion

The accuracy (percentage of correct classifications) obtained by each algorithm on each dataset are reported in Table 2. Bold numbers indicate the best performing algorithm. The bottom 2 lines show the average accuracy as well as the average rank obtained by each algorithm on all the datasets.

The results indicate that DEA managed to obtain the best average accuracy as well as the best average rank on the 25 datasets considered. In particular, the average rank obtained demonstrates the ability of DEA to perform consistently well regardless of the characteristics of the dataset and of the type of drifts encountered. This is an interesting property because it is often impossible to predict how the stream will evolve over time and thus, an algorithm which can deal with a very diversified set of environments could be useful as it wouldn't be possible to pick right from the beginning the algorithm which is the best suited for the whole dataset.

This good performance also confirms that using the local expertise of the BL as a selection criteria to decide which subset will be used for prediction should be considered as a way to improve the performances of an ensemble learning algorithm. Indeed, DEA over-performed the ensemble learning algorithms which rely on the latest performances to weight their BL (ADACC, DWM, AUE, ...) as well as a Majority Vote algorithm which simply ask all the algorithms to collaborate for prediction, independently of the observation received.

## 6     Conclusion

Learning on a data stream subject to concept drifts is a challenging task. The hidden underlying distribution on which the algorithm is trying to learn can change in many unexpected ways, requiring an algorithm which is capable of good performances regardless of the environment encountered.

In order to tackle this issue, we have proposed the Droplets Ensemble Algorithm (DEA), a novel algorithm which combines the properties of an instance base learning algorithm with the ones of an ensemble learning algorithm. It maintains into memory a set of hyper-spheres, each of which includes a pointer to the BL which is the most likely to obtain the best performance in the region of the feature space around that observation. When a new observation is received, it selects the BL which are likely to obtain the best performance in this region and use them for prediction.

The experiments carried on a set of 25 diversified datasets, reproducing a wide variety of drifts show that our algorithm is able to over-perform each of its base learners, a majority vote algorithm using the same base learners as well as SAM KNN (one of the currently best adaptive algorithm) by obtaining the best average accuracy and rank. These results indicate that our algorithm is well suited to be used as a general purposed algorithm for predicting on data streams with concept drifts and that taking into account the local expertise of each BL should be considered in order to improve the performances of an ensemble learning algorithm.

The algorithm can still be further improved and future work will focus on improving the efficiency of the search algorithm.

# References

1. Losing, V., Hammer, B., Wersing, H.: KNN classifier with self adjusting memory for heterogeneous concept drift. In: ICDM (2016)
2. Tavallaee, M., Bagheri, E., Lu, W., Ghorbani, A.A.: A detailed analysis of the KDD CUP 99 data set. In: Proceedings of the Second IEEE International Conference on Computational Intelligence in Security and Defense Applications, pp. 53–58 (2009)
3. Katakis, I., Tsoumakas, G., Vlahavas, I.: Tracking recurring contexts using ensemble classifiers: an application to email filtering. Knowl. Inf. Syst. **22**(3), 371–391 (2010)
4. Brzezinski, D., Stefanowski, J.: Reacting to different types of concept drift: the accuracy updated ensemble algorithm. IEEE Trans. Neural Netw. Learn. Syst. **25**(1), 81–94 (2014)
5. Elwell, R., Polikar, R.: Incremental learning of concept drift in nonstationary environments. IEEE Trans. Neural Netw. **22**(10), 1517–1531 (2011)
6. Kolter, J.Z., Maloof, M.A.: Dynamic weighted majority: a new ensemble method for tracking concept drift. In: Third IEEE International Conference on Data Mining, ICDM 2003, pp. 123–130 (2013)
7. Bifet, A., Holmes, G., Pfahringer, B.: Leveraging bagging for evolving data streams. In: Balcázar, J.L., Bonchi, F., Gionis, A., Sebag, M. (eds.) ECML PKDD 2010. LNCS, vol. 6321, pp. 135–150. Springer, Heidelberg (2010). doi:10.1007/978-3-642-15880-3_15
8. Katakis, I., Tsoumakas, G., Vlahavas, I.: An ensemble of classifiers for coping with recurring contexts in data streams. In: 18th European Conference on Artificial Intelligence, Patras, Greece. IOS Press (2008)
9. Bifet, A., Holmes, G., Pfahringer, B., Kirkby, R., Gavaldà, R.: New ensemble methods for evolving data streams. In: Proceedings of the 15th ACM SIGKDD International Conference on Knowledge Discovery and Data Mining - KDD 2009 (2009)
10. Oza, N., Russell, S.: Online bagging and boosting. In: Artificial Intelligence and Statistics 2001, pp. 105–112. Morgan Kaufmann (2001)
11. Jaber, G., Cornuéjols, A., Tarroux, P.: A new on-line learning method for coping with recurring concepts: the ADACC system. In: Lee, M., Hirose, A., Hou, Z.-G., Kil, R.M. (eds.) ICONIP 2013. LNCS, vol. 8227, pp. 595–604. Springer, Heidelberg (2013). doi:10.1007/978-3-642-42042-9_74
12. Bifet, A., Gavaldà, R.: Adaptive learning from evolving data streams. In: Adams, N.M., Robardet, C., Siebes, A., Boulicaut, J.-F. (eds.) IDA 2009. LNCS, vol. 5772, pp. 249–260. Springer, Heidelberg (2009). doi:10.1007/978-3-642-03915-7_22

13. Breiman, L., Friedman, J.H., Olshen, R.A., Stone, C.J.: Classification and Regression Trees. Wadsworth (1984)
14. Agrawal, R., Imielinski, T., Swami, A.: Database mining: a performance perspective. IEEE Trans. Knowl. Data Eng. **5**(6), 914–925 (1993)
15. Domingos, P., Hulten, G.: Mining high-speed data streams. In: Knowledge Discovery and Data Mining, pp. 71–80 (2000)

# Predictive Clustering Trees for Hierarchical Multi-Target Regression

Vanja Mileski[1,2]([⊠]), Sašo Džeroski[1,2], and Dragi Kocev[1,2]

[1] Department of Knowledge Technologies, Jožef Stefan Institute, Ljubljana, Slovenia
{vanja.mileski,saso.dzeroski,dragi.kocev}@ijs.si
[2] International Postgraduate School Jožef Stefan Institute, Ljubljana, Slovenia

**Abstract.** Multi-target regression (MTR) is the task of learning predictive models for problems with multiple continuous target variables. In this work, we introduce the task of hierarchical multi-target regression (HMTR), where these target variables are organized in a hierarchy. The hierarchy contains the target variables and has an aggregation function that defines the parent child relationships in the hierarchy. This information can be used by learning methods to obtain better predictive models. We then propose to extend the approach of predictive clustering trees for MTR towards addressing the task of HMTR. The information from the hierarchy is exploited by defining the variance function through a weighted Euclidean distance. We evaluate the proposed method on 4 practically relevant HMTR datasets. The results show that HMTR performs better than standard MTR. Finally, we illustrate the enhanced interpretability potential of PCTs for HMTR.

**Keywords:** Multi-target regression · Hierarchical multi-target regression · Interpretable models · Predictive clustering trees

## 1 Introduction

The task of building a model that is capable of making predictions is called predictive modeling. Supervised learning is the machine learning task of inferring a function from given training data. It is an area of machine learning that has been extensively researched. The goal in supervised learning is to learn, from a set of examples with known class, a function that outputs a prediction for the class of a previously unseen example. Regression models make predictions for continuous variables, e.g. housing prices, weather temperature and similar.

In this work, we are interested in predicting multiple continuous variables since many real-life problems have multiple outputs. Multi-target regression (MTR), also known in the literature as multi-output, multi-response or multivariate regression, tries to simultaneously predict multiple continuous target variables based on a set of input variables [1]. The methods addressing the MTR task can be categorized into two groups: local and global methods [2]. Local methods construct multiple models for each target variable separately and then

© Springer International Publishing AG 2017
N. Adams et al. (Eds.): IDA 2017, LNCS 10584, pp. 223–234, 2017.
DOI: 10.1007/978-3-319-68765-0_19

combine the predictions from each model, whereas global models construct only one model that outputs the predictions for all of the target variables. If a problem has multiple outputs, it is generally better to build a model that gives a prediction for all of the outputs, rather than one by one [1]. It has been shown that global methods perform better than local methods [3]. They have several advantages over local methods: (a) they can achieve higher predictive performance since they exploit the dependencies that exist between the output variables; (b) they can be more efficient if there are many outputs since building a separate model for each output will be slower, and (c) can produce smaller models than the combined size of the models created with local methods.

Many applications for MTR have been studied due to its applicability to a wide range of domains, including the assessment of vegetation condition, water quality, stock market selection, in chemometrics, to predict wind noise of vehicle components and gas tanks level prediction, to predict biophysical parameters, to perform channel estimation, etc. These applications of MTR also inherit the many challenges present when working with real-world applications, such as missing data and noise, but also provide the means to create the model considering the underlying relationships between the targets, and not only the relationships between the inputs. Global multi-target approaches therefore give a better representation and interpretability of the given problems, as well as simpler models with higher computational efficiency [3]. For a survey on the MTR task, we refer the reader to [4].

Many real-life objects tend to exist within organizational structures. For example, in education, students exist within a hierarchical social structure that can include family, peer group, classroom, grade level, school, school district, state and country [5]. Data collected about an individual is hierarchical, as all the observations are nested within individuals. While there are other methods to deal with this type of data, the assumptions relating to them are rigorous, whereas procedures relating to hierarchical modeling require fewer assumptions. The individuals in the hierarchy follow the hierarchy constraint - an individual that belongs to a given hierarchy node also belongs to all its supernodes [1]. If an example has multiple continuous outputs in a hierarchy that we want to predict, then the task is called hierarchical multi-target regression (HMTR).

In this work, we extend the predictive clustering trees (PCTs) [1,6] towards the HMTR task. They are a state-of-the-art interpretable global approach for different types of outputs and can efficiently learn models valid for the output structure as a whole [1,7,8]. One major reason why we consider PCTs beside their good predictive performance is their interpretability since the model is a tree. This representation is easily understood by people from different backgrounds and expertise [9].

We evaluate the proposed PCTs for HMTR on 4 practically relevant domains. The major goal of the comparison is to check whether the introduction of a hierarchical structure in the output space can improve the predictive power as well as the interpretability of the predictive models.

The remainder of this paper is organized as follows. First, we introduce the task of hierarchical multi-target regression (HMTR). We then propose a method for addressing the HMTR task, i.e., the predictive clustering framework. Next, we outline the experimental design used to evaluate the proposed method. Finally, we conclude and give directions for further work.

## 2    The Hierarchical Multi-Target Regression Task

The work presented in this paper concerns the learning of a model for hierarchical multi-target regression (HMTR). In accordance with Džeroski [10], where predictive modeling is defined for arbitrary types of input and output data, we define the HMTR task as follows:

**Given:**

- A description space $X$ that consists of tuples of values of continuous (or discrete) primitive data types, i.e. $\forall X_i \in X, X_i = (x_{i_1}, x_{i_2}, \ldots, x_{i_{N_d}})$, where $N_d$ is the number of descriptive variables,
- A target space $Z$, defined with a numeric variable hierarchy $(H, \leq_p)$, where $H$ is a set of numeric variables and $\leq_p$ is structured as a rooted tree representing the supervariable relationship $(\forall h_1, h_2 \in H : h_1 \leq_p h_2$ if and only if $h_1$ is a supervariable of $h_2$). The supervariables are the products of aggregate functions on their respective children (for example, sum, minimum, maximum, average...) i.e. $h_j = agg(h_i) \forall h_i \leq_p h_j$,
- A set of examples $E$, where each example is a pair of a tuple and a set, from the descriptive and target space, respectively, and each set satisfies the hierarchy constraint, i.e., $E = \{(X_i, Z_i) | X_i \in X, Z_i \subseteq H, h \in Z_i \implies \forall h' \leq_p h : h' \in Z_i, 1 \leq i \leq N_e\}$ and $N_e$ is the number of examples in $E(N_e = |E|)$, and
- A quality criterion $q$, which rewards models with high predictive accuracy and low complexity.

**Find:** A function $f : X \to R^H$ where $R^H$ are all of the variables from $H$ such that $f$ maximizes $q$ and $h \in f(x) \implies \forall h' \leq_p h : h' \in f(x)$ in order to satisfy the hierarchy constraint.

To the best of our knowledge, the task of HMTR as defined above has not been treated before by the research community. The HMTR task applies to problems whose target variables have hierarchical dependencies. The HMTR learning algorithm can exploit this information, which means that it can help in building a better predictive model. Such problems up until now could not have been solved, at least not in such a way that the hierarchy would be taken into account when learning the predictive model. The only solution for solving such problems until now was by using a MTR algorithm and simply ignoring the hierarchy of the output space.

The major advantage of HMTR over MTR is that HMTR is a broader task that encapsulates MTR, i.e., if we instantiate HMTR without a hierarchy weighting scheme, the results will be the same as using a MTR algorithm. Even further, one of the major gains of this design of the learning algorithm is the possibility to define the hierarchy aggregation function based on the task at hand. Most commonly the aggregate function of the supervariable is the sum of its children, however it can also be any other function, for example minimum, maximum, average and similar. In other words, this design provides an additional degree of freedom for the tree induction algorithm and facilitates its application to a wider range of practical tasks.

Let us illustrate the generality of the HMTR definition by deriving the task of hierarchical multi-label classification (HMLC) from the definition above. HMLC is a variant of classification where an example can have multiple labels at the same time and the labels are organized in a form of hierarchy. The presence or absence of a label for a given example can be presented as a boolean variable or as a binary 0/1 variable. Then, by instantiating the aggregation function as logical OR, we obtain the hierarchy constraint defined in [9], i.e., we define the task of HMLC.

## 3   Predictive Clustering Trees for Hierarchical Multi-Target Regression

The PCT framework views decision trees as a hierarchy of clusters where the root node corresponds to a cluster that contains all of the examples: The other nodes at the lower levels of the tree are smaller clusters partitioned from the nodes that are above them (parent nodes). The predictive clustering framework is implemented in the system CLUS [11].

The TDIDT algorithm (top-down induction of decision trees) is used for inducing the PCTs [12] as presented in Algorithm 1. It takes a set of examples $E$ as input and produces a tree as an output. The heuristic score $s$ that is used for selecting the best tests $b$ is the reduction of variance that is caused by the partitioning $p$ of the examples. With maximization of the variance reduction, we maximize the cluster homogeneity and improve the predictive performance of the model.

The main difference between the algorithm for learning PCTs and an algorithm for learning decision trees is that the former considers the variance function and the prototype function (that computes a label for each leaf) as parameters that can be instantiated for a given learning task. The PCTs have been instantiated for multi-target prediction [6,13], prediction of time series [14] and hierarchical multi-label classification (HMLC) [9]. We instantiate these two functions for the HMTR task as follows.

The variance is calculated using a distance function among the values of the variables. The proposed variance for HMTR of a set of examples $E$ is defined as the average squared distance between each node vector $L_i$ of the examples and the mean node vector $\bar{L}$ [9]:

$$Var(E) = \frac{1}{|E|} \cdot \sum_{E_i \in E} (d(L_i, \bar{L})^2) \tag{1}$$

**Algorithm 1.** The top-down induction algorithm for learning PCTs.

**Procedure** PCT
**Input:** A dataset $E$
**Output:** A predictive clustering tree
    $(b^*, s^*, p^*) = BestTest(E)$
    **if** $t^* \neq none$ **then**
        **for each** $E_i \in p^*$ **do**
            $tree_i = PCT(E_i)$
        **return** $node(b^*, \cup_i \{tree_i\})$
    **else**
        **return** $leaf(Prototype(E))$
**Procedure** BestTest
**Input:** A dataset $E$
**Output:** best test $b^*$, heuristic score $s^*$, induced partition, $p^*$ on the dataset $E$
    $(b^*, s^*, p^*) = (none, 0, \emptyset)$
    **for each** test $b$ **do**
        $p =$ induced partition by $b$ on $E$
        $s = Var(E) - \sum_{E_i \in p} \frac{|E_i|}{|E|} \cdot Var(E_i)$
        **if** $(s > s^*) \& Acceptable(b, p)$ **then**
            $(b^*, s^*, p^*) = (b, s, p)$
    **return** $(b^*, s^*, p^*)$

Note that we represent the output hierarchy as a vector of variables by traversing the hierarchy in *preorder* mode.

The target variables are normalized so that they would contribute equally to the overall score. As for the distance $d$, any distance can be essentially used. For the task of HMTR, we propose to use a weighted Euclidean distance:

$$d(L_1, L_2) = \sqrt{\sum_{l=1}^{|L|} w(n_l) \cdot (L_{1,l} - L_{2,l})^2} \tag{2}$$

where $L_{i,l}$ is the $l$-th component of the node vector $L_i$ of an instance $E_i$, $|L|$ is the node's vector size, and $w(n)$ are the node weights which decrease exponentially with the depth of the node in the hierarchy, e.g., as $w(n) = w_0^{depth(n)}$.

We provide the following example to better illustrate the calculation of the distance for HMTR. We consider the toy hierarchy depicted in Fig. 1 and two

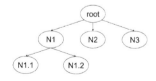

**Fig. 1.** An example of a toy hierarchy.

data examples $(X_1, Z_1)$ and $(X_2, Z_2)$ that use a vector representation for the numeric values of the nodes {root, N1, N1.1, N1.2, N2, N3} in that order. The aggregate function for the supervariables used in this example is *sum*. The examples have the following numeric variables for their outputs: $Z_1 = \{1.7, 1.1, 0.5, 0.6, 0.1, 0.5\}$ and $Z_2 = \{1.1, 0.3, 0.1, 0.2, 0.1, 0.7\}$. The weighted Euclidean distance is then calculated as follows:

$$
d_{w_0}(Z_1, Z_2) = d_{w_0}(\{1.7, 1.1, 0.5, 0.6, 0.1, 0.5\}, \{1.1, 0.3, 0.1, 0.2, 0.1, 0.7\})
$$

$$
= \sqrt{w_0^0(0.6)^2 + w_0^1(0.8)^2 + w_0^2(0.4)^2 + w_0^2(0.4)^2 + w_0^1(0)^2 + w_0^1(-0.2)^2} \quad (3)
$$

$$
= \sqrt{0.36w_0^0 + 0.68w_0^1 + 0.32w_0^2}
$$

Setting a specific value for $w_0$, $\frac{3}{4}$ for example, yields the following distance:

$$
d_{\frac{3}{4}}(Z_1, Z_2) = \sqrt{0.36\left(\frac{3}{4}\right)^0 + 0.68\left(\frac{3}{4}\right)^1 + 0.32\left(\frac{3}{4}\right)^2} = \sqrt{1.05} \approx 1.025 \quad (4)
$$

The prototype function used is averaging the values of the examples belonging to a given leaf. One needs to be careful when defining this prototype function, since it can break the hierarchy constraint if it is not compatible with the used hierarchy aggregation function. For example, averaging works fine when the aggregation function is *sum* or *avg*, while it may break the hierarchy constraint if the aggregation function is *min* or *max* (the value calculated by averaging the parents may not be the *min* or *max* of the averaged children).

Next, we analyse the computational complexity of PCTs for HMTR and compare it with the complexity of PCTs for MTR. Let us assume that the size of the training set is $N_e$, the number of descriptive attributes is $N_d$ out of which $N_c$ are continuous, the number of target attributes is $N_t$ and the number of supervariables is $N_s$. From the algorithm for induction of PCTs, we can note that sorting the $N_c$ numeric attributes is of the order of $\mathcal{O}(N_c N_e \log N_e)$ and $N_c = \mathcal{O}(N_d)$. Calculating the best split for multiple variables has the complexity order of $\mathcal{O}(N_t N_d N_e)$ and applying the split to the examples has a linear complexity, i.e. $\mathcal{O}(N_e)$. We assume that the tree is balanced, which means that the depth of the tree is the logarithm of the number of examples, i.e. $\log N_e$. With these calculations, the computational cost of inducing a single MTR tree is:

$$
\mathcal{O}(MTR) = \mathcal{O}(N_d N_e \log^2 N_e) + \mathcal{O}(N_t N_d N_e \log N_e) + \mathcal{O}(N_e \log N_e) \quad (5)
$$

For the HMTR algorithm, we also have the supervariables which in this case act like targets. This changes only the cost of calculating the best split to $\mathcal{O}((N_t + N_s)N_d N_e \log N_e)$. With this in mind, the order of complexity for the HMTR tree is very similar to the MTR and is given as:

$$
\mathcal{O}(HMTR) = \mathcal{O}(N_d N_e \log^2 N_e) + \mathcal{O}((N_t + N_s)N_d N_e \log N_e) + \mathcal{O}(N_e \log N_e) \quad (6)
$$

From the complexity analysis of single PCTs for HMTR, we can see that the HMTR algorithm has a higher computational complexity than the MTR algorithm. The increase, however, is linear with the number of targets from the introduced supervariables. We can see that the dominant elements in the computational costs are the first two in the parentheses, i.e. the one containing the second logarithmic power of the number of examples, and the one that is multiplied with the number of targets. The first element is $\mathcal{O}(N_d N_e \log^2 N_e)$, and the second is $\mathcal{O}(N_t N_d N_e \log N_e)$ or $\mathcal{O}((N_t + N_s)N_d N_e \log N_e)$ for MTR and HMTR respectively. If we compare the two terms, we can see that the first term is bigger than the second when $\log N_e > N_t$ for MTR and $\log N_e > (N_t + N_s)$ for HMTR. Let us explore the first case where the first term is smaller. This means that when comparing MTR and HMTR, HMTR will have greater computational cost, due to the addition of $N_s$. Let us now explore the second case where $\log N_e$ is greater. This will make the first term of the equation the major contributor to the cost of the algorithm. With this, the linear increase in the second terms of the equations for the HMTR task (i.e. the addition of $N_s$ in $\mathcal{O}((N_t + N_s)N_d N_e \log N_e)$) will be insignificant in this case, resulting in comparable performance between MTR and HMTR, on datasets with sufficiently large number of examples.

## 4  Experimental Design

We assess the effect of introducing a hierarchy in the output space by comparing the performance of PCTs for HMTR with the performance of PCTs for MTR. We estimate the predictive performance by using 10-fold cross-validation. We follow the recommendations from Borchani et al. [4] and adopt the average correlation coefficient ($aCC$) and average relative root mean squared error ($aRRMSE$) as evaluation measures. For a fair comparison, we calculate these errors only for the variables at the leafs of the hierarchy.

The parametrization of HMTR sets the weight parameter $w(n)$ for a node $n$. From the weights of the Euclidean distance, one might have already concluded that if we instantiate the algorithm such that $w = 1$, all of the nodes will have the same weight regardless of their depth in the hierarchy, thus giving little weight to the hierarchy. Lowering the weight $w$ from 1 downwards, we place increasingly greater importance on the hierarchy, meaning that a smaller weight $w$ assigns larger relative weights to the higher levels of the hierarchy in comparison to the deepest nodes in the hierarchy, which get the smallest weights.

For each application domain, the user needs to define the hierarchy for the output space of the problem (if it is not readily available) and are adequate aggregation function that will be applied in order to calculate the values for the parent nodes. The user also needs to choose the weighting parameter.

We perform the experimental evaluation of our method on 4 practically relevant datasets: **OSALES, MARSEXPRESS, ADNI** and **SIS**.

The task for the **OSALES** dataset is the prediction of online sales of products described with various product features. The dataset is from the Kaggle's Online

Product Sales competition in 2012 and the goal is to predict the monthly sales for 12 months of products. It has 639 examples, 413 features and 12 target variables that need to be predicted. The data and additional information are available at https://www.kaggle.com/c/online-sales.

The **MARSEXPRESS** dataset concerns the power consumption of the the the different systems of the European Space Agency's MARS Express spacecraft orbiting Mars. It includes context data as descriptive variables and 33 thermal power lines read-outs of the electric current (power consumption) in each thermal subsystem node. The data consists of 464 features describing the current and past operation of the satellite and 33 target features referring to the thermal power lines. Here, we consider a subsample of 20000 examples out of the over 2.6 million examples in the dataset. More information about the data can be found at https://kelvins.esa.int/mars-express-power-challenge/home/.

The **ADNI** dataset contains data from the Alzheimer's Disease Neuroimaging Initiative (ADNI) – a longitudinal study for developing clinical, genetic, imaging and biochemical biomarkers for the Alzheimer's disease (AD). The ADNI dataset consists of subjects with AD, as well as elderly controls and patients with mild cognitive impairment. The dataset consists of 659 examples [15]. We consider 10 descriptive features (APOE4 and PET imaging data) and 27 targets consisting of clinical scores that describe everyday cognition (eCOG), Montreal Cognitive Assessment (MoCA), Mini-Mental State Exam (MMSE) and Alzheimer's Disease Assessment Scale (ADAS13).

The **SIS** dataset refers to the New York City Social Indicators Survey (SIS) – a study on social problems and inequality in New York City for the year 2001. For the 229 descriptive features, we consider questionnaire answers on a variety of topics, while the 12 targets are the subjects' income from various sources. We consider data from 1501 subjects. More information is available at http://cupop.columbia.edu/research/research-areas/social-indicators-survey-sis.

For the four datasets, we consider the hierarchies presented in Fig. 2. The OSALES hierarchy combines the months into quarters (a period of 3 months),

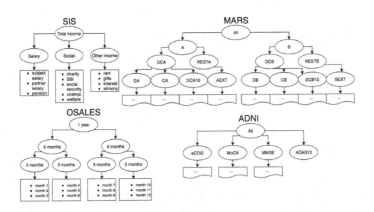

**Fig. 2.** The hierarchies for the output spaces of the four HMTR datasets.

then groups the quarters into semesters, and then to an entire year. In the
MARSEXPRESS dataset, the thermal power lines are grouped depending on
their location and subsystem, resulting in a hierarchy of depth 5. The ADNI
dataset before combining all of the target variables in the root node separates
them according to the four different assessment examinations. The SIS dataset
groups the target variables depending on the type of income: salary (and pen-
sion), social income and income from other sources.

## 5   Results and Discussion

In this section, we present and discuss the results obtained from the experimental
evaluation. We first compare the predictive performance of the proposed PCTs
for HMTR with the predictive performance of PCTs for MTR. We next illustrate
the enhanced interpretability of the PCTs for HMTR.

Table 1 gives a detailed overview of the performance of PCTs for HMTR
and PCTs for MTR as measured with $aRRMSE$ on the training set and on
unseen data (estimated by 10-fold cross-validation). First of all, it shows the
performance of PCTs for HMTR for different values of the parameter $w_0$. We
can note that the training error generally increases when decreasing the value
of $w_0$ (an exception is the SIS dataset for $w_0 = 0.25$). This means that putting
less weight on the more general concepts in the hierarchy guides the learning
algorithm towards learning better fitted descriptions of the data. However, we

**Table 1.** Train and test $aRRMSE$ errors of PCTs for HMTR and PCTs for MTR
with different weights $w_0$.

| Dataset | HMTR | | | MTR | |
|---|---|---|---|---|---|
| | Weight | Train | Test | Train | Test |
| OSALES | 0.75 | 0.445 | 0.840 | 0.422 | 0.878 |
| | 0.5 | 0.450 | 0.844 | | |
| | 0.25 | 0.476 | 0.842 | | |
| MARS | 0.75 | 0.453 | 0.836 | 0.449 | 0.839 |
| | 0.5 | 0.456 | 0.836 | | |
| | 0.25 | 0.472 | 0.841 | | |
| ADNI | 0.75 | 0.640 | 1.055 | 0.632 | 1.074 |
| | 0.5 | 0.648 | 1.051 | | |
| | 0.25 | 0.652 | 1.063 | | |
| SIS | 0.75 | 0.704 | 1.155 | 0.705 | 1.175 |
| | 0.5 | 0.720 | 1.124 | | |
| | 0.25 | 0.672 | 1.112 | | |

cannot make the same statement about the errors as estimated with 10-fold cross validation (the column test in the Table).

Furthermore, we observe that the PCTs for HMTR overfit less than PCTs for MTR. Namely, the training errors of PCTs for HMTR are always worse than the training errors of the PCTs for MTR, while this is reversed for the error estimated with 10-fold cross validation. This is due to the fact that with the hierarchy, the learning algorithm now is able to generalize better the data. Hence, PCTs for HMTR perform worse on the training set because the hierarchy prevents them from focusing on too specific concepts existing in the data.

Next, we can note that PCTs for HMTR have better predictive performance than PCTs for MTR across all datasets and almost all of the values for $w_0$ (an exception is the MARSEXPRESS dataset for $w_0 = 0.25$). Based on these results, the best values for the depth parameter $w_0$ per dataset are as follows. For the OSALES and MARSEXPRESS datasets, the best value for $w_0$ is 0.75. For the ADNI dataset, the best value is 0.5, while for the SIS dataset is 0.25.

We then compare the performance of PCTs for HMTR with the best values for the parameter $w_0$ with the performance of PCTs for MTR in Table 2. The performance is here measured with both $aCC$ and $aRRMSE$. For both evaluation

**Table 2.** Evaluation (test) and efficiency comparison of HMTR and MTR

| Dataset | HMTR | | MTR | |
|---------|------|--------|------|--------|
| | $aCC$ | $aRRMSE$ | $aCC$ | $aRRMSE$ |
| OSALES | 0.361 | 0.840 | 0.331 | 0.878 |
| MARS | 0.418 | 0.836 | 0.416 | 0.839 |
| ADNI | 0.081 | 1.051 | 0.074 | 1.074 |
| SIS | 0.014 | 1.112 | 0.010 | 1.175 |

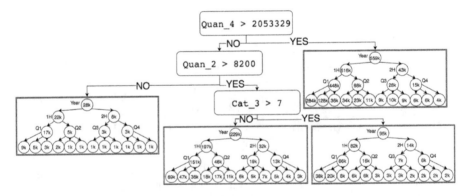

**Fig. 3.** A model for the OSALES dataset: A PCT for HMTR predicting the online sales of products by months (lowest part of the hierarchy), quarters, semesters and year (root node of the hierarchy).

measures, the same conclusions can be made: PCTs for HMTR outperform PCTs for MTR on the four datasets.

Finally, we demonstrate the enhanced interpretability of the PCTs by showing a heavily pruned PCT for HMTR learned on the OSALES dataset in Fig. 3. As it can be seen, the leafs of the PCT contain a prediction for the complete hierarchy, thus a domain expert can directly observe and comment on the more specific variables (lower in the hierarchy) and the more global variables (upper in the hierarchy, quarters, semesters and yearly). For example, if the attribute $QUAN\_4$ is higher than 2053329, from the predictions from the model, we can see that the sales of these products will be high overall, but will mostly take place in the first quarter of the year.

## 6  Conclusions

In this work, we introduce the task of hierarchical multi-target regression (HMTR). This structured output prediction task considers that a problem has multiple continuous target variables that can have hierarchical relationships among themselves. The formal task definition presented here unites all of the multi-target prediction tasks available.

We also propose a method for learning predictive clustering trees (PCTs) for the task of HMTR. Moreover, we propose to include the information about the hierarchy through adapting the Euclidean distance. The weighted Euclidean distance introduces weights to take into account the depth of the nodes in the hierarchy.

We evaluated the proposed method on 4 datasets and compared its performance to PCTs for MTR that do not exploit the hierarchical structure in the output space. The evaluation revealed that the PCTs for HMTR have better predictive performance than PCTs for MTR across all datasets. Furthermore, the PCTs for HMTR overfit less than PCTs for MTR. All in all, we showed that considering a hierarchical structure in the output space consisting of multiple continuous variables improves the performance.

We plan to extend this work along several directions. First, we will look more closely into the influence of various aggregation measures for different problems. We will also look into the weight parameter that the algorithm uses. As a general rule of thumb, $w = 3/4$ is a good starting point for most of the data sets, where higher values reduce the influence of the hierarchy and make the problem closer to standard MTR, and low values give more importance to the hierarchy than the target variables in the leaves. Finally, another direction to explore is the one of ensemble methods like random forests and bagging. We will investigate whether the performance improvement in HMTR carries over in the ensemble learning setting.

**Acknowledgments.** We acknowledge the financial support of the European Commission through the grants ICT-2013-612944 MAESTRA and ICT-2013-604102 HBP.

# References

1. Kocev, D., Vens, C., Struyf, J., Džeroski, S.: Tree ensembles for predicting structured outputs. Pattern Recogn. **46**(3), 817–833 (2013)
2. Bakır, G.H., Hofmann, T., Schölkopf, B., Smola, A.J., Taskar, B., Vishwanathan, S.V.N.: Predicting Structured Data. Neural Information Processing. The MIT Press, Cambridge (2007)
3. Kocev, D., Džeroski, S., White, M.D., Newell, G.R., Griffioen, P.: Using single- and multi-target regression trees and ensembles to model a compound index of vegetation condition. Ecol. Model. **220**(8), 1159–1168 (2009)
4. Borchani, H., Varando, G., Bielza, C., Larrañaga, P.: A survey on multi-output regression. Wiley Interdisc. Rev. Data Min. Knowl. Discov. **5**(5), 216–233 (2015)
5. Osborne, J.W.: The advantages of hierarchical linear modeling (2000)
6. Struyf, J., Džeroski, S.: Constraint based induction of multi-objective regression trees. In: Bonchi, F., Boulicaut, J.-F. (eds.) KDID 2005. LNCS, vol. 3933, pp. 222–233. Springer, Heidelberg (2006). doi:10.1007/11733492_13
7. Tsoumakas, G., Spyromitros-Xioufis, E., Vrekou, A., Vlahavas, I.: Multi-target regression via random linear target combinations. In: Calders, T., Esposito, F., Hüllermeier, E., Meo, R. (eds.) ECML PKDD 2014. LNCS, vol. 8726, pp. 225–240. Springer, Heidelberg (2014). doi:10.1007/978-3-662-44845-8_15
8. Spyromitros-Xioufis, E., Tsoumakas, G., Groves, W., Vlahavas, I.: Multi-target regression via input space expansion: treating targets as inputs. Mach. Learn. **104**(1), 55–98 (2016)
9. Vens, C., Struyf, J., Schietgat, L., Džeroski, S., Blockeel, H.: Decision trees for hierarchical multi-label classification. Mach. Learn. **73**(2), 185–214 (2008)
10. Džeroski, S.: Towards a general framework for data mining. In: Džeroski, S., Struyf, J. (eds.) KDID 2006. LNCS, vol. 4747, pp. 259–300. Springer, Heidelberg (2007). doi:10.1007/978-3-540-75549-4_16
11. Blockeel, H., Struyf, J.: Efficient algorithms for decision tree cross-validation. J. Mach. Learn. Res. **3**, 621–650 (2002)
12. Breiman, L., Friedman, J., Olshen, R., Stone, C.J.: Classification and Regression Trees. Chapman & Hall/CRC, Boca Raton (1984)
13. Madjarov, G., Kocev, D., Gjorgjevikj, D., Džeroski, S.: An extensive experimental comparison of methods for multi-label learning. Pattern Recogn. **45**(9), 3084–3104 (2012)
14. Slavkov, I., Gjorgjioski, V., Struyf, J., Džeroski, S.: Finding explained groups of time-course gene expression profiles with predictive clustering trees. Mol. BioSyst. **6**(4), 729–740 (2010)
15. Gamberger, D., Ženko, B., Mitelpunkt, A., Shachar, N., Lavrač, N.: Clusters of male and female alzheimers disease patients in the alzheimers disease neuroimaging initiative (adni) database. Brain Inf. **3**(3), 169–179 (2016)

# Identifying Novel Features from Specimen Data for the Prediction of Valuable Collection Trips

Nicky Nicolson[1,2](✉) and Allan Tucker[2]

[1] Biodiversity Informatics and Spatial Analysis, Royal Botanic Gardens, Kew,
Richmond, UK
n.nicolson@kew.org
[2] Department of Computer Science, Brunel University London, London, UK

**Abstract.** Primary biodiversity data provide "what, where, and when" data points: the assertion that a species occurred at a particular point in space and time. These are most valuable when associated with specimens stored in natural history museums and herbaria, which evidence the assertions with reference to a physical specimen. The research presented uses novel data-mining techniques to uncover two hidden dimensions in specimen data - *who* collected the specimens and *how* they were collected. A combination of unsupervised and supervised learning techniques are used, which establish two new entities: *collector* and *collection trip*. Features are defined against these higher order representations of the data, which support the use of the data to answer novel questions such as *which collection trips discover the most new species?* We explore the features by building classifiers to predict species discovery, and compare these with a baseline model grouped using collector team transcriptions derived from the raw specimen data. Preliminary results are promising and whilst the particular focus of this research was botanical specimens, the technique is equally applicable to datasets of field-collected specimens from other scientific domains.

**Keywords:** Data-mining · Clustering · Classification · Species discovery

## 1  Introduction

Biological specimens collected over hundreds of years and held in natural history museums and herbaria are a rich reference source with which to understand the natural world, and to analyse its changes over time. Estimates of the total number of specimens vary between 2.5–3 billion specimens globally [1]. Only a small percentage have associated digital data. Aggregation initiatives such as the Global Biodiversity Informatics Facility (GBIF) harvest and mobilise digital specimen data: at the time of writing (May 2017) the GBIF data portal includes information on 129,006,858 specimens. In order to aid the mobilization of the data, there has been an effort to develop standards regarding the representation of the data [2], and references to it [3]. These standards are important as due to the scale of the overall task, data have been digitised in a distributed fashion, at different rates and to different levels of completeness.

© Springer International Publishing AG 2017
N. Adams et al. (Eds.): IDA 2017, LNCS 10584, pp. 235–246, 2017.
DOI: 10.1007/978-3-319-68765-0_20

In addition to the structured data held on the specimens themselves, field collected specimens are often accompanied by a wealth of information about the collection site, habitat and associated species, logged in field books, which are also being digitised via literature digitisation initiatives.

Although plants are a comparatively well known group, and are well represented with digitised specimen data, species discovery is not yet complete, and approximately two thousand new species are described per year [4]. Not all species discovery is via field work: a significant proportion of species discovery is conducted from pre-existing specimens already lodged in institutional collections [5]. Estimates of the total number of plant species recognise the importance of species discovery from pre-existing collections and the use of collections data to plan species discovery in the field.

The application of intelligent data analysis techniques on the specimen data can help meet two key aims: *data mobilisation* by better utilising and curating the existing data, and finding efficiencies that will help the digitisation process, and *data understanding* by uncovering patterns that will help plan future scientific effort as research is conducted with specimens or in the field.

The novel data-mining techniques demonstrated here detect new entities (*collector* and *collection trip*) from the duplicated, incomplete and variably transcribed specimen datasets. These are used to draw together heterogonous data, which has been recorded in different places, to different standards, in order to build classification models to support and develop our understanding of a complex system - species discovery.

The remainder of the paper is structured as follows: a background section further introduces the nature of the specimen data available by defining terms and outlining the specimen collection process, methods describes a data-mining process to detect collector and collection trip entities from raw specimen data, defines a novel set of features using these new entities to group the raw specimen data, and describes the creation of classifiers using these and baseline data. Preliminary results of the data-mining and classification steps are shown, and ideas for further work are discussed.

## 2    Background and Definition of Terms

A *specimen* is a physical sample of biological material collected in the field. In botany, a collected sample may consist of multiple specimens, named *duplicates*. The *collecting team* is the team of collectors responsible for gathering and documenting the specimen, this may include multiple collectors, referred to by personal name. The *primary collector* is the first listed member of the *collecting team*, and controls the *recordnumber* - a number given to the specimen in the field, usually sequential and unique to the primary collector. Recordnumbers are locally managed, rather than centrally assigned. When duplicate specimens are collected, they are given the same recordnumber [6]. A *collection trip* is a circumscribed period of specimen collecting activity conducted by a particular primary collector, focussed on a particular place and time. An *itinerary* is a list of the collecting localities visited by a primary collector in a collection trip, which may be documented in a *field book*, cross referenced to specimens via the recordnumber.

An *institution* is the holder of specimens for long term storage and reference consultation, usually natural history museums or herbaria (botanically focussed institutions). Institutions may distribute duplicate specimens to external institutions, to form a globally distributed reference collection. *Digitisation* is the process of creating electronic records from the data held on the physical specimen, which may include *imaging* - the creation of a digital image of the specimen, and/or *georeferencing* - the process of determining a latitude/longitude pair from a textual description of the collecting locality. This is necessary due to the historic nature of the specimen collection effort, which pre-dates technologies such as hand-held global positioning systems. Duplicates are recognised as a source of data to speed the digitisation process [7]. The *collector name transcription* is the transcription of the collector names made when specimen data is read for digitisation. A single collector may have multiple varying collector name transcriptions, depending on the standards used in the different institutions, transcription errors and spelling mistakes. As the collector name transcription is necessary to identify duplicates [7], variability in this data element impedes efficient use of the global specimen dataset. *Aggregation* is the collation of digitised specimen records from many institutions into a single data repository, represented using a structured *data standard*.

A *type specimen* is the reference use of a specimen as the basis of a new species description, published in the academic literature. The reference to a specimen is made using the collector name and recordnumber [8]. The use of a specimen as a type specimen is indicated in its digital record. A *name author* is the author of a new species description, a person who may also act as a collector. A *career* is the complete body of work performed by one collector/name author. The subject focus of the career may be examined to determine if the person is a *specialist* (focussed on a particular taxonomic subset) or *generalist* (working across many different areas of taxonomy). Some generalists may be regional specialists, focussing on the plants of a particular geographical region.

Primary biodiversity data derived from specimens have many applications in research [1] including species description and discovery. The collector - who makes decisions in preparatory planning and in the field about what to collect - is obviously a major contributor to species discovery [9], and the collection trip has been recognised as a way to understand the accumulation of knowledge regarding the species found in a particular geographic area [10].

Differentiating collection trips based on the characteristics of the collector has also been proposed [10], including the differentiation between specialist and generalist collectors. Despite the scope for more advanced analyses of specimen data when differentiated and grouped by collector and/or collection trip, these entities are not formally managed - only the collecting team is a component of the main data standard used to share specimen data, which is supplied as a text transcription [2, 11]. Studies involving the grouping of specimen data by collector

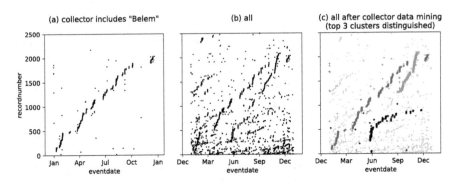

**Fig. 1.** Example specimen data from 1965, recordnumber less than 2500

and or collection trip have had to use manual specimen record allocation [9] and/or expert knowledge [10], which limits scope. This means that the sequential nature of recordnumbers has been minimally exploited to date - but they have been used to create itineraries, by cross-referencing by hand between specimen data and field books [12] as an aid to geo-referencing.

An example use of sequential recordnumber for a single collector is a test for a positive correlation - as a particular collector moves forward through time, their own personal sequential recordnumber increases (see Fig. 1(a)). Exploration of the data in this way can be useful to identify outliers (resulting from data transcription errors), but applications are limited due to the difficulty in initially identifying the set of specimens relating to a single primary collector, due to the variation in collector name transcriptions. Plotting a fuller corpus of specimen data (see Fig. 1(b) - a sample of points from specimens collected in a single year), shows some visually distinguishable elongated "clusters", each of which correspond to the set of specimens collected by a particular primary collector and labelled with their own sequential recordnumber, which ascends over time.

In this research, we propose the exploitation of the sequential recordnumber as a feature for clustering to detect the primary collector, thereby overcoming the variability encountered when using the un-standardised transcription of personal names. We employ a novel combination of data-mining techniques to detect these clusters, in order to identify higher order abstractions (collector and collection trip) from an incomplete raw specimen dataset. These abstractions are recognised in the domain, but are absent from digital datasets. We exploit the sequential nature of the recordnumber to cluster specimens as they were gathered over time, resulting in a grouping by primary collector. We then use the collector grouping to detect the collection trips made by that primary collector. These abstractions are used to group the data and to define features at the grouped level, these features are used to train classifiers to identify high-value collection trips which are relevant to species discovery.

# 3    Methods

The process described here was developed to allow visualisation of intermediate results at each stage, in order to allow an analyst to influence the design of the process. Visualisations were created as interactive scatter plots (as shown in Fig. 1), allowing the analyst to focus on particular areas of the data, and to examine the underlying specimen records.

The main specimen dataset was downloaded from the Global Biodiversity Informatics Facility, encompassing data generated from botanical specimens [11]. This large (59 million record) dataset was used for exploratory data analysis, and a subset representing the specimens collected since 1700 from a single political country (Brazil) was selected for data-mining. Brazil is recognised as a mega-diverse country [13], and Brazilian specimen data has been digitised and repatriated via the REFLORA project [14], meaning that a considerable amount of data is digitally available, from many different institutions. The subset of data used for data-mining contains 3493107 specimen records, which were collected between 1705 and 2016, and held in 132 different institutions. A biographical dataset was used as a data lookup, to check if collectors detected in the data-mining steps are also known to have authored new species. This is managed as part of the International Plant Names Index (IPNI), and contains the personal names and lifespan dates for those who have published new names since the start date for botanical nomenclature (1753) [4].

## 3.1    Data-Mining

**Preparation:** data are read from the data store and prepared for data-mining by making a `numeric feature-set` from `recordnumber` and `eventdate` (expressed as days since 1st January 1970). The details of the collector team transcription are quantified by extracting the primary collector name, standardising the order of recording of the name elements and deriving minimal textual features from the name to form a `lexical feature-set`. The first initial, and the first uppercase character, first lowercase character and last lowercase character from the first word (usually the surname) are extracted and converted to indicator variables where the value `1000` represents presence and `0` represents absence. A field for type status (`is_type`) is created and populated following the criteria used in [9].

The actual data mining process is composed of 4 steps, steps 1–3 identify collectors, step 4 examines the set of specimen data allocated to a particular collector to detect collection trips.

**Step 1 (cluster)** uses DBSCAN [15]. This clustering algorithm is used as via exploratory data analysis the data are observed to form elongated rather than spherical clusters, due to the use of the sequential `recordnumber` and `eventdate` features (along with the `lexical feature-set`). DBSCAN is configured to use a value of `300` for `epsilon` and `2` for `min_samples`. A low value of `min_samples` is used as the clustering results are computationally post-processed to lexically examine the collector names included within a cluster. Analyst examination of the data immediately after DBSCAN clustering shows that the primary collector names are so variably recorded that clusters contain multiple logical collector names. A pessimistic approach is taken, clusters are divided into multiple separate clusters if the lexical variation of the primary collector names included is too great (e.g. due to differing initials).

Expert analysis of the dataset after step 1 identified a common problem with many clusters, that the huge variation in the transcriptions of the primary collector names introduces variation into the lexical feature-set, and results in the assignment of logical collectors into separate clusters. When the data are examined using visualisation (an interactive scatter plot of eventdate against recordnumber, with the colour of the points determined by the cluster_id, these clusters show up as *interpolations*: an elongated stream of points flips between two or more different clusters, but the specimen data underlying is seen to have the same primary collector (transcribed in very different ways).

**Step 2 (classify)** uses a decision tree to detect sets of distinct clusters which are similar in terms of the numeric feature-set, but which differ in terms of the lexical feature-set (described as *interpolated* above). The classifier is trained on the numeric feature-set to predict the cluster identifier. Commonly confused classes are identified using the classifier, and these are considered candidates for joining after computational assessment for lexical similarity with respect to their primary collector names. Those with very similar names (as may result from differing transcriptions e.g. abbreviation to initials) are joined. As cluster manipulation will affect the extent of cluster interpolation, this is an iterative process, and is run for 10 iterations or until there are no more candidates for joining, whichever occurs first.

**Step 3 (join)** joins clusters to result in a grouping representing the career work of a single collector, so that all specimens collected by the same collector will be held in the same cluster. This is implemented in two stages: (i) the clusters output from step 2 are joined if their most frequently occurring first collector name is shared and all name variants in the cluster agree lexically and (ii) clusters are matched against an external bibliographic database of taxonomic name authors, those matching to the same bibliographic database record are joined. A unique identifier value for each collector is created (collector_id).

**Step 4 (detect collection trips)** subdivides the dataset by collector_id. The specimen data for each collector_id is passed into a DBSCAN clustering using minimal features - eventdate and recordnumber. A lower value of epsilon is used (90, in comparison to 300 used in the collector data-mining in step 1). The minimal value for min_samples (2) is retained in recognition of the gaps in the incomplete specimen dataset - a cluster of two specimen data points may indicate a trip which collected many specimens, only two of which are currently digitised. The clusters identified by each iteration of the DBSCAN process represent the collection trips for a single collector, a unique trip identifier is created and applied to the specimen dataset.

## 3.2   Data Abstraction: Creation of New Features

The results of the data mining are used to group the specimen data and to define metrics using these groupings. The identification of a collector allows the detection and elimination of specimen duplicates. Duplicates are defined as those specimens which share `recordnumber` and `collector_id`, all but the first occurrence of a particular `recordnumber`/`collector_id` pairing are flagged as duplicates and excluded from subsequent analyses.

Grouping the specimen data using the data-mined entity types (collector and collection trip) allows the definition of a new set of features. These are categorised as *temporal* (the start year of the grouping), the *scale* of the grouping (duration, the total number of specimens included and the range of recordnumbers allocated), the *rate* of accumulation (the slope of a line of best fit through the `eventdate` and `recordnumber` values), and the `correlation_score` of these points). The *character* of the grouping is defined in two ways - by creating a `specialist` flag, set if the grouping is more 60% composed of specimens from a single taxonomic family, and by creating a `nomenclaturalist` flag which is set if the collector is known to have also acted as a name author. The *experience* of the collector at a point in time is assessed by creating features for the total number of previous specimens collected and the total number of previous collection trips made. Finally, a feature is defined that will later be used as the class variable in classifiers: this encodes the *species discovery value* of the grouping, and is simply a flag indicating if the grouping contains material that was later used as a type specimen, representing a contribution towards species discovery. Data files containing these features are constructed, these are used as training data in the next step.

## 3.3   Development of a Classification Model Using the Results of Data-Mining and Data Abstraction

The feature-sets generated by the data-mining process are used to train classifiers to predict the species discovery value of the grouping. These are compared to a baseline grouping, derived without the data-mining process. The baseline is simply a grouping of the specimen data by the different values of the transcribed primary collector name, which was the source of the lexical feature-set used in the data-mining steps. Both the baseline and the data-mined datasets were downsampled to balance the binary class variable, as the samples for the positive class were far less frequent. A decision tree classifier was trained on the downsampled data, using 10-fold stratified cross-validation. Feature selection was also conducted to examine which of the features defined were the most indicative.

# 4   Results

**Data mining results:** DBSCAN identified 42096 clusters, lexically postprocessed to 51192 clusters (step 1); resolved via decision-tree classifier to 44768 clusters (step 2); joined to 19706 clusters representing collector entities (step 3). 79012 different collecting trips were identified (step 4). The raw specimen data underlying the entities recognised via data mining comprises 131582 unique collector team transcriptions and 41511 unique primary collector name transcriptions. 1127 (5.7%) of collectors and 3412 (4.3%) of trips collected specimens later labelled as type specimens.

The results of the collector data-mining process on the illustrative sample used in the background section are shown in Fig. 1(c). An alternative visualisation of the data, using textual rather than numeric attributes, demonstrates the grouping provided by the data-mining process, and the level of variation in the input data. A Sankey diagram represents flow. It is used here (see Fig. 2) to demonstrate how specimen records "flow" between the groupings established at each stage of the data mining process. The diagram compares three groupings - on the left the data are grouped using the baseline method (the text transcription of the primary collector name extracted from the collector team transcription). The central column shows the grouping as done by the data-mined collector

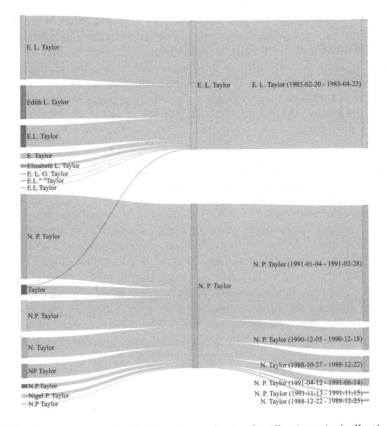

**Fig. 2.** Specimen grouping by (l-r) baseline, collector & collection trip (collection trips shown with start & end dates)

entities, and the rightmost column divides the specimen data still further, into collection trips (shown with start and end dates). The width of the connections between the groups in each of the columns is proportional to the number of specimens included. The diagram shows that the variation in the dataset is reduced as a result of the data-mining, and that specimens can be more meaningfully grouped by collector and/or collection trip.

The selected subset shows both the strengths and the weaknesses of the current data-mining technique - a strength is that the ambiguous name *Taylor* is split into two different collectors based on the context provided by the date and recordnumber values; a weakness is the over-enthusiastic grouping of what seem to be two distinct collectors (*Edith L Taylor* and *Elizabeth L Taylor*) in the top-most collector. The trip detection results should be considered preliminary - due to an incomplete dataset and immature trip data-mining process, many very small trip groupings are defined.

**Classification and feature selection results:** these were assessed by calculating the mean area under the receiver operator curve from the 10-fold cross-validated runs. Classification results (see Fig. 3) from the baseline and trip datasets are similar (73.00% and 71.22%), collector shows an improvement over these (77.92%), and the execution of feature selection on the trip and collector datasets improves each over the comprehensive feature-set (76.27% and 80.69% respectively).

Collector appears to perform better than trip; the similarity of the performance of the trip groupings to the baseline is likely to be due to the numbers of small trips detected in the data-mining process. The datasets derived from datamining were used to conduct feature selection using an exhaustive search strategy, scoring using the area under the receiver operator curve (ROC AUC) metric.

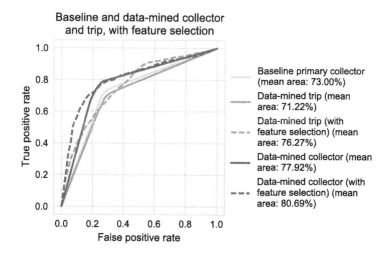

**Fig. 3.** Receiver-operator curves for classifiers trained on baseline & data-mined datasets

The features selected via this process were the temporal feature (start_year) and the two features encoding character (specialist and nomenclaturalist) for both the collector and trip groupings.

## 5   Discussion

When analysing the results of the data-mining via a simple classification task, collector has shown to perform well, and trip has similar results to the baseline. Re-examination of the methods shows that collector data-mining has a set of post-processing steps which validate and modify the entities, trip data-mining by contrast is rather immature and is perhaps more badly affected by the incomplete input data: when trying to subdivide data into trips, a time gap can be either due to a legitimate trip boundary or an artefact due to incomplete input data. Further work with the trip data-mining process to define a similar set of post-processing steps to be applied after DBSCAN clustering would likely improve this part of the method.

The immediate context for this research is data generated from field-collected scientific specimens. The particular focus of this research was botanical specimens, however the technique is equally applicable to datasets of field-collected specimens from other scientific domains. It is possible to define simple eligibility criteria for this kind of analysis: the specimen dataset must contain a string representation of the primary collector (or collecting team), a date of collection and a field-assigned recordnumber. Applying these eligibility criteria to the datasets available via GBIF shows that datasets comprising specimens from *icthyology*, *ornithology* and *mycology* meet the criteria for this kind of analysis.

The data-mining process created in this research can be generalized to the use of a *product* (specimen) dataset, to identify the *agent* responsible for its generation (collector), and to place the product within a *sequence of work* (a collection trip). This is possible as the product is identified by an *agent-managed sequence* (recordnumber) which ascends over time. We scanned the biodiversity (and related) domains for other examples of data generated via a similar process. Many of the datasets generated in the digital age have recognised the need for shared persistent identifiers across distributed datasets (e.g. the use of DOIs in publishing) and by implementing these have sidestepped the need for this kind of analysis. The examples selected represent data generation via digitisation of historic information, that which pre-dates easily accessible shared identifiers. Species names for plants are referenced using micro-citations, page level bibliographic references. There is an effort to standardise the authorship for these to enable trends analysis e.g. to detect changes in gender balance of the authors of plant names [16]. Page level microcitations can be seen as another representation of the *product/agent/sequence of work* data generation process: the product is a page-level microcitation, generated by an author, fitting into a bibliographic container (article or book) as a sequence of work. As page number is sequential and pages located in close proximity are likely to be authored by the same person, this dataset is a candidate for data-mining using this technique.

Feature selection on the grouped datasets result in the inclusion of the two character features - which indicate if a grouping is specialist (taxonomically focussed) and if the collector was also a nomenclaturalist (participating in the publication of new species) - these results support previous work which propose the specialist/generalist distinction as relevant [10]. Future work will implement the selective inclusion of features using the categorisations defined earlier, for example to define classification models applicable across temporal scales or to see how relevant different feature categories remain over time. There is scope for further data integration to expand the set of features used here by including data from bibliographic sources. An expanded feature-set should allow further advances towards understanding the process of species discovery and the people involved in it.

## 6   Conclusions

This paper proposes the application of data-mining techniques to specimen data in order to create higher-order data abstractions. These abstractions were used to define new features, which were tested by building classifiers to predict species discovery value. The input data are acknowledged as being incomplete - both in terms of the number of records available and the population of individual fields in a particular record. Specimen data are expensive to fully digitise, one of the aims of this research was to understand what could be done with a minimal dataset, with no dependence on expensive augmentation processes such as georeferencing.

The positive preliminary results shown here have impacts in two core areas - data mobilization and data understanding. In data mobilization we are able to suggest practical modifications to increase efficiency of the specimen digitisation process. Recognition of the data-mined collector and trip entities allows better integration of data from different sources, one element of the data-mining process - classification to predict collector from easily digitised features of specimen (`eventdate`, `recordnumber`) - has potential application as a tool to aid the transcription of specimen data by digitisation staff. We have advanced data understanding by demonstrating the potential for reshaping specimen data to define novel features. These were used to populate models which can detect subsets of particular value in species discovery.

This work has demonstrated that although incomplete and variably recorded, the aggregated specimen data now form a critical mass which support the development and application of alternative approaches towards data mobilization and data understanding.

## References

1. Chapman, A.D.: Uses of Primary Species-Occurrence Data. Global Biodiversity Information Facility, Copenhagen (2005)
2. Wieczorek, J., Bloom, D., Guralnick, R., Blum, S., Döring, M., Giovanni, R., Robertson, T., Vieglais, D.: Darwin core: an evolving community-developed biodiversity data standard. PLoS ONE **7**(1), e29715, January 2012

3. Güntsch, A., Hyam, R., Hagedorn, G., Chagnoux, S., Röpert, D., Casino, A., Droege, G., Glöckler, F., Gödderz, K., Groom, Q., Hoffmann, J., Holleman, A., Kempa, M., Koivula, H., Marhold, K., Nicolson, N., Smith, V.S., Triebel, D.: Actionable, long-term stable and semantic web compatible identifiers for access to biological collection objects. Database **2017**(1), January 2017

4. ipni.org: International Plant Names Index, http://www.ipni.org

5. Bebber, D.P., Carine, M.A., Wood, J.R.I., Wortley, A.H., Harris, D.J., Prance, G.T., Davidse, G., Paige, J., Pennington, T.D., Robson, N.K.B., Scotland, R.W.: Herbaria are a major frontier for species discovery. Proc. Nat. Acad. Sci. **107**, 22169–22171 (2010)

6. Bridson, D.M.: The Herbarium Handbook, 3rd edn. Royal Botanic Gardens, Kew (1998)

7. Tulig, M., Tarnowsky, N., Bevans, M., Kirchgessner, A., Thiers, B.M.: Increasing the efficiency of digitization workflows for herbarium specimens. ZooKeys **209**, 103–113 (2012)

8. Turland, N.: The code decoded a user's guide to the International code of nomenclature for algae, fungi, and plants. Koeltz Scientific Books, Königstein (2013)

9. Bebber, D.P., Carine, M.A., Davidse, G., Harris, D.J., Haston, E.M., Penn, M.G., Cafferty, S., Wood, J.R.I., Scotland, R.W.: Big hitting collectors make massive and disproportionate contribution to the discovery of plant species. Proc. Roy. Soc. London B: Biol. Sci. **279**(1736), 2269–2274 (2012)

10. Utteridge, T., de Kok, R.: Collecting strategies for large and taxonomically challenging taxa. In: Reconstructing the Tree of Life: Taxonomy and Systematics of Species Rich Taxa, pp. 297–304. CRC Press (2006)

11. GBIF.org: GBIF occurrence download (taxon: Tracheophyta, basis of record: specimen) 4 October 2016, http://doi.org/10.15468/dl.68z1mf

12. Smith, L., Smith, R.: Itinerary of William John Burchell in Brazil, 1825–1830. Phytologia **14**(8), 492–505 (1967)

13. Mittermeier, R.A.: Megadiversity: Earth's biologically wealthiest nations. Agrupacion Sierra Madre (1997)

14. REFLORA: REFLORA programme, http://reflora.jbrj.gov.br

15. Ester, M., Kriegel, H.P., Sander, J., Xu, X., et al.: A density-based algorithm for discovering clusters in large spatial databases with noise. In: KDD, vol. 96, pp. 226–231 (1996)

16. Lindon, H.L., Gardiner, L.M., Brady, A., Vorontsova, M.S.: Fewer than three percent of land plant species named by women: author gender over 260 years. Taxon **64**(2), 209–215 (2015)

# Adapting Supervised Classification Algorithms to Arbitrary Weak Label Scenarios

Miquel Perelló-Nieto[1(✉)], Raúl Santos-Rodríguez[1], and Jesús Cid-Sueiro[2]

[1] University of Bristol, Bristol, UK
{miquel.perellonieto,enrsr}@bristol.ac.uk
[2] Universidad Carlos III de Madrid, Leganés, Spain
jcid@tsc.uc3m.es

**Abstract.** In many real-world problems, labels are often weak, meaning that each instance is labelled as belonging to one of several candidate categories, at most one of them being true. Recent theoretical contributions have shown that it is possible to construct proper losses or classification calibrated losses for weakly labelled classification scenarios by means of a linear transformation of conventional proper or classification calibrated losses, respectively. However, how to translate these theoretical results into practice has not been explored yet. This paper discusses both the algorithmic design and the potential advantages of this approach, analyzing consistency and convexity issues arising in practical settings, and evaluating the behavior of such transformations under different types of weak labels.

**Keywords:** Weak labels · Noisy labels · Proper losses

## 1 Introduction

Most machine learning algorithms are grounded on two common assumptions: (1) a pool of annotated examples is available for training, and (2) the labelling process satisfies some nice statistical properties, e.g., balanced label proportions or statistical independence. However, in practice, real datasets pose major challenges regarding the quality of the labels, from label noise to partial supervision.

In the last decade several authors addressed these and similar tasks using different terminologies depending on the specific properties and assumptions of the scenarios at hand. For instance, learning from *partial labels* [5,6,11], *multiple labels* [10] and *ambiguous labels* [8], describe settings where each sample is labeled using a subset of classes, required to contain the true label. On the other hand, *crowd learning* [12,13] assumes that we may have access to multiple labels provided by a number of different annotators, that are bound to sometimes disagree when labelling the same example. A related problem, *noisy labels* [2,14],

JCS is supported by the TEC2014-52289-R project funded by the Spanish MEC. MPN and RSR are supported by the SPHERE IRC funded by the UK Engineering and Physical Sciences Research Council (EP/K031910/1).

© Springer International Publishing AG 2017
N. Adams et al. (Eds.): IDA 2017, LNCS 10584, pp. 247–259, 2017.
DOI: 10.1007/978-3-319-68765-0_21

restricts the setting to a unique label per sample, but in this case labels have some constant probability of being flipped. Finally, *superset learning* [9] generalizes many of the previous approaches and allows for each example to be linked to a subset of classes that contains the true outcome but may also encompass additional ones. In this paper, our interpretation of the weak label paradigm also accounts for each instance being labelled as belonging to one of several candidate categories, at most one of them being true, but we do not require the true label to be present.

In [3] we suggested a general procedure to transform a standard (i.e. fully-supervised) proper loss into a weak loss that is also proper, in the sense that posterior class probabilities can be estimated provided that the label mixing process is restricted to lie in certain linear subspace. Recently, in [4] we analyzed the conditions under which the true class can be inferred from weak labels. In this paper we built upon this previous theoretical results to describe a simple algorithmic approach to seamlessly adapt existing classification algorithms that are based on the empirical minimization of proper or classification calibrated losses, in such a way that they explicitly take into account the mixing process underlying the generation of the annotations. In short, the contributions of this work are twofold:

- We depict a transparent procedure for the machine learning practitioner to transform weak labels into what we refer to as *virtual labels*, that can then be used within standard out-of-the-shelf machine learning toolboxes, providing advice on practical implementation issues.
- We thoroughly test the approach studying realistic scenarios in which, (1) partial information might be available regarding the relationship between some true and weak labels and, (2) no information at all is revealed.

The remainder of the paper is organized as follows: the problem of learning from weak labels is formulated in Sect. 2. Some results on losses for weak labels are reviewed in Sect. 3, and the algorithmic design is detailed in Sect. 4. In Sect. 5 we analyse five common weakly supervised case studies. Finally, we state some conclusions in Sect. 6.

## 2 Formulation

### 2.1 Notation

Vectors are written in boldface, matrices in boldface capital and sets in calligraphic letters. For any integer $n$, $\mathbf{e}_i^n$ is a $n$-dimensional unit vector with all zero components apart from the $i$-th component which is equal to one, and $\mathbb{1}_n$ is a $n$-dimensional all-ones vector. Superindex $^\top$ denotes transposition. We will use $\Psi()$ to denote a loss based on weak labels (for brevity, "weak loss"), and $\tilde{\Psi}$ to losses based on the true class. The number of classes is $c$, and the number of possible weak label vectors is $d \leq 2^c$. $|\mathbf{z}|$ is the number of nonzero elements in $\mathbf{z}$. The set of all $d \times c$ matrices with stochastic columns is $\mathcal{M} = \{\mathbf{M} \in [0,1]^{d \times c} : \mathbf{M}^\top \mathbb{1}_d = \mathbb{1}_c\}$, and the simplex of $n$-dimensional probability vectors is $\mathcal{P}_n = \{\mathbf{p} \in [0,1]^n : \sum_{i=0}^{n-1} p_i = 1\}$.

## 2.2   Learning from Weak Labels

Let $\mathcal{X}$ be a sample space, $\mathcal{Y} = \{e_j^c, j = 0, 1, \ldots, c - 1\}$ a set of labels, and $\mathcal{Z} = \{b_1, \ldots, b_d\} \subset \{0, 1\}^c$ a set of weak or partial label vectors. Sample $(x, z) \in \mathcal{X} \times \mathcal{Z}$ is drawn from an unknown distribution $P$.

Weak label vector $z \in \mathcal{Z}$ is a noisy version of the actual true class $y \in \mathcal{Y}$. A common assumption [1,5,7,10] is that the true class is always present in $z$, i.e., $z_j = 1$ when $y_j = 1$, but this assumption is not required in our setting. We assume that $\mathcal{Z}$ contains only weak labels with nonzero probability (i.e. $P\{z = b\} > 0$ for any $b \in \mathcal{Z}$). The dependency between $z$ and $y$ is modelled through an arbitrary $d \times c$ conditional mixing probability matrix $M(x) \in \mathcal{M}$ with components

$$m_{ij}(x) = P\{z = b_i | y_j = 1, x\} \tag{1}$$

where $b_i \in \mathcal{Z}$ is the $i$-th element of $\mathcal{Z}$. Defining posterior probability vectors $p(x)$ and $\eta(x)$ with components $p_i = P\{z = b_i | x\}$ and $\eta_j = P\{y = e_j^c | x\}$, we can write $p(x) = M(x)\eta(x)$. In general, the dependency with $x$ will be omitted and we will write, for instance, $p = M\eta$. The mixing matrix could depend on $x$, though a constant mixing matrix is a common assumption [1,7,10,13], as well as the statistical independence of the incorrect labels [1,7,10]. Assuming a constant matrix is not required in our analysis. Any property derived for $M$ can be extended to a property that must be satisfied by $M(x)$ for all $x$.

The goal is to infer $y$ given $x$ without knowing model $P$. To do so, a set of i.i.d. weakly labelled samples, $\mathcal{S} = \{(x_k, z_k), k = 1, \ldots, K\} \sim P$ is available. True classes $y_k$ are not observed. In this paper we are interested in algorithms based on the minimization of an empirical risk

$$\hat{R}_\Psi(\mathcal{S}) = \sum_{k=1}^{K} \Psi(z_k, f(x_k)) \tag{2}$$

where $\Psi(z, f)$ is a weak loss function that takes a weak label (instead of the true label) as an argument, and $f(x)$ is a scoring function. The class prediction is computed through some function $\text{pred}(x) \in \text{argmax}_i\{f_i(x)\}$.

# 3   Transforming a Conventional Loss into a Weak Loss

This section summarizes some of the theoretical results in [3,4] that have practical implications on the design of algorithms for weakly labeled datasets.

## 3.1   Virtual Labels

We will consider weak loss functions that can be computed as linear combinations of conventional loss functions for clean labels. Defining the vector representation of the weak loss as $\Psi(f) = (\Psi(b_0, f), \ldots \Psi(b_{d-1}, f))$, we construct losses following

$$\Psi(f) = \tilde{Y}^\top \tilde{\Psi}(f) \tag{3}$$

where $\tilde{\boldsymbol{\Psi}}$ is a vector representation of a conventional loss $\tilde{\Psi}(\mathbf{y}, \mathbf{f})$, that is, $\tilde{\boldsymbol{\Psi}}(\mathbf{f}) = \left( \tilde{\Psi}(\mathbf{e}_0^c, \mathbf{f}), \ldots \tilde{\Psi}(\mathbf{e}_{c-1}^c, \mathbf{f}) \right)$ and $\tilde{\mathbf{Y}}$ is a weight matrix.

Note that the weak loss for a weak label $\mathbf{b}_i$ can be written as

$$\Psi(\mathbf{f}, \mathbf{b}_i) = \tilde{\mathbf{y}}_i^\mathsf{T} \tilde{\boldsymbol{\Psi}}(\mathbf{f}) \tag{4}$$

where $\tilde{\mathbf{y}}_i$ is the $i$-th column of $\tilde{\mathbf{Y}}$. By comparing this expression with the loss for a clean label $\mathbf{y}$

$$\tilde{\Psi}(\mathbf{f}, \mathbf{y}) = \mathbf{y}^\mathsf{T} \tilde{\boldsymbol{\Psi}}(\mathbf{f}) \tag{5}$$

we can interpret the $i$-th column of $\tilde{\mathbf{Y}}$ as a *virtual* label vector. The weak loss can thus be computed by replacing the true label by the virtual label corresponding to the observed weak label. For this reason we will call $\tilde{\mathbf{Y}}$ a *virtual label matrix*.

### 3.2   Properness and Classification Calibration

Linear transformations in the form (3) have been studied in [4]. In particular, the authors studied the conditions on the virtual matrix that guarantee that the weak loss is **M**-proper or **M**-classification calibrated.

A loss function is said to be **M**-proper if the minimizer of the expected loss is the true posterior class probability, i.e. $\boldsymbol{\eta} \in \arg\min_{\mathbf{f}} \mathbb{E}_\mathbf{z} \{\Psi(\mathbf{z}, \mathbf{f})\}$, where $\boldsymbol{\eta}$ is the probability vector with components $\eta_j = P\{y_j = 1\}$. The loss is strictly proper if $\boldsymbol{\eta}$ is the unique minimizer. If posterior class probability estimates are not required and the main goal is to minimize classification error, classification calibration can be enough. We say that a weak loss is **M**-classification calibrated (or **M**-CC) if $\mathbf{f}^* \in \arg\min_{\mathbf{f}} \mathbb{E}_\mathbf{z} \{\Psi(\mathbf{z}, \mathbf{f})\}$ satisfies $(\eta_i > \max_{j \neq c} \eta_j \Rightarrow f_i^* > \max_{j \neq c} f_j^*)$, that is, a class with maximum a posteriori probability is also a class with the highest score value in $\mathbf{f}$.

**Example.** The difference between a proper and a classification calibrated loss is illustrated in Fig. 1, which represents the probability simplex for a 3-class problem. Every point in the triangle is a probability vector $\mathbf{p} = (p_0, p_1, p_2)$. If the true posterior class probability for a given input $\mathbf{x}$ is given by point $\boldsymbol{\eta}$, the minimizer of the risk associated with a proper loss should be exactly $\boldsymbol{\eta}$. Classification calibration is less restrictive as the minimizer of the risk associated to a CC loss can be any point in the lighter region around $(0, 0, 1)$. Intuitively, the choice of a good virtual matrix for a given mixing matrix $\mathbf{M}$ is less restrictive for an **M**-CC loss than for an **M**-proper loss. This is formalized below.

### 3.3   Constructing a Weak Loss When the Mixing Matrix is Known

If $\mathbf{M}$ is know, we might be interested in systematic procedures to design an **M**-proper or **M**-CC loss. The main result is summarized in the following theorems.

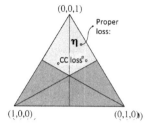

**Fig. 1.** Location of the minimizers of a proper loss (a single point) or a classification calibrated loss (the region around $\eta$, for a true probability $\eta$).

**Theorem 1 ([4]).** *Given a strictly proper loss $\tilde{\Psi}(\mathbf{f}, \mathbf{y})$ and a virtual label matrix $\tilde{\mathbf{Y}}$, the weak loss $\boldsymbol{\Psi}(\mathbf{f}) = \tilde{\mathbf{Y}}^{\mathsf{T}}\tilde{\boldsymbol{\Psi}}(\mathbf{f})$ is strictly $\mathcal{M}$-proper, for any $\mathbf{M} \in \mathcal{M}$ such that $\tilde{\mathbf{Y}}\mathbf{M} = \mathbf{I}$.*

The theorem states that any left inverse of the mixing matrix $\mathbf{M}$ is a valid virtual label matrix. The analogous result for CC losses is the following.

**Theorem 2 ([4]).** *Given a CC loss $\tilde{\Psi}(\mathbf{f}, \mathbf{y})$ and a matrix $\tilde{\mathbf{Y}}$, the weak loss $\boldsymbol{\Psi}(\mathbf{f}) = \tilde{\mathbf{Y}}^{\mathsf{T}}\tilde{\boldsymbol{\Psi}}(\mathbf{f})$ is $\mathbf{M}$-classification calibrated, for any $\mathbf{M}$ such that $\tilde{\mathbf{Y}}\mathbf{M} = \lambda\mathbf{I} + \mathbb{1}_c\mathbf{v}^{\mathsf{T}}$, for some $\lambda \in \mathbb{R}^+$ and $\mathbf{v} \in \mathbb{R}^c$.*

As a consequence, any matrix in the form $\tilde{\mathbf{Y}} = (\lambda\mathbf{I} + \mathbb{1}_c\mathbf{v}^{\mathsf{T}})\tilde{\mathbf{Y}}_0$, where $\tilde{\mathbf{Y}}_0$ is an arbitrary left inverse of the mixing matrix, is a valid virtual label matrix to construct a classification-calibrated loss. Note that, parameters $\lambda$ and $\mathbf{v}$ provide more degrees of freedom to construct a CC loss with respect to a proper loss.

### 3.4  Constructing a Weak Loss When the Mixing Matrix is Unknown

If the mixing matrix is unknown, the classification problem becomes unresolvable, as the mixing matrix can not be inferred from a weakly labelled dataset. For instance, without any additional side information, $\mathbf{M} = \mathbf{I}$ is not distinguishable from any random permutation of its rows or columns. However, Theorems 1 and 2 show that a weak loss $\boldsymbol{\Psi}(\mathbf{f}) = \tilde{\mathbf{Y}}^{\mathsf{T}}\tilde{\boldsymbol{\Psi}}(\mathbf{f})$ can be proper for a large set of different mixing matrices. Under some conditions, loss functions can be constructed to be $\mathbf{M}$-proper or $\mathbf{M}$-cc over pre-specified sets of mixing matrices. In particular, we can derive generic proper losses that are admissible under some general independence assumptions on the mixing process.

**Case 1: Losses for Quasi Independent Labels.** Consider the conditional probability model given by

$$P(\mathbf{z}|\mathbf{y} = \mathbf{e}_m^c) = \begin{bmatrix} z_m\beta_{m,|\mathbf{z}|} & |\mathbf{z}| < c \\ 0 & |\mathbf{z}| = c \text{ or } |\mathbf{z}| = 0 \end{bmatrix} \tag{6}$$

where coefficients $\beta_{m,n}$ satisfy the linear constraint

$$\sum_{n=1}^{c} \binom{c-1}{n-1} \beta_{m,n} = 1 \tag{7}$$

In [4], it is shown that, for some particular values of coefficient $\beta_{m,|z|}$, this model is almost equivalent to the case where the observation of a class in the weak label vector does not depend on all other classes, but only on the true class (the model would be equivalent in the probability of observing all classes would be nonzero). However, it can be shown that the virtual label matrix $\tilde{\mathbf{Y}}$ with virtual label vectors

$$\tilde{y}_j = \begin{bmatrix} 1 & z_j = 1 \\ -\frac{|\mathbf{z}|-1}{c-|\mathbf{z}|} & z_j = 0 \end{bmatrix} \tag{8}$$

(the case $|\mathbf{z}| = c$ is ignored), is admissible for the quasi independent model, no matter what the specific values of parameter $\beta_{m,n}$ are.

**Case 2: Classification Calibrated Losses for Independent Labels.** Another interesting choice for the virtual label matrix is to take $\tilde{\mathbf{y}}_i = \mathbf{b}_i$, i.e. taking the virtual label vectors equal to the weak label vectors. It can be shown that, though a loss based on this matrix is not **M**-proper, it is **M-CC** for the independent label model

$$P(\mathbf{z}|\mathbf{y} = \mathbf{e}_m^c) = \alpha^{z_m}(1-\alpha)^{1-z_m}\beta^{|\mathbf{z}|-1}(1-\beta)^{c-|\mathbf{z}|} \tag{9}$$

## 4    Algorithmic Design

Building upon the results in the previous Section, we now discuss the implementation of specific algorithms for weak labels, analyzing consistency and convexity issues arising in practical settings.

### 4.1    Replacing True Labels with Virtual Labels

The analogy between Eqs. (4) and (5) suggests that we can easily transform any conventional algorithm for clean labels into an algorithm minimizing a weak loss, by simply replacing true label vectors by the virtual label vectors corresponding to the observed weak labels. In practice, this is not always the case. Some specific implementations of algorithms for the minimization of some losses for clean labels do not work when the target vector has not the conventional form of all zeros but a single one.

For instance, assume the logistic regression model, where the scoring function is given by

$$f_i = \frac{\exp(\mathbf{w}_i^\mathsf{T}\mathbf{x})}{\sum_{k=0}^{c-1} \exp(\mathbf{w}_j^\mathsf{T}\mathbf{x})} \tag{10}$$

The cross entropy is a common choice of a proper loss for this model, and is given by $\tilde{\Psi}(\mathbf{f}) = -\log(\mathbf{f})$ (where the log is the natural logarithm and it is computed component-wise). The gradient of the weak loss (4) with respect to the model weights is given by

$$g(\mathbf{w}_0, \ldots, \mathbf{w}_{c-1}) = \mathbf{x}^\mathsf{T}(b_i \cdot \mathbf{f} - \tilde{\mathbf{y}}_i) \qquad (11)$$

where the coefficient $b_i = \mathbb{1}^\mathsf{T}\tilde{\mathbf{y}}_i$ is equal to one in a clean label case, so it is ignored in usual implementations of gradient based learning rules for the cross entropy. It is, however, required for learning from weak losses.

## 4.2  Consistency

When the loss function $\tilde{\Psi}$ is not upper bounded, the implementation of an $\mathbf{M}$ proper loss may present consistency issues. For instance, for a logistic model with a cross entropy loss function, we have

$$\tilde{\Psi}(\mathbf{f}, \tilde{\mathbf{y}}_i) = \tilde{\mathbf{y}}_i^\mathsf{T}\mathbf{W}\mathbf{x} - (\mathbb{1}^\mathsf{T}\tilde{\mathbf{y}}_i)\log\left(\sum_{j=0}^{c-1}\exp(\mathbf{w}_j^\mathsf{T}\mathbf{x})\right) \qquad (12)$$

If the virtual labels contain negative components, it is not difficult to show that, for some values of $\mathbf{x}$, the loss is not bounded below and, for some datasets, the minimum empirical risk is $-\infty$. Thus, the minimizer of the empirical risk may not converge to the true posterior class probabilities.

This problem can be resolved by taking into account that $|\tilde{\Psi}(\mathbf{f}, \tilde{\mathbf{y}}_i)| \geq -\lambda\|\mathbf{W}\| - \max_i\|\tilde{\mathbf{y}}_i\|\log(C)$, where $\lambda = 2\max_i\|\tilde{\mathbf{y}}_i\|\max_{\mathbf{x}\in\mathcal{S}}\|\mathbf{x}\|$. Thus, the regularized loss $\tilde{\Psi}_\lambda(\mathbf{f}, \tilde{\mathbf{y}}_i) = \tilde{\Psi}_\lambda(\mathbf{f}, \tilde{\mathbf{y}}_i) + \lambda\|\mathbf{W}\|$ is bounded below. In order to avoid these inconsistency issues, we have used the Brier score (square loss) $\tilde{\Psi}_\lambda(\mathbf{f}, \tilde{\mathbf{y}}_i) = \|\mathbf{y}_i - \mathbf{f}\|^2$, which is known to be proper, in our experiments.

## 4.3  Convexity

It can be show that, if the loss $\tilde{\Psi}(\mathbf{f}, \tilde{\mathbf{y}}_i)$ is $\mathbf{M}$-proper, its conditional expectation is a proper loss, which is necessarily a convex function of the true posterior, $\boldsymbol{\eta}$. This is illustrated in Fig. 2, which shows the contour plots of the expected loss for the Brier score (left) and Cross Entropy (center) for a given mixing matrix and a true posterior $\boldsymbol{\eta} = (0.35, 0.2, 0.45)$ on the 3 class probability simplex. The right plot shows the conditional expected value of the Optimistic Superset Loss [9], which consists on replacing the weak label by a tentative single-class label, selected as the class with the highest score among the candidate labels. Note that this loss is not convex and may have several local minima.

Note that, despite the conditional loss being a convex function of the score, the weak loss might be a non convex function of the model parameters, depending on the type of virtual label matrix. Consider, for instance, a parametric score function $\mathbf{f}_\mathbf{w}(\mathbf{x})$ and assume that the conventional loss $\tilde{\Psi}(\mathbf{f}_\mathbf{w}(\mathbf{x}), \mathbf{y})$ is convex on

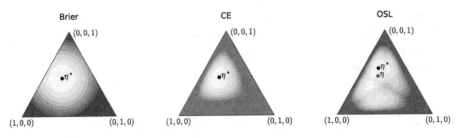

**Fig. 2.** Conditional expected loss over the 3 class probability simplex for a true posterior $\eta = (0.35, 0.2, 0.45)$, for three different losses: Brier score (left), Cross Entropy (center) and Optimistic Superset Loss (right). $\eta$ is the true posterior, and $\eta^*$ is the minimizer. For the Brier score and the CE score, as expected, $\eta$ and $\eta^*$ coincide.

the model parameters $\mathbf{w}$, for any target vector $\mathbf{y} \in \mathcal{Y}$ (as, for instance, in logistic regression). If the virtual label contains negative values, then the weak loss $\tilde{\mathbf{Y}}\mathbf{\Psi}$ is in general not convex. This is a major difficulty in order to preserve convexity and properness as, if $\tilde{\mathbf{Y}}$ is a left inverse of a nonnegative probability matrix, it usually contains negative components. Fortunately, we can always construct a CC-loss with the appropriate selection of the virtual label matrix.

**Theorem 3.** *If $\tilde{\psi}(\mathbf{f}_\mathbf{w}(\mathbf{x}), \mathbf{y})$ is a convex function of $\mathbf{w}$ for any $\mathbf{y}$ and $\tilde{\mathbf{Y}}\tilde{\mathbf{\Psi}}(\mathbf{f})$ is $\mathbf{M}$-proper, then the weak loss given by $\mathbf{\Psi}(\mathbf{f}) = \tilde{\mathbf{Y}}'\tilde{\mathbf{\Psi}}(\mathbf{f})$ is $\mathbf{M}$-CC and convex in $\mathbf{w}$, where*

$$\tilde{\mathbf{Y}}' = \tilde{\mathbf{Y}} - \mathbf{m}\mathbb{1}_d^\intercal \tag{13}$$

*where $\mathbf{m}$ is the row-wise minimum of matrix $\mathbf{M}$*

The proof is a direct consequence of: (1) $\tilde{\mathbf{Y}}'$ satisfies the conditions in Theorem 3, (so the weak loss is CC), and (2) $\tilde{\mathbf{Y}}'$ has nonnegative components, so the weak loss is a conic combination of convex functions, so it is convex.

### 4.4   Selection of the Virtual Matrix

Note that, in general (if the rank of $\mathbf{M}$ is higher than $c$) the set of admissible virtual label matrices for a given mixing matrix is infinite. However, this does not mean that the choice of $\tilde{\mathbf{Y}}$ in this set is irrelevant. Different virtual label matrices show different performances. Though we have no theoretical evidence, we have experimentally found that a good choice for $\mathbf{Y}$ is the Moore-Penrose pseudoinverse $\tilde{\mathbf{Y}} = (\mathbf{M}^T\mathbf{M})^{-1}\mathbf{M}^\intercal$ (we assume $\mathbf{M}$ is not rank-deficient, otherwise the classification problem would be severely degenerated by the mixing process).

## 5   Experiments

In this section, we demonstrate the use of our framework on real-world data. We show how the proposed method can effectively transform existing classification algorithms so that they incorporate the available information regarding the label quality, thereby providing robust solutions to weakly supervised data that is not addressed when using out-of-the-box machine learning techniques.

**Datasets.** We tested the models on 31 real-world datasets from `openml.org`. From the available datasets, we chose those with number of classes between 3 and 20, less than 11,000 instances and no missing values. Then, we sorted all the filtered datasets by highest impact and manually selected the final subset. Before training, all the categorical features were transformed into binary features using a one-hot encoding. Finally, every feature was standardised with mean zero and standard deviation one. Table 1 contains a summary of the datasets.

**Table 1.** Summary of the 31 datasets used in the experiments

| Name | Size | Features | Classes | Name | Size | Features | Classes |
|---|---|---|---|---|---|---|---|
| GesturePhaseSegme | 9873 | 32 | 5 | flags | 194 | 120 | 8 |
| JapaneseVowels | 9961 | 14 | 9 | glass | 214 | 9 | 7 |
| abalone | 4177 | 10 | 3 | iris | 150 | 4 | 3 |
| analcatdata_dmft | 797 | 21 | 6 | mfeat-zernike | 2000 | 47 | 10 |
| autoUniv-au6-1000 | 1000 | 44 | 8 | page-blocks | 5473 | 10 | 5 |
| autoUniv-au6-750 | 750 | 44 | 8 | pendigits | 10992 | 16 | 10 |
| autoUniv-au7-1100 | 1100 | 18 | 5 | prnn_fglass | 214 | 9 | 6 |
| autoUniv-au7-500 | 500 | 18 | 5 | satimage | 6430 | 36 | 6 |
| balance-scale | 625 | 4 | 3 | segment | 2310 | 19 | 7 |
| car | 1728 | 21 | 4 | vehicle | 846 | 18 | 4 |
| cardiotocography | 2126 | 35 | 10 | visualizing_liv | 130 | 27 | 5 |
| collins | 500 | 35 | 15 | vowel | 990 | 27 | 11 |
| confidence | 72 | 3 | 6 | wine | 178 | 13 | 3 |
| diggle_table_a2 | 310 | 8 | 9 | yeast | 1484 | 8 | 10 |
| ecoli | 336 | 7 | 8 | zoo | 101 | 31 | 7 |
| fl2000 | 67 | 16 | 5 | | | | |

As all datasets have only one true label per sample, we needed to artificially weaken the labels. We simulated five common scenarios by generating different random sets of mixing matrices $\mathbf{M}$:

1. **Noisy:** Each sample has only one label. The weak label is the true label with probability $\alpha$. The rest of the classes are equally probable. This represents scenarios where, in order to reduce annotation costs, the labels are bound to contain mistakes.
2. **Random_noise:** Each sample has only one label. The weak label is the true label with at least probability $\alpha$. The rest of the classes have probability $\beta_m$, initially drawn from a uniform distribution. The value of $\alpha$ defines the degree of supervision, from fully supervised to a completely random mixing matrix $\mathbf{M}$.

3. **IPL:** Each sample can have multiple labels. In this case the true label has a probability $\alpha$ of appearing, while other labels are present with a certain probability $\beta$. This scenario occurs in complex classification tasks where multiple annotators label the samples but there is not known ground truth.
4. **Quasi_IPL:** Each sample can have multiple labels. In this case the true label is always present but other labels may appear with certain probability $\beta$. This scenario could be applied to the identification of an animal subspecies given that we know its taxonomic parents.
5. **Random_weak:** Each sample can have multiple labels. In this case all the possible weak labels may appear with a uniform probability. However, the correct label is present with a probability of at least $\alpha$.

The amount of noise in the aforementioned scenarios increases with the value of $\alpha$ and decreases with the value of $\beta$. In all the tested scenarios we constrained both parameters to sum to one.

**Models.** In all cases, we compare the performance of our proposed methods with three baselines. First, we include the results of using the set of clean labels without added noise (`Superv`). This method gives us a lower bound on the error rate. Second, we show the expected error if the weak labels are used without any consideration (`Weak`). Third, we compare our approaches with optimistic superset loss (OLS) [9], as this is a popular technique to deal with weak labels (`OSL`).

As for our models, we explore three different scenarios. If we know the mixing matrix $\mathbf{M}$, we obtain the new virtual labels following the approach suggested in Sect. 3.3 (`Mproper`). If we do not know $\mathbf{M}$, and we assume a quasi Independent mixing matrix $\mathbf{M}$ we use the method suggested in Sect. 3.4 (`qIPL`). Finally, the alternative assumption of an Independent mixing matrix suggested in Sect. 3.4 is equivalent to the baseline `Weak`.

**Implementation.** In order to evaluate the models we used the framework Keras that adds an abstraction layer on top of TensorFlow and Theano. Keras allows an easy specification of the loss function and automatic differentiation. In our case we used a Brier loss that is suitable for the virtual labels. All the models were trained with full batch gradient descent with a fixed learning rate of 1.0 for 40 epochs and with 10 times 10-fold cross-validation. For each dataset and mixing matrix $\mathbf{M}$ we trained a Logistic Regression (LR), and a Feed-forward Neural Network (FNN) with two layers of 200 rectified linear units. Although all the comparisons on this section are focused on LR, the results of the FNN are shown in the second row of Fig. 3 to illustrate the applicability of our method with richer hypothesis classes. All implementations are publicly available[1].

---

[1] https://github.com/Orieus/WeakLabelModel/.

**Results.** Figure 3 shows the error rate of every model and mixing matrix **M** averaged over the 31 datasets, increasing the noise level from left to right; LR on top and FNN at the bottom. Although the mean error rate over several datasets is not fully informative, we performed a one tailed Wilcoxon rank sum test for each pair of models and every type of mixing matrix. The null hypothesis is that the models follow the same distribution. All the conclusions are extracted from the statistical tests while the figures are only here to provide a visual intuition. Fig. 3a presents the case where the weak labels are similar to the true labels. Here the only significant comparison (p-value of 0.02) confirms that Superv outperforms OSL. In Fig. 3b for both the Independent Partial Labels (IPL) and quasi Independent Partial Labels (qIPL) mixing matrices, all methods surpass OSL (p-values of $10^{-5}$ and $10^{-4}$, respectively). In Fig. 3c Weak and qIPL are still not statistically different but all the rest are (p-values under $10^{-3}$). The IPL mixing matrix with this level of noise is a degenerate case, as the model that uses the known **M** can not retrieve the original labels. Figure 3d shows how for higher noise levels, the differences between the models are highlighted and are easier to rank (p-values under $10^{-3}$ except for pairwise comparison between Weak and qIPL models which has a p-value of 0.31). Similar conclusions can be extracted from the second row of Fig. 3 for more complex FNNs.

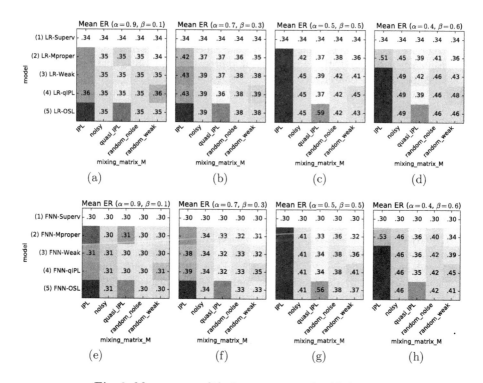

**Fig. 3.** Mean cross-validation error rates for 31 datasets.

As expected, if the random process that generated the noise is known, the best approach is to use the `Mproper` method. For that reason, when possible, it is advisable to try to estimate the mixing matrix from a clean set with true labels. When the mixing process is unknown, the results show that assuming a quasi independent or independent mixing process will help in most of the cases, achieving the best results when the prior assumption is true. With respect to the poor performance of the `OSL` on the proposed scenarios, we hypothesise that this method makes important assumptions about particular correlations between classes with respect to the input space.

# 6    Conclusion

In this paper we built upon previous theoretical results to describe a simple approach to adapt existing classifiers that are based on the empirical minimization of proper or classification calibrated losses, in such a way that they explicitly incorporate the available information regarding the label quality. Furthermore, we show that the constructed weak loss achieves similar and in some cases surpasses the performance of a state-of-the-art method on a variety of scenarios.

# References

1. Ambroise, C., Denoeux, T., Govaert, G., Smets, P.: Learning from an imprecise teacher: probabilistic and evidential approaches. Appl. Stochast. Models Data Anal. **1**, 100–105 (2001)
2. Angluin, D., Laird, P.: Learning from noisy examples. Mach. Learn. **2**(4), 343–370 (1988)
3. Cid-Sueiro, J.: Proper losses for learning from partial labels. In: Advances in Neural Information Processing Systems, vol. 25, pp. 1574–1582 (2012)
4. Cid-Sueiro, J., García-García, D., Santos-Rodríguez, R.: Consistency of losses for learning from weak labels. In: Calders, T., Esposito, F., Hüllermeier, E., Meo, R. (eds.) ECML PKDD 2014. LNCS, vol. 8724, pp. 197–210. Springer, Heidelberg (2014). doi:10.1007/978-3-662-44848-9_13
5. Cour, T., Sapp, B., Taskar, B.: Learning from partial labels. J. Mach. Learn. Res. **12**, 1225–1261 (2011)
6. Grandvalet, Y.: Logistic regression for partial labels. In: 9th Information Processing and Management of Uncertainty in Knowledge-based System, pp. 1935–1941 (2002)
7. Grandvalet, Y., Bengio, Y.: Learning from partial labels with minimum entropy. Centre Universitaire de recherche en analyse des organizations (2004)
8. Hüllermeier, E., Beringer, J.: Learning from ambiguously labeled examples. Intell. Data Anal. **10**(5), 419–439 (2006)
9. Hüllermeier, E., Cheng, W.: Superset learning based on generalized loss minimization. In: European Conference on Machine Learning and Knowledge Discovery in Databases, pp. 260–275 (2015)
10. Jin, R., Ghahramani, Z.: Learning with multiple labels. In: Advances in Neural Information Processing Systems, vol. 15, pp. 897–904 (2002)
11. Nguyen, N., Caruana, R.: Classification with partial labels. In: SIGKDD International Conference on Knowledge Discovery and Data Mining, pp. 551–559 (2008)

12. Ni, Y., McVicar, M., Santos-Rodriguez, R., De Bie, T.: Understanding effects of subjectivity in measuring chord estimation accuracy. IEEE Trans. Audio Speech Lang. Process. **21**(12), 2607–2615 (2013)
13. Raykar, V.C., Yu, S., Zhao, L.H., Valadez, G.H., Florin, C., Bogoni, L., Moy, L.: Learning from crowds. J. Mach. Learn. Res. **99**, 1297–1322 (2010)
14. van Rooyen, B., Menon, A.K., Williamson, R.C.: Learning with symmetric label noise: the importance of being unhinged. In: Advances in Neural Information Processing Systems, pp. 10–18 (2015)

# The Combination of Decision in Crowds When the Number of Reliable Annotator Is Scarce

Agus Budi Raharjo[(✉)] and Mohamed Quafafou

Aix-Marseille University, CNRS, LSIS UMR 7296, 13397 Marseille, France
agus-budi.raharjo@etu.univ-amu.fr, mohamed.quafafou@univ-amu.fr

**Abstract.** Crowdsourcing appears as one of cheap and fast solutions of distributed labor networks. Since the workers have various expertise levels, several approaches to measure annotators reliability have been addressed. There is a condition when annotators who give random answer are abundance and few number of expert is available Therefore, we proposed an iterative algorithm in crowds problem when it is hard to find expert annotators by selecting expert annotator based on EM-Bayesian algorithm, Entropy Measure, and Condorcet Jury's Theorem. Experimental results using eight datasets show the best performance of our proposed algorithm compared to previous approaches.

**Keywords:** Annotator reliability · EM algorithm · Crowdsourcing

## 1 Introduction

Within the rise of the sharing era, crowdsourcing appears as a task that is solved by a group of annotators online, such as InnoCentive[1], Amazon Mechanical Turk[2], etc. [1]. It is considered as a cheap and fast solution for distributed labor networks [2]. In another way, since the workers have various expertise level, it is difficult to measure their reliability. From this context, a lot of literatures addressed the problem of learning from different annotators. For example in [3], authors categorize worker reliability estimation techniques as follows: (1) before a rater can rate any instance (before the start), (2) during the process (dynamic reliability), (3) after the process is finished (static reliability). As an example of the first category, authors in [4] conducted a small test before the real annotation process is started. Although, it is difficult to select the appropriate test data that represent the real problem.

For dynamic category, it was used in [3,5–8] by considering worker availability and instance allocation during the crowdsourcing process. By applying this, the number of worker involved will be minimized and it leads to a lower cost. But, the quality of estimation result will be limited by the user involved [9]. In this case, static estimation is important to be applied in order to get better

---

[1] https://www.innocentive.com.
[2] https://www.mturk.com.

© Springer International Publishing AG 2017
N. Adams et al. (Eds.): IDA 2017, LNCS 10584, pp. 260–271, 2017.
DOI: 10.1007/978-3-319-68765-0_22

results. Static worker estimation tends to eliminate spammer decision by using several approaches. For example in [10], authors propose a spammer score to rank annotator reliability. In addition, authors in [11] extended crowds problem in supervised learning into multidimensional domain. Another study focused on how to handle biased annotator by considering statistical difference between the labeling qualities of the two classes [12]. It is important to mention that there is no fully reliable annotator in certain situations [7]. Hence, a condition in which a task has few or no reliable worker has to be considered.

In this paper we propose a new algorithm that selects expert voter based on estimating sensitivity and specificity of each worker using EM algorithm [13]. Also, we use the Condorcet Jury's Theorem (CJT) [14] to handle the situation where the number of reliable worker is rare or even does not exist. We applied our method in eight UCI datasets [15] with several experimental conditions based on different expert availability and evaluated the result according to Area Under the Curve (AUC) where our method shows a better results compared to previous mentioned works. The rest of the paper is organized as follows. Section 2 discuss related work to our approach. The formulation of our algorithm detailed in Sect. 3. The experimental results are presented in Sect. 4. Finally, we conclude and present some future work in Sect. 5.

## 2 Related Work

The EM algorithm has been widely used in literatures, due to its high performance in estimating latent variables [16–18]. In this section, we provide a brief description about the previous approaches use EM based method in order to verify annotators prediction and use it in annotator selection, also we discuss the studies that used the CJT theorem to decide the number of reliable worker.

### 2.1 EM-Bayesian Approaches

Authors in [18] applied the Bayesian algorithm within the EM algorithm in their method for binary problem, by estimating the probability of sensitivity and specificity for each annotator. Where sensitivity is defined as the probability of predicting a positive value. On the contrary, specificity is the probability that annotator predicts a negative value correctly. Also in their study, EM was used as iterative procedure to compute the maximum-likelihood and calculate the probability parameters. In [19], authors extended the previous work by considering the instance vector to validate annotator prediction .

In [10], authors improve their work and propose a spammer score measure (SpEM) which been used in ranking the annotators. Although the SpEM algorithm provides a good performance, it takes a lot of computation time since it calculates all annotators for each EM iteration. Therefore, [20] proposed ExpertS algorithm which is a discrete version of SpEM that use an Entropy measure [21] to divide a set of annotators in order to computes only the selected annotators for each EM iteration. Entropy will calculate the similarity between class distribution of annotator prediction and the reference distribution. However, ExpertS requires the total number of class distribution in true label. So The application of this algorithm is difficult to find in real world case, due to the absent of true labels value. For this reason our work enhances the earlier discussed approaches and propose a new method to define an annotator as a reference distribution instead of true label information.

## 2.2    Condorcet's Jury Theorem (CJT)

CJT is a public decision theorem used in the case where there is two options of solution [14]. This theorem can be described by the following example, suppose we need to make a decision from two available options, and we have a group of independent voter where each one has probability knowledge value $p$. the theorem will only consider the voter who give correct answer with high knowledge probability ($p_{correct} > 0.5$). if there is no voter that satisfy $p_{correct} > 0.5$, then CJT selects a voter with the highest $p_{correct}$.

Recently, this theorem was used in several disciplines, such as in social choice and welfare [22], medical imaging [23], artificial intelligence [24, 25]. On the contrary, some limitations in CJT have been addressed. For example, authors in [26] mentioned that this theorem is only suitable for binary case. While in [27], authors concluded that the number of selected annotators must be odd for not reaching equal probability option. We have been influenced by this theorem, and we apply it to decide the number of annotators that should be involved in decision making.

# 3    The Combination of Reliable Annotator (CRA)

In this section, first we describe the basic notions used in binary decision problem which is our interest in this study, then we give a detailed explanation of the proposed algorithm, the Combination of Reliable Annotator (CRA). As shown in Fig. 1, our approach can be summarized into four steps: (1) selecting reference annotator by using EM-Bayesian algorithm, (2) ranking and dividing annotators in descending order by considering entropy measure, (3) calculating spammer score by considering EM and CJT for each group, (4) selecting consistent reliable candidate. Selecting reference annotator is important in our model to estimate the potential of annotators and explore the maximal entropy score. Where this reference will be used later on to calculate the similarity of class distribution.

## 3.1    Problem Formulation and Modeling

Crowd sourcing scenario consist of a set of $N$ independent instances which contains $Z$ unknown true labels. in which instances are rated by a set of $T$ annotators. $T$ annotator prediction for $N$ instances can be noted as follow:

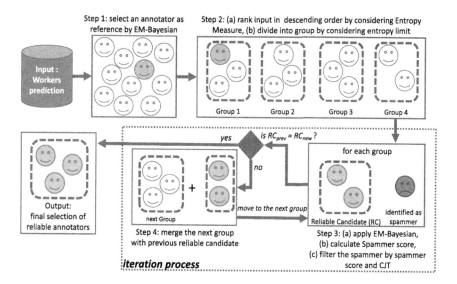

**Fig. 1.** The overview of proposed approach.

$D = \{y_i^1, y_i^2, ..., y_i^t, ..., y_i^T\}_{i=1}^N$ such that $y_i^t \in \{0, 1\}$ and $y_i^t$ is independent annotator prediction for each instance. In this case, Log likelihood model in [18] is applied to find the probability of annotators giving the correct answer.For the reason that there is a missing parameters in this mentioned model, EM algorithm has been applied to optimize the global parameter. These missing parameters can be summarized as follow : $\theta = \{\alpha^1, \beta^1, ..., \alpha^t, \beta^t, ..., \alpha^T, \beta^T, p\}$. $\alpha_i$ which represent sensitivity for $i^{th}$ instance can be expressed by as $P(y_i = 1|z_i = 1)$ where $z_i$ is a true label of instance $i$. In another way, specificity is expressed by $\beta_i := P(y_i = 0|z_i = 0)$. And $p = P(z = 1)$ is the probability of positive class.

By applying $\theta$ in $D$ the probability can be represented as follow: $P(D|\theta) = \prod_{i=1}^N P(y_i^1, ..., y_i^T|\theta)$. From which, The log likelihood model can be formulated by Eq. 1.

$$\log P(D|\theta) = \sum_{i=1}^N \log \sum_{y_i=0}^1 \prod_{t=1}^T P(y_i^t|z_i, \theta).P(z_i|\theta) \tag{1}$$

By giving $a_i = \prod_{t=1}^T P(y_i^t|z_i = 1, \alpha^t) = \prod_{t=1}^T (\alpha^t)^{y_i^t}(1 - \alpha^t)^{1-y_i^t})$ and $b_i = \prod_{t=1}^T P(y_i^t|z_i = 0, \beta^t) = \prod_{t=1}^T (\beta^t)^{y_i^t}(1 - \beta^t)^{1-y_i^t})$, Eq. 1 can be written as follow:

$$\log P(D|\theta) = \sum_{i=1}^N \log(a_i p + b_i(1 - p)) \tag{2}$$

EM algorithm is used in Eq. 2 in order to estimate maximum-a-posteriori parameters, which consist of E-step and M-step. Where in E-step, The probability of true positive label of $i^{th}$ instance is estimated by $\mu_i = P(z_i = 1|y_i^1, ..., y_i^T, \theta)$. By Considering Bayes theorem for $P(z|y)$, $\mu_i$ can be expressed as in Eq. 3.

$$\mu_i \propto P(y_i^1, ..., y_i^T | z_i = 1, \theta) P(z_i = 1 | \theta) = \frac{a_i p}{a_i p + b_i (1 - p)} \tag{3}$$

By applying Eqs. 2 and 3, the conditional expectation for $D$ and $\theta$ is computed as below:

$$\mathbb{E}\{\log P(D, z | \theta)\} = \sum_{i=1}^{N} \mu_i \log(a_i p) + (1 - \mu_i) \log(b_i (1 - p)) \tag{4}$$

While in M-step, the $\theta$ parameters in formula 4 are maximized as follow:

$$\alpha^t = \frac{\sum_{i=1}^{N} \mu_i y_i^t}{\sum_{i=1}^{N} \mu_i} \qquad \beta^t = \frac{\sum_{i=1}^{N} (1 - \mu_i)(1 - y_i^t)}{\sum_{i=1}^{N} (1 - \mu_i)} \qquad p = \frac{\sum_{i=1}^{N} \mu_i}{N} \tag{5}$$

In general, the same process is repeated until convergence. While in initial state, $\mu_i$ value is set to $1/T \sum_{t=1}^{T} y_i^t$, $\alpha^t$ and $\beta^t$ are also set to zero. Once the optimal $\alpha$ and $\beta$ are obtained, spammer score is applied as in [10]. All the previous discussed method have been used in our framework in step 1. The maximal spammer score is considered as an entropy reference distribution which will be used further in step 2. The Eqs. 1–5 are also implemented in step 3 to obtain the reliable annotator(see Sect. 3.3).

## 3.2   Annotator Discretization

Entropy score is used to measure class distribution similarity according to the maximal entropy $w$ [21]. In the work of [20], authors select true label distribution as the maximal entropy. While in real world problem, this distribution is difficult to get. Therefore, an annotator is selected in our framework as a maximal entropy. Then, we measure annotator entropy score compared to maximal entropy distribution (see Fig. 1 step 2) and rank it in descending order. Suppose that $H^t$ is an entropy of each annotator, $C$ is the number of classes, $f$ is the frequency of each class distribution, and $w_i$ is the maximal entropy of class $i$. Thus, entropy score of each annotator can be computed as follow:

$$H^t(N, f_1, ..., f_c, ..., f_C) = \sum_{i=1}^{C} \frac{\lambda_i (1 - \lambda_i)}{(-2w_i + 1)\lambda_i + w_i^2} \tag{6}$$

where $\lambda_i = (N f_i + 1)/(N + C)$ and $f_i = |y^t = c|/N$.

$g_1$ is the first group of annotator that has higher $H$ value compared to the limit score $\epsilon_H$ (see Eq. 7).

$$g_1 = \{t \in T | H^t > \epsilon_H, \epsilon_H \in [0 : +\infty]\} \tag{7}$$

Based on $|g_1|$, the next annotators is divided into smaller group. In this paper, we define the value of $\epsilon_H$ by multiply spammer threshold $\epsilon_S$ with the maximal entropy distribution.

### 3.3    The Selection of Reliable Annotator

In crowds problem, workers can be either reliable, spammer, biased, or malicious annotators. Reliable worker is an annotator who gives a prediction that is near to true label. On the contrary, malicious worker gives a false answer, while spammer provides a random prediction that the sum of its sensitivity and specificity is near to one. Biased is described by an annotator that has a large sensitivity and specificity value difference. By using the model proposed in [18,28], Malicious and biased annotators can be ignored due to the ability of EM algorithm to reverse their annotations. Authors in [10] define a spammer score to measure the spam degree of annotator based on $\alpha$ and $\beta$ estimation. For example, the spammer score of $t^{th}$ annotator is expressed as follow:

$$S^t = (\alpha^t + \beta^t - 1)^2 \tag{8}$$

Based on Eq. 8, the annotator who has $S^t$ value close to 0 is considered as a spammer, While reliable annotator has a $S^t$ value close to 1. In [20] work, reliable annotator is computed as follow:

$$RC = \{t \in A | S^t > \epsilon_S\} \tag{9}$$

where $RC$ is a reliable candidate who satisfy spammer threshold $\epsilon_S$ and $\epsilon_S \in [0, 1]$. In Fig. 1 step 3, the final reliable candidate $RC_f$ is calculated as follow:

$$RC_f = RC - \{t \in RC | S^t \leq \epsilon_S\} \tag{10}$$

It is important to mention here that $RC$ is hard to find in certain situations. Our contribution is shown in Fig. 1 step 3.c where we filter reliable annotator by considering CJT. In the case when there is no annotator has higher $S^t$ value than $\epsilon_S$, we select only the one with the highest value of $S^t$. By considering Eq. 9, in our proposed method, $RC$ selection can be formulated as follow:

$$RC = \begin{cases} S^t > \epsilon_S & \text{if } |S^t > \epsilon_S| > 0 \\ \max(S^t) & \text{otherwise} \end{cases} \tag{11}$$

Our approach can be explained by Algorithm 1. Line 1 describes the input value $D$ as the combination of $T$ annotator predictions for $N$ instances. In the same line, $\alpha, \beta, \mu$, and threshold $\epsilon_S$ are declared. The estimation of maximal entropy is calculated in lines 2 and 3. Then it will be used in lines 4 and 5 for the computation of entropy measure, also for the division of $T$ annotators into several group. After defining the first group, we extract the reliable candidate $RC_{new}$ in line 7. Then, the final reliable candidate $RC_f$ is extracted from the process in line 8–12. Finally, majority voting of $RC_f$ is applied in line 13 to produce a set of recommended decision.

## 4    Experiments

A series of experiments are performed on eight different datasets to evaluate our algorithm. In the following section, we discuss the dataset used, the protocol, and then the results.

---

**Algorithm 1.** The Combination of Reliable Annotator

---

1: Set D as input, $\alpha^t = 0, \beta^t = 0, \epsilon_S, k = 1, \mu_i = 1/T \sum_{t=1}^{T} y_i^t, RC_{prev} = \emptyset$.
2: Apply EM-Bayesian algorithm for $T$ annotators by considering equations 3 and 4 in e-step and equation 5 in m-step.
3: Calculate $S^t$ with the help of equation 8 and define $w = \max(S^t)$.
4: $\epsilon_H = \epsilon_S * H^w$
5: Calculate $H^t$ according to equation 6 and select $g_k$ by applying equation 7.
6: Apply EM-Bayesian algorithm for $g_k$.
7: Select $RC_{new}$ by the help of equations 9, 10, and 11.
8: **while** $RC_{prev} \neq RC_{new}$ and $g_{k+1} \neq \emptyset$ **do**
9:    Set $RC_{prev} = RC_{new}$ and $k = k + 1$.
10:    Move to a new $g_k$ and apply EM-Bayesian algorithm.
11:    Select $RC_{new}$ by the help of equations 9,10, and 11.
12: **end while**
13: Calculate the final decision with the majority voting of $RC_f$.

---

### 4.1 Data Description and Protocol

Eight datasets from UCI repository [15] are used as follow: SPECT Heart, Haberman's Survival (Haberman), Vertebral Column (Vertebral), Liver Disorders (Bupa), Ionosphere, Blood Transfusion Service Center (Blood) [29], Pima Indians Diabetes (Diabetes), and Spambase dataset. Spectheart is the smallest data with 267 instances and 22 attributes, while Spambase has the largest size of data with 4601 instances and 57 attributes. The ratios of negative and positive labels are between 32%:68% (Vertebral) to 76%:24% (Blood).

All instances were applied as unlabeled data and were predicted by 100 simulated annotators. The experiments were conducted by changing the number of reliable annotator from zero to five experts. It means that if there are $k$ experts, the number of spammers in this situation will be $100 - k$ annotators. Based on earlier experiment [10], reliable annotator is defined as an annotator that has sensitivity and specificity between 0.65 and 0.85. In another experiment [20], expert is simulated with a sensitivity and specificity values between 0.75 and 0.95. Based to our knowledge, there is no specific range used to define the expertise of annotator. Our inter-

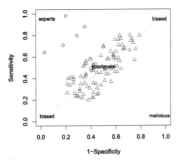

**Fig. 2.** Generated annotators for Diabetes dataset according to their sensitivity and (1-specificity).

est is to study the case where there is no fully reliable annotator. For this purpose, we set the range value of sensitivity and specificity between 0.6 to 1 which will cover more heterogeneous state. On the contrary, the spammer is simulated as annotator that satisfies $S^t \leq 0.04$ and avoids biased condition. The example of 5 experts and 95 spammers representation for Diabetes dataset is shown in Fig. 2. We evaluated our proposed algorithm by calculating the Area Under

Curve (AUC) value and the selected annotator within its computation time. CRA is compared with ExpertS [20], SpEM [10], and majority voting (MV). All the algorithms were implemented in R software[3] and were performed on an Intel core i7 2.50 GHz with 16 GB of memory. CRA implementation is available on GitHub[4].

## 4.2   Experimental Results

Figure 3 shows the AUC score comparison for four algorithms on six annotator availability situations. X-axis represents the number of reliable annotator, while y-axis represents AUC score with $AUC \in [0,1]$ and $AUC \in \mathbb{R}$, where the best algorithm has the highest value of AUC. We can notice from the results that: first, there is an increase in AUC score whenever the number of reliable annotators is higher. Second, CRA shows a high AUC score in Haberman, Blood, Diabetes, and Spambase where there is only one or two reliable annotators considered as expert, which means that CRA is able to detect the annotator that has the high AUC score estimation. Finally, we can observe that the best AUC average is given by CRA. In another way, SpEM has the second rank in the six datasets, while ExpertS has a better result compared to SpEM in Haberman and Ionosphere. Moreover, MV provides average scores between 0.5 to 0.7 in all the used dataset, which means that the result does not change significantly because the percentage of reliable annotator is small.

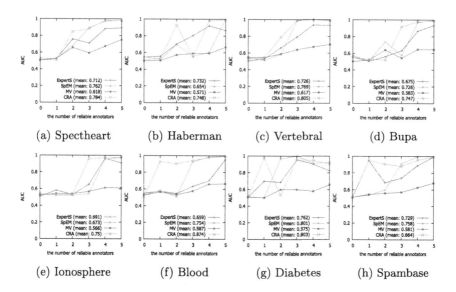

**Fig. 3.** Comparison of AUC score for eight datasets (higher value is better).

---

[3] www.r-project.org.
[4] https://github.com/agusbudi/CRA.

Table 1 shows the comparison between the ratio of expert annotator and spammers within the number of real expert. It is figured out that CRA chooses minimal number of annotators among the others, due to the implementation of CJT. This behavior leads CRA to minimize getting an answer from a spammer, while ExpertS and SpEM tend to select a lot of annotators. Although, ExpertS selects smaller number of annotators than SpEM in general. It prevents CRA to select at least one correct expert in certain case of all datasets, except in Table 1. For the whole data, we can conclude the relation that whenever the number of real expert increase there is a decrease in spammer selection.

**Table 1.** The number of real expert, selected expert, and spam annotators comparison.

(a) Spectheart

| real expert | selected expert:spammer | | |
|---|---|---|---|
| | ExpertS | SpEM | CRA |
| 1 | 0:17 | 1:55 | 0:9 |
| 2 | 2:22 | 2:55 | 2:1 |
| 3 | 3:13 | 3:49 | 3:2 |
| 4 | 4:10 | 4:36 | 4:0 |
| 5 | 5:13 | 5:34 | 5:1 |

(b) Haberman

| real expert | selected expert:spammer | | |
|---|---|---|---|
| | ExpertS | SpEM | CRA |
| 1 | 0:26 | 1:58 | 0:4 |
| 2 | 2:18 | 2:50 | 2:1 |
| 3 | 3:11 | 1:57 | 0:5 |
| 4 | 2:3 | 3:62 | 4:0 |
| 5 | 5:11 | 5:35 | 5:0 |

(c) Vertebral

| real expert | selected expert:spammer | | |
|---|---|---|---|
| | ExpertS | SpEM | CRA |
| 1 | 1:23 | 0:61 | 0:4 |
| 2 | 0:30 | 2:64 | 2:3 |
| 3 | 3:12 | 3:29 | 2:0 |
| 4 | 4:9 | 4:22 | 4:0 |
| 5 | 5:8 | 5:28 | 5:0 |

(d) Bupa

| real expert | selected expert:spammer | | |
|---|---|---|---|
| | ExpertS | SpEM | CRA |
| 1 | 0:24 | 1:56 | 0:6 |
| 2 | 0:30 | 2:46 | 0:3 |
| 3 | 1:25 | 2:59 | 2:1 |
| 4 | 4:7 | 4:27 | 4:0 |
| 5 | 5:4 | 5:36 | 5:0 |

(e) Ionosphere

| real expert | selected expert:spammer | | |
|---|---|---|---|
| | ExpertS | SpEM | CRA |
| 1 | 1:42 | 0:45 | 0:3 |
| 2 | 1:55 | 0:32 | 0:1 |
| 3 | 3:31 | 1:24 | 3:0 |
| 4 | 4:2 | 4:10 | 4:0 |
| 5 | 5:7 | 5:13 | 5:0 |

(f) Blood

| real expert | selected expert:spammer | | |
|---|---|---|---|
| | ExpertS | SpEM | CRA |
| 1 | 0:23 | 1:55 | 0:1 |
| 2 | 0:9 | 0:53 | 0:1 |
| 3 | 3:40 | 3:34 | 3:0 |
| 4 | 4:40 | 4:33 | 4:0 |
| 5 | 5:42 | 5:22 | 5:0 |

(g) Diabetes

| real expert | selected expert:spammer | | |
|---|---|---|---|
| | ExpertS | SpEM | CRA |
| 1 | 1:34 | 0:53 | 1:0 |
| 2 | 2:57 | 2:0 | 0:1 |
| 3 | 3:0 | 3:32 | 3:0 |
| 4 | 4:2 | 4:26 | 4:0 |
| 5 | 5:4 | 5:40 | 4:0 |

(h) Spambase

| real expert | selected expert:spammer | | |
|---|---|---|---|
| | ExpertS | SpEM | CRA |
| 1 | 1:4 | 1:55 | 1:0 |
| 2 | 2:4 | 2:59 | 2:0 |
| 3 | 3:3 | 3:15 | 3:0 |
| 4 | 4:3 | 4:5 | 4:0 |
| 5 | 5:0 | 5:4 | 5:0 |

It is also important to consider algorithms performance by comparing the computation time. Table 2 shows the comparison of computation time (in second), where the better algorithm performance has the lower time value. SpEM needs a lot of time in the seven datasets, except in Ionosphere. Also, it shows that CRA gives the higher performance in the seven dataset compared to ExpertS and SpEM.

**Table 2.** Comparison of computation time for eight datasets (lower value is better).

(a) SpectHeart

| | ExpertS | SpEM | MV | CRA |
|---|---|---|---|---|
| 0 | 29 | 162.27 | 0.01 | 215.94 |
| 1 | 34.33 | 161.2 | 0.02 | 15.24 |
| 2 | 26.11 | 181.29 | 0.02 | 207.64 |
| 3 | 13.87 | 236.91 | 0.02 | 7.99 |
| 4 | 10.98 | 174.33 | 0.01 | 5.82 |
| 5 | 5.78 | 141.03 | 0.02 | 3.53 |
| mean | 20.01 | 176.17 | 0.02 | 76.03 |

(b) Haberman

| | ExpertS | SpEM | MV | CRA |
|---|---|---|---|---|
| 0 | 2.2 | 166.41 | 0.11 | 2.36 |
| 1 | 2.32 | 144.44 | 0.06 | 2.32 |
| 2 | 2.23 | 156.07 | 0.02 | 1.28 |
| 3 | 2.24 | 153.28 | 1 | 1.6 |
| 4 | 2.77 | 149.97 | 0.41 | 1.58 |
| 5 | 1.38 | 215.44 | 0.11 | 1.2 |
| mean | 2.19 | 164.27 | 0.29 | 1.72 |

(c) Vertebral

| | ExpertS | SpEM | MV | CRA |
|---|---|---|---|---|
| 0 | 3.22 | 306.02 | 0.03 | 2.9 |
| 1 | 3.62 | 208.66 | 0.02 | 2.46 |
| 2 | 4.59 | 188.98 | 0.01 | 5.39 |
| 3 | 2.22 | 305.16 | 0.02 | 1.99 |
| 4 | 1.64 | 188.75 | 0.02 | 1.64 |
| 5 | 1.66 | 186.26 | 0.01 | 1.78 |
| mean | 2.83 | 230.64 | 0.02 | 2.69 |

(d) Bupa

| | ExpertS | SpEM | MV | CRA |
|---|---|---|---|---|
| 0 | 6.03 | 280.8 | 0.02 | 3.84 |
| 1 | 3.97 | 189.46 | 0.01 | 3.84 |
| 2 | 9.88 | 202.98 | 0.01 | 5.52 |
| 3 | 5.38 | 217.21 | 0.02 | 3.63 |
| 4 | 2.22 | 194.77 | 0.0001 | 2.06 |
| 5 | 2.11 | 134.22 | 0.02 | 1.91 |
| mean | 4.93 | 203.24 | 0.01 | 3.47 |

(e) Ionosphere

| | ExpertS | SpEM | MV | CRA |
|---|---|---|---|---|
| 0 | 93.77 | 136.14 | 0 | 62.42 |
| 1 | 143.25 | 175.17 | 0.01 | 83.34 |
| 2 | 48.29 | 162.99 | 0.01 | 108.27 |
| 3 | 1355.91 | 140.72 | 0.01 | 495.45 |
| 4 | 2241.22 | 148.69 | 0.02 | 1339.71 |
| 5 | 13.94 | 146.27 | 0.02 | 9.03 |
| mean | 649.4 | 151.66 | 0.01 | 349.7 |

(f) Blood

| | ExpertS | SpEM | MV | CRA |
|---|---|---|---|---|
| 0 | 5.69 | 448.18 | 0.03 | 1.4 |
| 1 | 17.74 | 574.02 | 0.04 | 19.67 |
| 2 | 25.34 | 86.35 | 0.04 | 25.1 |
| 3 | 5.21 | 969.01 | 0.02 | 2.46 |
| 4 | 5.11 | 781.61 | 0.03 | 2.3 |
| 5 | 2.77 | 559.18 | 0.03 | 1.96 |
| mean | 10.31 | 569.73 | 0.03 | 8.82 |

(g) Diabetes

| | ExpertS | SpEM | MV | CRA |
|---|---|---|---|---|
| 0 | 18.38 | 116.41 | 0.03 | 12.73 |
| 1 | 386.71 | 112.88 | 0.05 | 15.76 |
| 2 | 20.82 | 972.46 | 0.03 | 24.05 |
| 3 | 4.81 | 486.68 | 0.02 | 4.09 |
| 4 | 5.02 | 742.24 | 0.03 | 3.72 |
| 5 | 4.6 | 698.48 | 0.03 | 3.83 |
| mean | 73.39 | 521.53 | 0.03 | 10.70 |

(h) Spambase

| | ExpertS | SpEM | MV | CRA |
|---|---|---|---|---|
| 0 | 2546.5 | 627.82 | 0.2 | 1369.63 |
| 1 | 208.14 | 638.33 | 0.21 | 181.2 |
| 2 | 1387.96 | 639.31 | 0.21 | 222.05 |
| 3 | 809.35 | 11449.61 | 0.2 | 451.95 |
| 4 | 2667.55 | 45923.77 | 0.19 | 190.55 |
| 5 | 2349.98 | 8057.47 | 0.22 | 78.21 |
| mean | 1661.58 | 11222.72 | 0.21 | 415.60 |

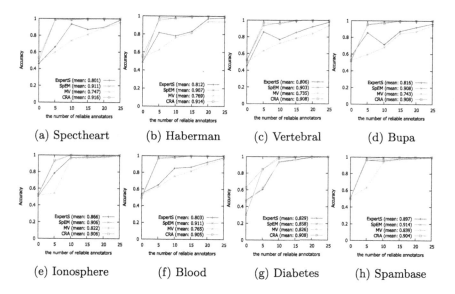

**Fig. 4.** Comparison of accuracy value for eight datasets (higher value is better).

In addition, accuracy evaluation was conducted to observe how the higher number of selected expert can minimize the prediction error. This experiment applied by using ExpertS, SpEM, MV, and CRA within 0, 5, 10, 15, 20, and 25 reliable annotators. We used these different numbers in order to show the validity of our proposed algorithm in handling not only small amount of experts but also in normal crowd cases. Figure 4 shows that CRA provides the best average values of accuracy in four datasets. In the other way, SpEM gives a better results in two datasets, while two other datasets show that CRA and SpEM have the same highest scores. ExpertS shows the good accuracy score in Ionosphere and Spambase. Based on MV accuracy score, the higher number of reliable annotator increase the value of accuracy. Therefore, CRA and SpEM provide the similar results when there are five or more experts given 100 annotators.

## 5 Conclusion

Reliability worker estimation is a challenging problem especially in the crowd-sourcing context, as the online occupation has beyond the regional borders. The fact that there is a few number of expert annotator in certain cases will affects crowds results, since some online workers give random answer to get the rewards. In this paper, the combination of reliable annotators (CRA) has been presented based on EM-Bayesian algorithm, entropy measure, and Condorcet Jury's Theorem. Experimental results using eight binary datasets have shown the good performance of the proposed method. Furthermore, applying CRA decrease the possibility to select spammers when reliable worker not clearly detected.

In future work, we will extend our algorithm by considering multiclass problem. We will also adapt our recent works on dynamic reliability worker estimation case for the efficient cost of crowdsourcing purposes.

# References

1. Estellés-Arolas, E., González-Ladrón-De-Guevara, F.: Towards an integrated crowdsourcing definition. J. Inf. Sci. **38**(2), 189–200 (2012)
2. Howe, J.: Crowdsourcing: How the Power of the Crowd is Driving the Future of Business. Business books. Random House Business (2008)
3. Tarasov, A., Delany, S.J., Namee, B.M.: Dynamic estimation of worker reliability in crowdsourcing for regression tasks: making it work. Expert Syst. Appl. **41**(14), 6190–6210 (2014)
4. Heer, J., Bostock, M.: Crowdsourcing graphical perception: using mechanical turk to assess visualization design. In: Proceedings of the SIGCHI Conference on Human Factors in Computing Systems, CHI 2010, pp. 203–212. ACM, New York (2010)
5. Ho, C.J., Jabbari, S., Vaughan, J.W.: Adaptive task assignment for crowdsourced classification. In: Proceedings of the 30th International Conference on International Conference on Machine Learning, ICML2013, vol. 28, pp. I-534-I-542, JMLR.org (2013)
6. Boutsis, I., Kalogeraki, V.: On task assignment for real-time reliable crowdsourcing. In: 2014 IEEE 34th International Conference on Distributed Computing Systems, pp. 1–10, June 2014
7. Moayedikia, A., Ong, K.L., Boo, Y.L., Yeoh, W.: Bee colony based worker reliability estimation algorithm in microtask crowdsourcing. In: 2016 15th IEEE International Conference on Machine Learning and Applications (ICMLA), pp. 713–717, December 2016
8. Dekel, O., Gentile, C., Sridharan, K.: Selective sampling and active learning from single and multiple teachers. J. Mach. Learn. Res. **13**(Sep), 2655–2697 (2012)
9. Downs, J.S., Holbrook, M.B., Sheng, S., Cranor, L.F.: Are your participants gaming the system? Screening mechanical turk workers. In: Proceedings of the SIGCHI Conference on Human Factors in Computing Systems, CHI 2010, pp. 2399–2402. ACM, New York (2010)
10. Raykar, V.C., Yu, S.: Eliminating spammers and ranking annotators for crowd-sourced labeling tasks. J. Mach. Learn. Res. **13**(1), 491–518 (2012)
11. Hernández-González, J., Inza, I., Lozano, J.A.: Multidimensional learning from crowds: usefulness and application of expertise detection. Int. J. Intell. Syst. **30**(3), 326–354 (2015)
12. Zhang, J., Sheng, V.S., Li, Q., Wu, J., Wu, X.: Consensus algorithms for biased labeling in crowdsourcing. Inf. Sci. **382–383**, 254–273 (2017)
13. Dempster, A.P., Laird, N.M., Rubin, D.B.: Maximum likelihood from incomplete data via the em algorithm. J. Roy. Stat. Soc. B **39**(1), 1–38 (1977)
14. Condorcet, M.d.: Essai sur l'application de l'analyse à la probabilité des décisions rendues à la pluralité des voix (1785)
15. Lichman, M.: UCI machine learning repository (2013)
16. Whitehill, J., Ruvolo, P., Wu, T., Bergsma, J., Movellan, J.: Whose vote should count more: Optimal integration of labels from labelers of unknown expertise. In: Proceedings of the 22nd International Conference on Neural Information Processing Systems, NIPS 2009, USA, Curran Associates Inc., pp. 2035–2043 (2009)

17. Welinder, P., Perona, P.: Online crowdsourcing: Rating annotators and obtaining cost-effective labels. In: 2010 IEEE Computer Society Conference on Computer Vision and Pattern Recognition - Workshops, pp. 25–32, June 2010

18. Raykar, V.C., Yu, S., Zhao, L.H., Valadez, G.H., Florin, C., Bogoni, L., Moy, L.: Learning from crowds. J. Mach. Learn. Res. **11**, 1297–1322 (2010)

19. Yan, Y., Fung, G., Schmidt, M., Hermosillo, G., Bogoni, L., Moy, L., Dy, J.G.: Modeling annotator expertise: learning when everyone knows a bit of something. In: In Proceedings of the Thirteenth International Conference on Artificial Intelligence and Statistics (AISTATS, 2010), pp. 932–939 (2010)

20. Wolley, C., Quafafou, M.: Scalable experts selection when learning from noisy labelers. In: 12th International Conference on Machine Learning and Applications (ICMLA) Poster Session (2013)

21. Zighed, D.A., Ritschard, G., Marcellin, S.: Asymmetric and sample size sensitive entropy measures for supervised learning. In: Ras Z.W., Tsay L.S. (eds.) Advances in Intelligent Information Systems. Studies in Computational Intelligence, vol 265, pp. 27–42. Springer, Heidelberg (2010)

22. Peleg, B., Zamir, S.: Extending the condorcet jury theorem to a general dependent jury. Soc. Choice Welfare **39**(1), 91–125 (2012)

23. Gottlieb, K., Hussain, F.: Voting for image scoring and assessment (visa) - theory and application of a 2 + 1 reader algorithm to improve accuracy of imaging endpoints in clinical trials. BMC Med. Imaging **15**(1), 6 (2015)

24. Xia, L.: Quantitative extensions of the condorcet jury theorem with strategic agents. In: Proceedings of the Thirtieth AAAI Conference on Artificial Intelligence, February 12–17, 2016, Phoenix, Arizona, USA, pp. 644–650 (2016)

25. Jain, B.J.: Condorcet's jury theorem for consensus clustering. CoRR abs/1604.07711 (2016)

26. Gehrlein, W.V.: Condorcet's paradox and the likelihood of its occurrence: different perspectives on balanced preferences*. Theor. Decis. **52**(2), 171–199 (2002)

27. Peyton, H.: Group choice and individual judgements, pp. 181–200. Cambridge University Press, Cambridge (1997)

28. Dawid, A.P.: A.M.S.: Maximum likelihood estimation of observer error-rates using the em algorithm. J. Roy. Stat. Soc.: Ser. C (Appl. Stat.) **28**(1), 20–28 (1979)

29. Yeh, I.C., Yang, K.J., Ting, T.M.: Knowledge discovery on RFM model using Bernoulli sequence. Expert Syst. Appl. **36**(3(Part 2)), 5866–5871 (2009)

# Estimating Sequence Similarity from Contig Sets

Petr Ryšavý[⊠] and Filip Železný

Department of Computer Science, Faculty of Electrical Engineering,
Czech Technical University in Prague, Prague, Czech Republic
{petr.rysavy,zelezny}@fel.cvut.cz

**Abstract.** A key task in computational biology is to determine mutual similarity of two genomic sequences. Current bio-technologies are usually not able to determine the full sequential content of a genome from biological material, and rather produce a set of large substrings (*contigs*) whose order and relative mutual positions within the genome are unknown. Here we design a function estimating the sequential similarity (in terms of the inverse Levenshtein distance) of two genomes, given their respective contig-sets. Our approach consists of two steps, based respectively on an adaptation of the tractable Smith-Waterman local alignment algorithm, and a problem reduction to the weighted interval scheduling problem soluble efficiently with dynamic programming. In hierarchical-clustering experiments with Influenza and Hepatitis genomes, our approach outperforms the standard baseline where only the longest contigs are compared. For high-coverage settings, it also outperforms estimates produced by the recent method [8] that avoids contig construction completely.

## 1   Introduction

A key task in computational biology is to determine mutual similarity of two genomic sequences. This is important for purposes such as database retrieval of similar genomes or for construction of phylogenetic trees by means of hierarchical genome clustering.

Current laboratory devices that identify the sequential content of genomes from biological material, so called *sequencers*, are not able to read sequences in their entirety (e.g. billions of nucleotides). Rather, they read small (e.g. tens or hundreds of nucleotides) subsequences, called *reads*, at random positions of the target sequence. The number of such reads is set high enough so that most of the nucleotides in the sequence are covered by multiple reads. Thus most of the reads overlap with some other reads.

Mutual read overlaps then allow guiding the in-silico reconstruction of the target sequence. For example, a possible assembly of reads $\{\mathsf{AGGC}, \mathsf{TGGA}, \mathsf{GCT}\}$ is $\mathsf{AGGCTGGA}$, and another possible assembly of the same reads is $\mathsf{GCTGGAGGC}$. These two solutions correspond to two different Hamiltonian paths in the *overlap graph* where vertices and directed edges represent reads and non-empty suffix-prefix overlaps, respectively. By an Occam-razor heuristic, the former (shorter)

© Springer International Publishing AG 2017
N. Adams et al. (Eds.): IDA 2017, LNCS 10584, pp. 272–283, 2017.
DOI: 10.1007/978-3-319-68765-0_23

assembly would be considered a more likely estimate of the target sequence. This motivates the generally accepted formulation of the assembly task: given a bag (multiset) of reads such that their overlap graph is connected, find the *shortest superstring* of all the reads.

A straightforward way to measure similarity of two sequences represented by their read bags would be to assemble each bag first and then compute the sequential similarity of the results. The problem, however, is that the shortest superstring problem is NP-hard. In previous work [8] we tackled this problem by proposing a similarity function computable directly on the input read sets thus avoiding the intractable assembly task.

Here we address a different problem, namely that popular assembly algorithms such as [2,7,11,13,14] are typically not able to produce a single putative sequence but rather yield a set of so called *contigs*. Contigs are maximal, mutually non-overlapping sequences which can be assembled from reads. Their ordering and mutual distance within the target sequence cannot be determined from the read bags alone. This has statistical reasons (the original sequence may not be completely covered by the reads, and the overlap graph may be disconnected) as well as more principal reasons. The latter are best illustrated by considering single-letter repeats longer than any read. Such repeat regions simply cannot be reconstructed, and their left and right neighborhoods need to be treated separately. There are however other sequential patterns more complex than single-letter repeats that also incur a break-down into contigs.

With additional requirements on laboratory equipment, the ordering of contigs and pairwise distances of neighboring contigs can be determined through a process called *scaffolding* (cf. Fig. 1). We consider this process unnecessary if the objective is just to estimate genome similarity, and enter the study with the hypothesis that the latter can be estimated directly from the contig sets.

On a superficial level, the problem considered here seems technically similar to the one we studied in [8]. Indeed, in both cases, we have two sets of substrings of the respective two sequences and want to determine the similarity of the latter two. However, contig sets do not satisfy some important properties of read sets (constant and short read length, roughly uniform distribution on the target sequence) and call for different methods. Briefly, the similarity (or, more precisely *dissimilarity* measure) in [8] was based on a one-to-one matching of reads between the respective read bags and an adaptation of the Monge-Elkan similarity principle. On the contrary, in the present study, we need to assume M-to-N matching between various-size contigs. Our present strategy consists of a tractable heuristic algorithm adapting the Smith-Waterman local alignment algorithm, and a subsequent problem reduction to the weighted interval scheduling problem which is solved efficiently through dynamic programming.

In the next section, we develop the novel dissimilarity measure. In Sect. 3 we test it empirically in hierarchical clustering tasks with 12 influenza virus genomes and 81 hepatitis strains, in comparison to a range of alternative approaches. Section 4 concludes the paper.

## 2    Dissimilarity Measure

Let $A$ be a string over an alphabet $\Sigma$. Here we consider $\Sigma = \{A, C, G, T\}$. Then $\mathsf{dist}(A, B)$ denotes the *Levenshtein distance* [5] between strings $A$ and $B$. The Levenshtein distance is one of the most common string distance measures that is used not only in bioinformatics. It measures the number of insertions, deletions and substitutions that are needed to convert one string into the other. It can be viewed as a simplified model that shows how many mutations had to be introduced during evolution. It holds that

$$\mathsf{dist}(A, B) \leq \max\{|A|, |B|\}, \tag{1}$$

where $|\cdot|$ denotes the length of $A$ when $A$ is a string. In our approach, we will use the post-normalized Levenshtein distance defined as

$$\overline{\mathsf{dist}}(A, B) = \frac{\mathsf{dist}(A, B)}{\max\{|A|, |B|\}}. \tag{2}$$

There are other approaches [6] to normalize the Levenshtein distance dealing with the fact that $\overline{\mathsf{dist}}$ does not satisfy the triangle inequality. However, here we adhere to (2).

A *read bag* $R_A$ is a multiset of $|R_A|$ substrings of length $l \ll |A|$ sampled with replacement from a distribution on all $|A| - l + 1$ substrings of length $l$ of $A$. *Coverage* $\alpha$ of $R_A$ is defined as

$$\alpha = l \frac{|R_A|}{|A|}. \tag{3}$$

Informally, the coverage $\alpha$ indicates the average number of reads covering a particular position in $A$. In other words, coverage is a ratio between the amount of data produced by the sequencing machine and the true DNA content. We assume that all genomes are sequenced with the same coverage.

*Assembly* is the process of reconstruction of string $A$ from $R_A$. For reasons explained earlier, the complete reconstruction is usually impossible, and a set of *contigs* $C_A$ is obtained instead. Contigs are mutually non-overlapping approximate substrings of $A$ that are longer than $l$. The adjective *approximate* means that contigs usually contain errors incurred by the assembly process. The relationships between the genome carrier DNA, reads, contigs and scaffolds is illustrated in Fig. 1.

Our goal is to propose a dissimilarity measure $\mathsf{Dist}(C_A, C_B)$ that approximates $\mathsf{dist}(A, B)$. The plan is briefly as follows. First, given any pair $(a \in C_A, b \in C_B)$ we estimate their most likely overlap in the unknown optimal alignment of $A$ and $B$, assuming they do overlap. Note that here the term 'overlap' is with respect to the mutual positioning of $a$ and $b$ in the *alignment* of $A$ and $B$, so the parts of $a$ and $b$ deemed to overlap may, in general, include insertions and mismatches. Second, the $|C_A||C_B|$ estimates resulting from the latter for all pairs of contigs are filtered using further (heuristic) constraints imposed on M-to-N

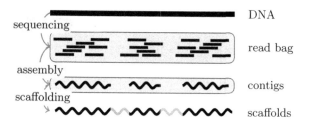

**Fig. 1.** Types of strings that are connected with the sequencing and assembly process.

contig matching; here we mainly want to filter out those pairs of contigs which likely *do not* overlap in the optimal alignment of $A$ and $B$. Lastly, the resulting set of hypothesized contig overlaps is used to estimate the distance between $A$ and $B$. These steps are described in the following three subsections, respectively.

## 2.1   Estimating Overlaps for Contig Pairs

Consider two contigs $a \in C_A$ and $b \in C_B$. In the optimal alignment of $A$ and $B$, there are several options how $a$ and $b$ can be positioned with respect to each other, assuming they have an overlap. Let $P(a)$ $(S(a))$ be the set of all prefixes (suffixes) longer[1] than 20 of string $a$, and let $c \in a$ denote that $c$ is a substring of $a$. The first option is that a suffix of one contig overlaps with a prefix of the other contig, which leads to the following set of overlap candidates: $U(a,b) = \{(c,d) \mid c \in S(a), d \in P(b)\}$. The second option is that one contig matches a substring of the other contig, which gives $V(a,b) = \{(a,d) \mid d \in b\}$. The set of candidate pairs is then the symmetric closure of the union of both options: $S(a,b) = U(a,b) \cup U(b,a) \cup V(a,b) \cup V(b,a)$. The following function then yields our estimate of the most likely mutual overlap of $a$ and $b$, assuming that the latter overlap at all.

$$\mathsf{overlap}(a,b) = \underset{(c,d)\in S(a,b)}{\arg\min}\ \overline{\mathrm{dist}}(c,d). \tag{4}$$

To calculate $\mathsf{overlap}(a,b)$ we modify the Smith-Waterman local alignment algorithm [12]. For this purpose we maintain an array `dist` storing the Levenshtein distance of suffixes of prefixes of $a$ and $b$. Two other arrays `lenA` and `lenB` store lengths of matching substrings. When filling each cell we decide the value based on (2). The complete algorithm that we use is described in Algorithm 1. It is a heuristic algorithm; we have not been able to design an exact algorithm with time-complexity of $\mathcal{O}(|a||b|)$ solving (4).[2] An execution of the algorithm is exemplified in Table 1.

---

[1] The threshold of 20 nucleotides is set to prevent very short random overlaps.
[2] While the problem is polynomial, the time complexity of the brute-force solution incurs a polynomial of order 4 rendering it unusable even on small datasets.

---

**Algorithm 1.** Pseudocode for the heuristic used for finding overlap$(a, b)$

---

function overlap$(a,b)$

$\quad dist, lenA, lenB \leftarrow$ 2D arrays of zeros of size $(|a| + 1) \times (|b| + 1)$

$\quad$ **for** $i \in \{1, 2, \ldots, |a|\}$ **do**                                          ▷ for each row

$\quad\quad$ **for** $j \in \{1, 2, \ldots, |b|\}$ **do**                                   ▷ for each column

5:$\quad\quad\quad$ choose option that leads to lowest $\frac{d}{\max\{lA,lB\}}$:

$\quad\quad\quad\quad$ **gap in** $a$: $d \leftarrow dist[i, j-1]+1; lA \leftarrow lenA[i, j-1]; lB \leftarrow lenB[i, j-1]+1$

$\quad\quad\quad\quad$ **gap in** $b$: $d \leftarrow dist[i-1, j]+1; lA \leftarrow lenA[i-1, j]+1; lB \leftarrow lenB[i-1, j]$

$\quad\quad\quad\quad$ **(mis)match:** $d \leftarrow dist[i-1, j-1] + (a[i-1] \neq b[j-1]);$

$\quad\quad\quad\quad\quad\quad lA \leftarrow lenA[i-1, j-1] + 1; lB \leftarrow lenB[i-1, j-1] + 1$

$\quad\quad\quad$ $dist[i, j] \leftarrow d, lenA[i, j] \leftarrow lA, lenB[i, j] \leftarrow lB$

$\quad\quad$ **end for**

$\quad\quad$ CHECKOPTIMUM$(i, |b|)$

10:$\quad$ **end for**

$\quad$ **for** $j \in \{1, 2, \ldots, |b| - 1\}$ **do** CHECKOPTIMUM$(|a|, j)$ **end for**

end function

function CHECKOPTIMUM$(i, j)$

$\quad$ **if** $\frac{dist[i,j]}{\max\{lenA[i,j], lenB[i,j]\}}$ is the smallest & $\min\{lenA[i, j], lenB[i, j]\} \geq 20$ **then**

15:$\quad\quad$ the new optimum is located at $(i, j)$, store it

$\quad$ **end if**

end function

---

**Table 1.** An illustration of Algorithm 1 showing the three involved data matrices for inputs $a = $ CATG, $b = $ AAGC and minimum overlap 2. The entries yielding the minimal value of the criterion are marked in boldface, producing (ATG, AAG) as the detected overlap.

| | *dist* matrix | | | | | *lenA* matrix | | | | | *lenB* matrix | | | |
|---|---|---|---|---|---|---|---|---|---|---|---|---|---|---|---|
| | | A | A | G | C | | | A | A | G | C | | | A | A | G | C |
| | 0 | 1 | 2 | 3 | 4 | | 0 | 1 | 2 | 3 | 4 | | 0 | 1 | 2 | 3 | 4 |
| 0 | 0 | 0 | 0 | 0 | 0 | 0 | 0 | 0 | 0 | 0 | 0 | 0 | 0 | 0 | 0 | 0 | 0 |
| C 1 | 0 | 1 | 1 | 1 | 0 | C 1 | 0 | 1 | 1 | 1 | 1 | C 1 | 0 | 0 | 0 | 0 | 1 |
| A 2 | 0 | 0 | 1 | 2 | 1 | A 2 | 0 | 1 | 1 | 1 | 2 | A 2 | 0 | 1 | 2 | 3 | 1 |
| T 3 | 0 | 1 | 1 | 2 | 2 | T 3 | 0 | 2 | 2 | 2 | 3 | T 3 | 0 | 1 | 2 | 3 | 1 |
| G 4 | 0 | 2 | 2 | **1** | 2 | G 4 | 0 | 3 | 3 | **3** | 3 | G 4 | 0 | 1 | 2 | **3** | 4 |

A technical remark is in order. Prior to executing the above algorithm, we need to pre-process the contig sets for reasons irrelevant to the algorithmic principles described. In particular, because contigs represent DNA, we do not know which strand they come from. The two strands of DNA contain the same sequence; however, pairs of symbols A and T and C and G are switched. The complementary strands are being read from the opposite ends, and this direction is fixed due to the chemical structure of DNA molecule. Therefore if we calculate overlap of $a$ and $b$, we do not know whether to match $a$ and $b$ or $a$ with reversed complement[3] of $b$. To deal with this, we simply expand one (but not both, for obvious reasons) of the two contig sets by the reversed complements of all its elements.

## 2.2  Estimating Overlaps for Contig Sets

The procedure from the previous section can be used to yield the most likely overlap for each possible pair of contigs $a \in C_A$, $b \in C_B$. Of course, not all such $|C_A||C_B|$ pairs actually overlap in the unknown optimal alignment of $A$ and $B$. To filter the overlap candidates towards a smaller, more plausible set, we adhere to the following rules. (1) For a given $a \in C_A$, we should only pick elements from $\{\,\mathsf{overlap}(a, b) \mid b \in C_B\,\}$ which do not overlap between themselves, (2) the resulting overlap pairs should minimize the sum of the $\overline{\mathsf{dist}}$ values.

Note that the two selection rules are only a heuristic. Not even rule (1) is dictated strictly by biological principles. Indeed, it may, in fact, happen that two contigs $b, b' \in C_B$ map to the same contig (or its substring) $a \in C_A$ in a way making $b$ and $b'$ overlap.[4]

The application of the two selection rules reduces to the *weighted interval scheduling* problem defined in [4]. In this problem, we have $n$ tasks, each of a value $v_t$ (for $t = 1, 2, \ldots, n$) and a starting and finishing time. Our goal is to select a subset of non-overlapping tasks that maximizes the sum of the selected task values.

Weighted interval scheduling can be solved in $\mathcal{O}(n \log n)$ time by a simple dynamic programming algorithm. We pass the tasks ordered by the finishing time, and for each task we have two options — to include it or not. If $S_t$ is an optimal solution for all tasks up to a task $t$ (in the sorted order), then $S_{t+1}$ is the maximum of $S_t$ and $v_{t+1} + S_{t'}$, where $t'$ is the task with the largest finishing time, which is smaller than the starting time of $t + 1$.

In our case the starting and finishing times represent location of $c$ in $a$. The value we assign to a pair $(c, d)$ is given by

$$v_{(c,d)} = \frac{1}{\overline{\mathsf{dist}}(c, d)} = \frac{\max\{|c|, |d|\}}{\mathsf{dist}(c, d)}. \tag{5}$$

Note that this value can be computed since Algorithm 1 has maintained for each potential match $(c, d)$ the values of $\mathsf{dist}(c, d)$, $|c|$, $|d|$.

To sum up, the procedure described accepts $a \in C_A$ and $C_B$, and produces a set we denote $o(a, C_B)$ which is selected from the initial overlap candidates, i.e.

$$o(a, C_B) \subseteq \{\,\mathsf{overlap}(a, b) \mid b \in C_B\,\}.$$

---

[3] For example for string ACCGGATT its reversed complement is AATCCGGT.

[4] Often long substrings, called repeats, occur multiple times in a DNA sequence. Assembly algorithms may identify a repeat as a single contig or as two contigs based on the number of mutations.

Further, overloading the overlap functor for contig sets, we denote

$$\mathsf{overlap}(C_A, C_B) = \bigcup_{a \in C_A} o(a, C_B).$$

Note that the above function is not symmetric and this fact will be dealt with in the next section.

## 2.3   Combining the Results

Having the filtered set of suspected overlaps, we first define the pre-distance $d(C_A, C_B)$ of contig sets $C_A$, $C_B$ as the sum of the distances associated with the individual overlaps

$$d(C_A, C_B) = \sum_{(c,d) \in \mathsf{overlap}(C_A, C_B)} \mathsf{dist}(c, d).$$

Since $\mathsf{overlap}(C_A, C_B)$ is not symmetric, neither is $d(C_A, C_B)$ and the final measure $\mathsf{Dist}(C_A, C_B)$ averages $d(C_A, C_B)$ and $d(C_B, C_A)$. Furthermore, it normalizes the scale.

$$\mathsf{Dist}(C_A, C_B) = Z \frac{d(C_A, C_B) + d(C_B, C_A)}{2}, \tag{6}$$

where $Z$ is a normalizing factor

$$Z = \frac{l \max\{|R_A|, |R_B|\}}{\alpha \sum_{(c,d) \in \mathsf{overlap}(C_A, C_B)} \max\{|c|, |d|\}}$$

dividing by the maximum distance that all matching substrings can have (the sum in the denominator) and multiplying by the maximum distance that $A$ and $B$ can have (i.e. $\max\{|A|, |B|\}$). For the latter, it estimates sequence lengths $|A|, |B|$ from (3).

## 3   Experimental Evaluation

Here we test the dissimilarity measure (6) on data in comparative experiments where the distance is used to infer phylogenetic trees through hierarchical clustering.

We compare methods for estimating the Levenshtein distance $\mathsf{dist}(A, B)$ for DNA sequences. The methods include (i) the reference distance $\mathsf{dist}(A, B)$ (ground truth), (ii) our newly proposed dissimilarity measure (6), (iii) measure $\mathsf{Dist}_{\mathsf{MESSGq}}$ taken from [8,9], and applied directly on read-sets without contig assembly (iv) distance of two longest contigs, which represents the 'standard' state of the art option, (v) a trivial baseline method estimating $\mathsf{dist}(A, B)$ as $\max\{|R_A|, |R_B|\}$ based on (1). The methods were implemented in Java maximizing shared code. For assembly, we used current official implementations of five common assembly algorithms, namely ABySS [11], Edena [2], SSAKE [13], SPAdes [7] and Velvet [14].

We measured

(i) the Pearson's correlation coefficient showing the similarity of the distance matrices produced by the respective methods to the true distance matrix,

(ii) the Fowlkes-Mallows index [1] that measures the similarity between trees produced based on distance estimates and the reference tree defined by $\mathrm{dist}(A, B)$. The Fowlkes-Mallows index shows how much two hierarchical clusterings differ in structure. Both hierarchical trees are first cut into $k$ clusters for $k = 2, 3, \ldots, n-1$. Then clusterings are compared based on the number of common objects among each pair of clusters. In this way, we obtain a set of values $B_k$ that shows distances of the trees at various levels.

(iii) assembly time (if applicable) and the distance matrix calculation time,

(iv) how many times was distance calculation successful (i.e. assembly produced at least one contig).

For hierarchical clustering, we used the neighbor-joining algorithm [10].

The testing data contain two datasets. The *influenza* dataset[5] contains 12 influenza virus genome sequences plus an outgroup sequence. The *hepatitis* dataset contains 81 hepatitis C strains from the ENA repository.

To generate read sets from the genomes, we employ two strategies. One is 'idealized', based on error-free sampling and consecutive contig generation, denoted as $\mathrm{opt}_\theta$ below. The other one simulates closely real-life erroneous sequencing conducted by the Illumina technology, and is facilitated by the ART [3] program.

For the influenza dataset, we use the idealized option. For each position we calculated coverage, and we produced nine simulated assembly results $\mathrm{opt}_\theta$ for $\theta = 10, 20, \ldots, 90$. To generate $\mathrm{opt}_\theta$ we chose the highest number $\beta$ such that at least $\theta$ percent of nucleotides are covered in $\beta$ or more reads. The longest substrings of such nucleotides formed individual contigs. We sampled the influenza dataset for choices with a high range of coverage and read length[6]. For sampling reads from hepatitis dataset we used ART [3] program to simulate Illumina sequencing for $(\alpha, l) \in \{10, 30, 50\} \times \{30, 70, 100\}$.

The main experimental results are shown in Table 2, which shows the average results on both datasets. For averaging on influenza dataset we exclude three most extreme values of coverage and read length. Table 2 contains results only for the best assembly algorithm and the worst one. The columns of the table show the Pearson's correlation coefficient, run time and the Fowlkes-Mallows index for selected levels. Figure 4 shows plot of Pearson's correlation coefficient on coverage for influenza dataset. Figures 2 and 3 show the Fowlkes-Mallows index. Figure 5 shows the dependency of correlation on percent of nucleotides in simulated assembly, i.e. on parameter $\theta$ of $\mathrm{opt}_\theta$. For clarity, the figures do not plot

---

[5] AF389115, AF389119, AY260942, AY260945, AY260949, AY260955, CY011131, CY011135, CY011143, HE584750, J02147, K00423 and outgroup AM050555. The genomes are available at http://www.ebi.ac.uk/ena/data/view/<accession>.

[6] $(\alpha, l) \in \{0.1, 0.3, 0.5, 0.7, 1, 1.5, 2, 2.5, 3, 4, 5, 7, 10, 15, 20, 30, 40, 50, 70, 100\} \times \{3, 5, 10, 15, 20, 25, 30, 40, 50, 70, 100, 150, 200, 500\}$.

**Table 2.** Average runtime, Pearson's correlation coefficient between distance matrices and Fowlkes-Mallows index for $k = 4$ and $k = 8$. The 'reference' method calculates distances from the original sequences. We show only assembly algorithms that gave the highest and the lowest correlation.

| Dataset | Method | Finished | assem. ms | distances ms | corr. | $B_4$ | $B_8$ |
|---|---|---|---|---|---|---|---|
| Influenza | Reference | 112/112 | 0 | 2,518 | 1 | 1 | 1 |
| | $\max(|R_A|, |R_B|)$ | 112/112 | 0 | 184 | .801 | .66 | .32 |
| | Dist$_{\text{MESSGq}}$ | 112/112 | 0 | 43,553 | .966 | 1 | .97 |
| | Longest contig Velvet | 110/112 | 392 | 101 | .569 | .46 | .23 |
| | Longest contig SPAdes | 43/112 | 12,461 | 2,127 | .751 | .71 | .56 |
| | Longest contig opt$_{90}$ | 112/112 | 0 | 1,208 | .666 | .63 | .43 |
| | Dist$(C_A, C_B)$ SSAKE | 67/112 | 2,115 | 17,483 | .949 | .98 | .87 |
| | Dist$(C_A, C_B)$ SPAdes | 43/112 | 12,461 | 20,968 | .975 | .99 | .95 |
| | Dist$(C_A, C_B)$ opt$_{90}$ | 112/112 | 0 | 22,239 | .987 | 1 | .98 |
| Hepatitis | Reference | 9/9 | 0 | 2,145,104 | 1 | 1 | 1 |
| | $\max(|R_A|, |R_B|)$ | 9/9 | 0 | 7,738 | .181 | .72 | .83 |
| | Dist$_{\text{MESSGq}}$ | 9/9 | 0 | 701,726 | .897 | 1 | .98 |
| | Longest contig Velvet | 9/9 | 22,860 | 3,447 | .234 | .93 | .54 |
| | Longest contig SPAdes | 9/9 | 103,683 | 1,872,233 | .591 | .95 | .84 |
| | Dist$(C_A, C_B)$ SSAKE | 9/9 | 96,446 | 29,465,436 | .916 | 1 | .9 |
| | Dist$(C_A, C_B)$ Velvet | 9/9 | 22,860 | 28,186,784 | .966 | 1 | .98 |

all the methodological combinations; they are all included in the supplementary material.[7]

From the results, we see that the proposed method Dist$(C_A, C_B)$ gives results of quality comparable to Dist$_{\text{MESSGq}}$. For low coverage data Dist$_{\text{MESSGq}}$ is more successful because it does not need assembled contigs. On the opposite for high coverage data, Dist$(C_A, C_B)$ produces results with higher correlation because it has more data available. Both methods are better than the baseline method $\max(|R_A|, |R_B|)$ and also than the simple approach that considers only the longest contig. Figure 5 indicates that the proposed method gives good estimates even if assembly identified only a fraction of the original sequence.

Dist$_{\text{MESSGq}}$ is faster than the proposed method. On the opposite Dist$(C_A, C_B)$ is approximately $10\times$ slower than the reference method. Dist$(C_A, C_B)$ and the reference method have to fill dynamic programming tables of approximately same sizes and therefore their run times differ only by a multiplicative constant, as the proposed method has to fill entries in three tables instead of one.

---

[7] A sample implementation and supplementary material are available on https://github.com/petrrysavy/ida2017.

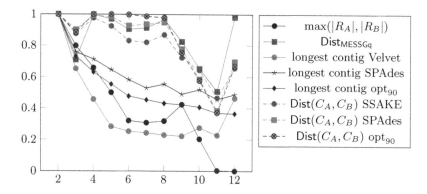

**Fig. 2.** Plot of average Fowlkes-Mallows index $B_k$ versus $k$ on influenza dataset. The index compares trees generated by the neighbor-joining algorithm. The tree is compared with the tree generated from the original sequences. If all values are equal to 1, the structures of the trees are the same.

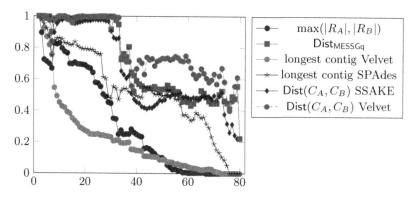

**Fig. 3.** Plot of average Fowlkes-Mallows index $B_k$ versus $k$ on hepatitis dataset. The index compares trees generated by the neighbor-joining algorithm.

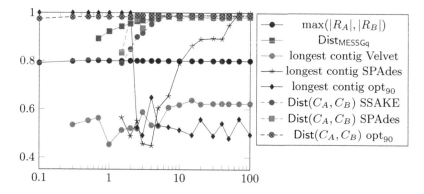

**Fig. 4.** Plot of average Pearson's correlation coefficient for several choices of coverage values on influenza dataset.

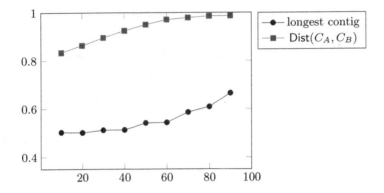

**Fig. 5.** Plot of average Pearson's correlation coefficient on $\theta$ parameter for simulated assembly $\mathrm{opt}_\theta$ on influenza dataset.

## 4    Conclusion and Future Work

We have proposed and evaluated a method for estimating Levenshtein distance between two sequences from partially assembled data. In experiments, our approach was better in terms of Pearson's correlation coefficient than the one of [9] for higher coverage data. The proposed method was significantly better than the trivial estimates and a straightforward solution where we use only the longest contig to represent the assembly.

The most promising goal for the follow-up work is to combine the method proposed in this paper with the one of [9] in order to get advantages of both. Method [9] is significantly faster and it produces better results for low coverage data. Instead of assembling whole contigs, we may assemble only a few neighboring reads in order to balance time needed for assembly together with distance estimation time and estimate accuracy.

**Acknowledgment.** This work was supported by the Grant Agency of the Czech Technical University in Prague, grant No. SGS17/189/OHK3/3T/13. Access to computing and storage facilities owned by parties and projects contributing to the National Grid Infrastructure MetaCentrum, provided under the programme "Projects of Large Research, Development, and Innovations Infrastructures" (CESNET LM2015042), is greatly appreciated.

## References

1. Fowlkes, E.B., Mallows, C.L.: A method for comparing two hierarchical clusterings. J. Am. Stat. Assoc. **78**(383), 553–569 (1983)
2. Hernandez, D., et al.: De novo bacterial genome sequencing: millions of very short reads assembled on a desktop computer. Genome Res. **18**(5), 802–809 (2008)
3. Huang, W., Li, L., Myers, J.R., Marth, G.T.: ART: a next-generation sequencing read simulator. Bioinformatics **28**(4), 593–594 (2012)

4. Kleinberg, J., Tardos, E.: Algorithm Design. Addison-Wesley Longman Publishing Co., Inc., Boston (2005)
5. Levenshtein, V.I.: Binary codes capable of correcting deletions, insertions, and reversals. Sov. Phys. Dokl. **10**(8) (1966)
6. Marzal, A., Vidal, E.: Computation of normalized edit distance and applications. IEEE Trans. Pattern Anal. Mach. Intell. **15**(9), 926–932 (1993)
7. Nurk, S., Bankevich, A., Antipov, D., Gurevich, A., Korobeynikov, A., Lapidus, A., Prjibelsky, A., Pyshkin, A., Sirotkin, A., Sirotkin, Y., Stepanauskas, R., McLean, J., Lasken, R., Clingenpeel, S.R., Woyke, T., Tesler, G., Alekseyev, M.A., Pevzner, P.A.: Assembling genomes and mini-metagenomes from highly chimeric reads. In: Deng, M., Jiang, R., Sun, F., Zhang, X. (eds.) RECOMB 2013. LNCS, vol. 7821, pp. 158–170. Springer, Heidelberg (2013). doi:10.1007/978-3-642-37195-0_13
8. Ryšavý, P., Železný, F.: Estimating sequence similarity from read sets for clustering sequencing data. In: Boström, H., Knobbe, A., Soares, C., Papapetrou, P. (eds.) IDA 2016. LNCS, vol. 9897, pp. 204–214. Springer, Cham (2016). doi:10.1007/978-3-319-46349-0_18
9. Ryšavý, P., Železný, F.: Estimating Sequence Similarity from Read Sets for Clustering Next-Generation Sequencing Data (preprint, 2017), http://arxiv.org/abs/1705.06125
10. Saitou, N., Nei, M.: The neighbor-joining method: a new method for reconstructing phylogenetic trees. Mol. Biol. Evol. **4**(4), 406–425 (1987)
11. Simpson, J.T., et al.: ABySS: a parallel assembler for short read sequence data. Genome Res. **9**(6), 1117–1123 (2009)
12. Smith, T.F., Waterman, M.S.: Identification of common molecular subsequences. J. Mol. Biol. **147**(1), 195–197 (1981)
13. Warren, R.L., et al.: Assembling millions of short DNA sequences using SSAKE. Bioinformatics **23**(4), 500–501 (2007)
14. Zerbino, D.R., Birney, E.: Velvet: algorithms for de novo short read assembly using de Bruijn graphs. Genome Res. **18**(5), 821–829 (2008)

# Computational Topology Techniques for Characterizing Time-Series Data

Nicole Sanderson[✉], Elliott Shugerman, Samantha Molnar, James D. Meiss, and Elizabeth Bradley

University of Colorado, Boulder, CO, USA
nicole.sanderson@colorado.edu

**Abstract.** Topological data analysis (TDA), while abstract, allows a characterization of time-series data obtained from nonlinear and complex dynamical systems. Though it is surprising that such an abstract measure of structure—counting pieces and holes—could be useful for real-world data, TDA lets us compare different systems, and even do membership testing or change-point detection. However, TDA is computationally expensive and involves a number of free parameters. This complexity can be obviated by coarse-graining, using a construct called the witness complex. The parametric dependence gives rise to the concept of persistent homology: how shape changes with scale. Its results allow us to distinguish time-series data from different systems—e.g., the same note played on different musical instruments.

## 1 Introduction

Topology gives perhaps the roughest characterization of shape, distinguishing sets that cannot be transformed into one another by continuous maps [16]. The Betti numbers $\beta_k$, for instance, count the number of $k$-dimensional "holes" in a set: $\beta_0$ is the number of components, $\beta_1$ the number of one-dimensional holes, $\beta_2$ the number of trapped volumes, etc. Of course, measures that are this abstract can miss much of what is meant by "structure," but topology's roughness can also be a virtue in that it eliminates distinctions due to unimportant distortions. This makes it potentially quite useful for the purposes of classification, change-point detection, and other data-analysis tasks[1].

Applying these ideas to real-world data is an interesting challenge: how should one compute the number of holes in a set if one only has samples of that set, for instance, let alone if those samples are noisy? The field of topological data analysis (TDA) [13,24] addresses these challenges by building *simplicial complexes* from the data—filling in the gaps between the samples by adding line segments,

---

[1] The work we describe in this paper calls upon areas of mathematics—including dynamical systems, topology and persistent homology—that may not be commonly used in the data-analysis community. As a full explanation of these would require several textbook length treatments, we content ourselves with discussing how these ideas can be applied, leaving the details of the theory to references.

© Springer International Publishing AG 2017
N. Adams et al. (Eds.): IDA 2017, LNCS 10584, pp. 284–296, 2017.
DOI: 10.1007/978-3-319-68765-0_24

triangular faces, etc.—and computing the ranks of the homology groups of those complexes. These kinds of techniques, which we describe in more depth in Sect. 2, have been used to characterize and describe many kinds of data, ranging from molecular structure [23] to sensor networks [5].

As one would imagine, the computational cost of working with a simplicial complex built from thousands or millions of data points can be prohibitive. In Sect. 2 we describe one way, the *witness complex*, to coarse-grain this procedure by downsampling the data. Surprisingly, one can obtain the correct topology of the underlying set from such an approximation if the samples satisfy some denseness constraints [1]. The success of this coarse-graining procedure requires not only careful mathematics, but also good choices for a number of free parameters—a challenge that can be addressed using *persistence* [7,18], an approach that is based on the notion that any topological property of physical interest should be (relatively) independent of parameter choices in the associated algorithms. This, too, is described in Sect. 2.

In this paper, we focus on time-series measurements from dynamical systems, with the ultimate goal of detecting bifurcations in the dynamics—change-point detection, in the parlance of other fields. Pioneering work in this area was done by Muldoon et al. [15], who computed Euler characteristics and Betti numbers of embedded trajectories. The scalar nature of many time-series datasets poses another challenge here. Though it is all very well to think about computing the topology of a state-space trajectory from samples of that trajectory, in experimental practice it is rarely possible to measure every state variable of a dynamical system; often, only a single quantity is measured, which may or may not be a state variable—e.g., the trace in Fig. 1(a), a time series recorded from a piano. The state space of this system is vast: vibration modes of every string, the movement of the sounding board, etc. Though each quantity is critical to the dynamics, we cannot hope to measure all of them. Delay reconstruction [2] lets one reassemble the underlying dynamics—up to smooth coordinate change, ideally—from a single stream of data. The coordinates of each point in such a reconstruction are a set of time $\tau$ delayed measurements $x(t)$: from a discrete time series $\{x_t\}_{t=1}^{N}$, one constructs a sequence of vectors $\{\mathbf{x}_t\}_{t=d_E\tau}^{N}$ where $\mathbf{x}_t = (x_t, x_{t-\tau}, \ldots, x_{t-(d_E-1)\tau})$ that trace out a trajectory in a $d_E$-dimensional reconstruction space. An example is shown in Fig. 1(b). Because the reconstruction process preserves the topology—but not the *geometry*—of the dynamics, a delay reconstruction can look very different than the true dynamics. Even so, this result means that if we can compute the topology of the reconstruction, we can assert that the results hold for the underlying dynamics, whose state variables we do not know and have not measured. In other words, the topology of a delay reconstruction can be useful in identifying and distinguishing different systems, even if we only have incomplete measurements of their state variables, and even though the reconstructed dynamics do not have the same geometry as the originals.

Like the witness-complex methodology, delay reconstruction has free parameters. A reconstruction is only *guaranteed* to have the correct topology—that is,

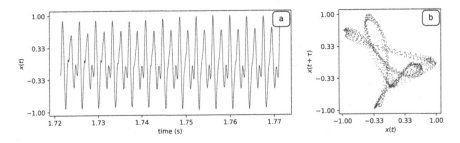

**Fig. 1.** A short segment (45 ms) of a recording of middle C ($\mathfrak{f} = 261.62$ Hz) played on a Yamaha upright piano, recorded at 44100 Hz sample rate using a Sony ICD-PX312 digital voice recorder: (a) time series data (b) two-dimensional delay reconstruction using $\tau = \frac{1}{\mathfrak{f}\pi}$.

to be an "embedding"—if the delay $\tau$ and the dimension $d_E$ are chosen properly. Since we are using the topology as a distinguishing characteristic, that correctness is potentially critical here. There are theoretical guidelines and constraints regarding both parameters, but they are not useful in practice. For real data and finite-precision arithmetic, one must fall back on heuristics to estimate values for these parameters [9,14], a procedure that is subjective and sometimes quite difficult. However, it is possible to compute the coarse-grained topology of 2D reconstructions like the one in Fig. 1(b) even though they are not true embeddings [11]. This is a major advantage not only because it sidesteps a difficult parameter estimation step, but also because it reduces the computational complexity of all analyses that one subsequently performs on the reconstruction.

This combination of ideas—a coarse-grained topological analysis of an incomplete delay reconstruction of scalar time-series data—allows us to identify, characterize, and compare dynamical systems efficiently and correctly, as well as to distinguish different ones. This advance can bring topology into the practice of data analysis, as we demonstrate using real-world data from a number of musical instruments.

## 2    Topological Data Analysis

There has been a great deal of work on change-point detection in data streams, including a number of good papers in past IDA symposia (e.g., [4]). Most of the associated techniques—queueing theory, decision trees, Bayesian techniques, information-theoretic methods, clustering, regression, and Markov models and classifiers (see, e.g., [6,10,17,22])—are based on statistics, though frequency analysis can also play a useful role. Though these approaches have the advantages of speed and noise immunity, they also have some potential shortcomings. If the regimes are dynamically different but the operative distributions have the same shapes, for example, these methods may not distinguish between them. They implicitly assume that it is safe to aggregate information, which raises complex issues regarding the window size of the calculation. Most of these techniques also assume that the underlying system is linear. If the data come from a nonstationary but deterministic nonlinear dynamical system—a common situation—all of these techniques can fail. Our premise is that computational topology can be useful in such situations; the challenge is that it can be quite expensive.

The foundation of TDA is the construction of a simplicial complex to describe the underlying manifold of which the data are a (perhaps noisy) sample: that is, to reconstruct the solid object of which the points are samples. A simplicial complex is, loosely speaking, a triangulation. The data points are the vertices, edges—one-simplices—join those vertices, two-simplices cover the faces, and so on. Abstractly, a $k$ simplex is an ordered list $\sigma = \{x_1, x_2, \ldots, x_{k+1}\}$ of $k+1$ vertices. The mathematical challenge is to connect the data points in geometrically meaningful ways. Any such solution involves some choice of scale $\epsilon$: a discrete set of points is only an approximate representation of a continuous shape and is accurate only up to some spatial scale. This is both a problem and an advantage: one can glean useful information from investigations of how the shape changes with $\epsilon$ [19]. While topology has many notions of shape, the most amenable to computation is homology, which determines the Betti numbers mentioned in Sect. 1. Computing these as a function of $\epsilon$ is the fundamental idea of persistent homology, as discussed further below.

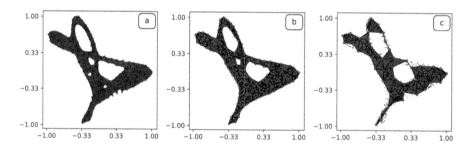

**Fig. 2.** Different simplicial complexes built from the data set in Fig. 1(b) with $\epsilon = 0.073$: (a) a Čech complex, with all 2000 points used as vertices; (b) and (c) witness complexes with $\ell = 200$ and $\ell = 50$, respectively—i.e., with 1/10th and 1/40th of the points used as landmarks. Complexes (a), (b) and (c) contain 2770627, 3938, and 93 triangles, respectively.

There are many ways to build a complex. In a Čech complex, there is an edge between two vertices if the two balls of radius $\frac{\epsilon}{2}$ centered at the vertices intersect; here the selection of $\epsilon$ fixes the scale. Similarly, three vertices in a Čech complex are linked by a two-simplex if the corresponding three $\frac{\epsilon}{2}$-balls have a common intersection, and so on. Figure 2(a) shows a Čech complex constructed from the points in Fig. 1(b). The distance checks involved in building such a complex— between all pairs of points, all triples, etc.—are computationally impractical for large data sets. There are many other ways to build simplicial complexes, including the $\alpha$-complex [8], the Vietoris-Rips complex [12], or even building a complex based on a cubical grid [13]. All of these approaches have major shortcomings for practical purposes: high computational cost, poor accuracy, and/or inapplicability in more than two dimensions.

An intriguing alternative is to coarse-grain the complex, employing a subset of the data points as vertices and using the rest to how to fill in the gaps. One way to do this is a *witness complex* [21], which is determined by the time-series data, $W$ (the witness set) and a smaller, associated set $L$—the landmarks, which form the vertices of the complex. Key elements of this process are the selection of appropriate landmarks, typically a subset of $W$, and a choice of a *witness relation* $R(W,L) \subset W \times L$, which determines how the simplices tile the landmarks: a point $w \in W$ is a witness to an abstract simplex $\sigma \in 2^L$ whenever $\{w\} \times \sigma \subset R(W,L)$. One connects two landmarks with an edge if they share at least one witness—this is a one-simplex. Similarly, if three landmarks have a common witness, they form a two-simplex, and so on. (This is similar to the Čech complex, except that not every point is a vertex.)

There are many ways to define what it means to share a witness. Informally, the rationale is that one wants to "fill in" the spaces between the vertices in the complex if there is at least one witness in the corresponding region. Following this reasoning, we could classify a witness $w_i \in W$ as shared by landmarks $l_j, l_k \in L$ if $-\epsilon < |l_j - w_i| - |l_k - w_i| < \epsilon$—that is, if it is roughly equidistant to both of them—and add an edge to the complex if we find such a witness. If the set of witnesses included a point that were shared between three landmarks, we would add a face to the complex, and so on. That particular definition is problematic, however: it classifies an $\epsilon$-equidistant witness as shared even if it is on the opposite side of the data set from the two landmarks. To address this, we add a distance constraint to the witness relation, classifying a witness $w_i \in W$ as shared between two landmarks $l_j, l_k \in L$ if both are within $\epsilon$ of being the closest landmark to $w_i$, i.e., if $\max(|l_j - w_i|, |l_k - w_i|) < \min_m |l_m - w_i| + \epsilon$:

Input: $\{x_t\}_{t=1}^N$, discrete $\mathbb{R}$-valued time series
**delay coordinate reconstruction,** $\{\mathbf{x}_t\}_{t=d_E\tau}^N$
select landmarks $L = \{l_i\}_{i=1}^\ell \subseteq \{\mathbf{x}_t\}_{t=d_E\tau}^N$
compute pair-wise distances $D_{ij} = |l_i - \mathbf{x}_j|$
**for** $\epsilon \in (\epsilon_{\min}, \epsilon_{\max}, \epsilon_{\text{step}})$ : (build witness complex, $\mathcal{W}^\epsilon$)
    **for** $\mathbf{x}_t \in X$:
        d $= |L - \mathbf{x}_t| + \epsilon$
        **for** $(l_i, l_j) \in L$ :   (check for edges)
            **if** $|l_i - \mathbf{x}_t|, |l_j - \mathbf{x}_t| < $ d :
                $\{l_i, l_j\} \in \mathcal{W}^\epsilon$
        **for** $(l_i, l_j, l_k) \in L$ :   (check for triangles)
            **if** $|l_i - \mathbf{x}_t|, |l_j - \mathbf{x}_t|, |l_k - \mathbf{x}_t| < $ d :
                $\{l_i, l_j, l_k\} \in \mathcal{W}^\epsilon$
            :   (check for higher dimensional simplices)
Output: $\{\mathcal{W}^\epsilon\}_{\epsilon_{\min}}^{\epsilon_{\max}}$, series of witness complexes for specified $\epsilon$ range

Figure 2(b) shows a witness complex constructed in this manner from the data of Fig. 2(a), with one-tenth of the points chosen as landmarks. The computation involved is much faster—an order of magnitude less than required for Fig. 2(a). The scale factor $\epsilon$ and the number $\ell$ of landmarks have critical implications for the correctness and complexity of this approach, as discussed further below[2].

Every simplicial complex has an associated set of homology groups, which depend upon the structure of the underlying manifold: whether or not it is connected, how many holes it has, etc. This is a potentially useful way to characterize and distinguish different regimes in data streams. An advantage of homology over homotopy or some other more complete topological theory is that it can be reduced to linear algebra [16]. Algorithms to compute homology depend on computing the null space and range of matrices that map simplices to their boundaries [13]. The computational complexity of these algorithms scales badly, though—both with the number of vertices in the complex and with the dimension of the underlying manifold. In view of this, the parsimonious nature of the witness complex is a major advantage. However, an overly parsimonious complex, or one that contains spurious simplices, may not capture the structure correctly.

The parsimony tradeoff plays out in the choices of both of the free parameters in this method. Figure 2(b) and (c) demonstrate the effects of changing the number of vertices $\ell$ in the witness complex. With 200 vertices, the complex effectively captures the three largest holes in the delay reconstruction; if $\ell$ is lowered to 50, the complex is too coarse to capture the smallest of these holes. In general, increasing $\ell$ will improve the match of the complex to the data, but it will also increase the computational effort required to build and work with that structure. Reference [11] explores the accuracy end of this tradeoff; the computational complexity angle is covered in the later sections of this paper.

The other free parameter in the process, the scale factor $\epsilon$, plays a subtler and more interesting role. When $\epsilon$ is very small, as in Fig. 3(a), very few witnesses are shared and the complex is very sparse. As $\epsilon$ grows, more and more witnesses fall into the broadening regions that qualify them as shared, so more simplices appear in the complex, fleshing out the structure of the sampled manifold. There is a limit to this, however. When $\epsilon$ approaches the diameter of the point cloud, the witness complex will be fully connected, which obscures the native structure of the sampled set; well before that, simplices appear that do not reflect the true structure of the data. One effective way to track all of this is the $\beta_1$ *persistence diagram* of [7], which plots the $\epsilon$ value at which each hole appears in the complex, $\epsilon_B$, on the horizontal axis and the value $\epsilon_D$ at which it disappears from the complex on the vertical axis[3]. A persistence diagram for the piano data, for example—part (d) of Fig. 3—shows a cluster of holes that are born and die

---

[2] Landmark choice is another issue. There are a number of ways to do this; here, we evenly space the landmarks across the data.

[3] Choosing the range and increment for $\epsilon$ in such a plot requires some experimentation; in this paper, we use $\epsilon_{step} = 20$ and $\epsilon_{max}$ set for each instrument when the first 20-dimensional simplex is witnessed. This is a good compromise between effectiveness and efficiency for the data sets that we studied.

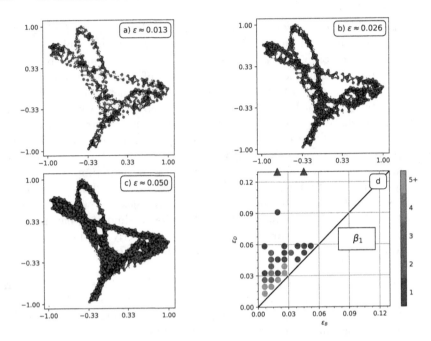

**Fig. 3.** The effects of the scale parameter $\epsilon$: (a)–(c) show witness complexes with $\ell = 200$ and different $\epsilon$ values. (d) shows a $\beta_1$ persistence diagram computed across a range of $\epsilon$ values. Each point in (d) represents a hole in the complex; its $x$ and $y$ coordinates show the $\epsilon$ values at which that hole appears and disappears, respectively. Holes that persist beyond the upper $\epsilon_{max} \approx 0.13$ are shown with triangles.

before $\epsilon = 0.06$. These represent small voids in the data. The three points near the top left of Fig. 3(d) represent holes that are highly persistent.

## 3   Persistent Homology and Membership Testing

Cycles are critical elements of the dynamical structure of many systems, and thus useful in distinguishing one system from another. A chaotic attractor, for example, is typically densely covered by unstable periodic orbits, and those orbits provide a formal "signature" of the corresponding system [3]. Topologically, a cycle is simply a hole, of any shape or size, in the state-space trajectory of the system. The persistent homology methods described above include some aspects of geometry, though, which makes the relationship between holes and cycles not completely simple. Musical instruments are an appealing testbed for exploring these issues. Of course, one can study the harmonic structure of a note from an instrument, or any other time series, using frequency analysis or wavelet transforms. Because delay reconstruction transforms time into space, it not only reveals which frequencies are present at which points in the signal, as well as their amplitudes. These reconstructions also bring out subtler features;

any deviation from purely elliptical shape, for instance—or the kind of "winding" that appears on Fig. 1(b)—signals the presence of another signal and also gives some indication of its amplitude and relative frequency.

Topological data analysis brings out those kinds of features quite naturally. The structure of the persistence diagrams for the same note played on two different musical instruments, for instance, is radically different, as shown in Fig. 4. The witness complex of a clarinet playing the A above middle C contains seven holes for $\epsilon < 0.05$. Six of these holes die before $\epsilon = 0.06$; they are represented by the points in the lower left corner of the persistence diagram. The other hole in the complex triangulates the large loop in the center of the reconstruction of Fig. 4(a). This hole, which remains open until the end of the $\epsilon$ range of the calculation, is represented by the triangle in the top left corner of the persistence diagram in Fig. 4(c). The viol reconstruction in Fig. 4(b), on the other hand, contains over twelve holes that are born at low $\epsilon$ values, including many short-lived features depicted in the lower left of the persistence diagram. By $\epsilon = 0.10$, only four holes remain open. The smallest of these four features, which closes up around $\epsilon = 0.23$, is represented by the point in the top left of Fig. 4(d). The three other holes, which remain open to the end of the $\epsilon$ range of the calculation, are represented by the colored triangles at the top left of Fig. 4(d).

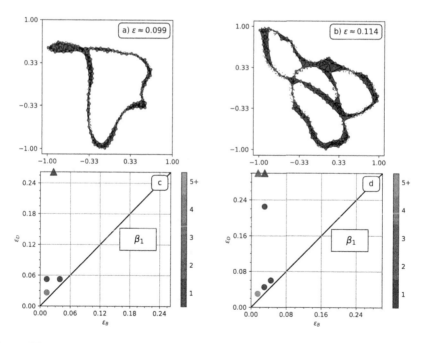

**Fig. 4.** Witness complexes for A440 ($\mathfrak{f} = 440\,\text{Hz}$) played on a (a) clarinet and (b) viol constructed using 2000 witnesses and 100 landmarks for $\epsilon = 0.099, 0.114$ respectively. Beneath each are the corresponding $\beta_1$ persistence diagrams. The delay reconstructions are for approximately 0.05 s each, with $\tau = \frac{1}{\mathfrak{f}\pi}$ s. (Color figure online)

The patterns in these persistence diagrams—the number of highly persistent holes and short-lived features, and the $\epsilon$ values at which they appear and disappear—suggest that computational topology can be an effective way to distinguish between musical instruments. To test this more broadly, we built a pair of simple classifiers that work with *persistent rank functions* (PRFs), cumulative functions on $\mathbb{R}^{2+}$ that report the number and location of points in a persistence diagram [20]. We trained each classifier on 25 disjoint 0.05 s windows from recordings of the corresponding instrument. This involved computing the persistent homology for each instance, then computing the mean, $\overline{\beta^1}$, and standard deviation, $\sigma$ of the set of corresponding PRFs. The test set comprised 50 0.05 s windows, 25 from each instrument; for each of these samples, we computed the $L^2$ distance between the PRF of the sample and the mean $\overline{\beta^1}$ for each instrument. If that distance was below $k\sigma$ for some threshold parameter $k$, we assigned membership in the corresponding instrument class. The receiver operating characteristic (ROC) curves in the top row of Fig. 5 plot the true positive rates versus the false positive rates for the PRF classifiers. The clarinet classifier achieves a true positive rate 70% around $k = 0.5$, and 100% when $k = 1$. The false positive rate remains near 0% up through $k = 5$, demonstrating a broad range of threshold values $k$ for which the PRF classifier will successfully assign membership in the clarinet class to most clarinet tones—and non-membership to most viol tones. The viol classifier achieves a true positive rate near 70% by $k = 1$ and over 90% by $k = 2$, maintaining a false positive rate below 50% for all $k$ up to 2.5.

As a comparison, we built a pair of FFT-based classifiers, whose results are shown in the bottom row of Fig. 5, training and testing them on the same samples used for the PRF-based classifiers. The feature vector in this case was a set of 2000 logarithmically spaced values between 10 Hz and 10,000 Hz from the power spectrum of the signal. As in the PRF-based classifier, we computed the mean and standard deviation of this set of feature vectors, classifying a sample as a viol or clarinet if its $L^2$ distance to the corresponding mean feature vector was less than $k\sigma$. As is clear from the shapes of the ROC curves, the PRF-based classifiers outperformed the FFT-based classifiers. The FFT-based clarinet classifier achieves 70% and 20% true and false positive rates, respectively, around $k = 0.5$. Above that threshold, the false positive rate rapidly catches up to the true positive rate, making the classifier equally likely to correctly classify a clarinet as a clarinet as it is to erroneously classify a viol as a clarinet. The ROC curve for the FFT-based viol membership classifier is even closer to the diagonal: it will correctly classify a viol as a viol only slightly more often than it will erroneously classify a clarinet as a viol, for any parameter value $0 < k < 5$.

Clarinets and viols produce very different sounds, of course, so distinguishing between them is not a hugely challenging task. A more interesting challenge is to compare two pianos. As shown in Fig. 6, persistence diagrams of the same note played on an upright piano and a grand piano are notably different: the former has a single long-lived hole—the fundamental tone of the note—while the latter has *two*, perhaps reflecting the greater sonic richness of the instrument.

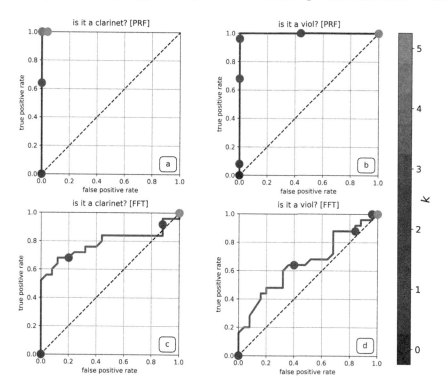

**Fig. 5.** ROC curves for a persistent homology-based classifier (top) and an FFT-based classifier (bottom) for clarinet (left) and viol (right) membership testing. Color bar indicates the threshold parameter value $0 < k < 5$. (Color figure online)

Persistence diagrams for both pianos contain many short-lived holes at low $\epsilon$ values, which also speaks to a notable variance in the volumes of the resonating frequencies.

Table 1 shows the runtime and memory costs involved in the construction of some of the complexes mentioned in this paper. These numbers make it quite clear why the parsimonious nature of the witness complex is so useful: using *all* of the points as landmarks is computationally prohibitive. And that parsimony, surprisingly, does not come at the expense of accuracy, as long as the samples satisfy some denseness constraints [1]. Nonetheless, this is still a lot of computational effort; the membership test process described here involves building the complex, computing the homology, repeating those calculations across a range of $\epsilon$ values, and perhaps computing a persistent rank function from the results. The associated runtime and memory costs will worsen with increasing $\epsilon$, and with the size of the data set, so computational topology is not the first choice technique for every IDA application. However, it can work when statistical- and frequency-based techniques do not.

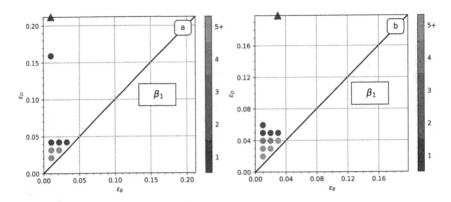

**Fig. 6.** Persistence diagrams for A440 on two different pianos. (a): a Steinway grand piano; (b): a Baldwin upright piano.

**Table 1.** The computational and memory costs involved in constructing different simplicial complexes from the 2000-point reconstruction of Fig. 1(b) with $\epsilon = 0.073$ on an Ubuntu Linux machine with an Intel Core i5 1.70 GHz 4 CPU and 12 GB of memory.

| Number of landmarks | Runtime (s) | Number of two-simplices | Memory usage |
|---|---|---|---|
| 2000 (Čech) | 59.1 | 2,770,627 | 9.3 MB |
| 200 | 3.9 | 3,938 | 0.9 MB |
| 50 | 0.8 | 93 | < 0.1 MB |

## 4    Conclusion

We have shown that persistent homology can successfully distinguish musical instruments using witness complexes built from two-dimensional delay reconstructions for a single note. This approach does not rely on the linearity or data aggregation of many traditional membership-testing techniques; moreover, topological data analysis can outperform these traditional methods. Though the associated computations are not cheap, the reduction in model order and the parsimony of the witness complex greatly reduce the associated computational costs.

Persistent homology calculations—on any type of simplicial complex—work by blending geometry into topology via a scale parameter $\epsilon$. Their leverage derives from the patterns that one observes upon varying $\epsilon$, which are presented here in the form of persistence diagrams. The witness complex uses the scale parameter to obtain its natural parsimony. While the specific form of the witness relation used in this paper is a good start, it can still create holes where none "should" exist, and vice versa. Better witness relations—factoring in the temporal ordering and/or the forward images of the witnesses, or the curvature of those paths—will be needed to address those issues. This is particularly important in the context of the kinds of reduced-order models that we use here to further control the computational complexity. An incomplete delay reconstruction is a projection of a high-dimensional structure onto a lower-dimensional manifold: an action that can collapse holes, or create false ones. Changing $\tau$ also alters the geometry of a delay reconstruction. Understanding the interplay of geometry and topology in an incomplete embedding, and the way in which the witness relation exposes that structure, will be key to bringing topological data analysis into the practice of intelligent data analysis.

# References

1. Alexander, Z., Bradley, E., Meiss, J., Sanderson, N.: Simplicial multivalued maps and the witness complex for dynamical analysis of time series. SIAM J. Appl. Dyn. Sys. **14**, 1278–1307 (2015)
2. Bradley, E., Kantz, H.: Nonlinear time-series analysis revisited. Chaos **25**(9), 097610 (2015)
3. Cvitanovic, P.: Invariant measurement of strange sets in terms of circles. Phys. Rev. Lett. **61**, 2729–2732 (1988)
4. Dasu, T., Krishnan, S., Lin, D., Venkatasubramanian, S., Yi, K.: Change (detection) you can believe in: finding distributional shifts in data streams. In: Adams, N.M., Robardet, C., Siebes, A., Boulicaut, J.-F. (eds.) IDA 2009. LNCS, vol. 5772, pp. 21–34. Springer, Heidelberg (2009). doi:10.1007/978-3-642-03915-7_3
5. de Silva, V., Ghrist, R.: Coverage in sensor networks via persistent homology. Alg. Geom. Topology **7**(1), 339–358 (2007)
6. Domingos, P.: A general framework for mining massive data streams. In: Proceedings Interface 2006 (2006)
7. Edelsbrunner, H., Letscher, D., Zomorodian, A.: Topological persistence and simplification. Disc. Comp. Geom. **28**, 511–533 (2002)
8. Edelsbrunner, H., Mücke, E.: Three-dimensional alpha shapes. ACM Trans. Graph. **13**, 43–72 (1994)
9. Fraser, A., Swinney, H.: Independent coordinates for strange attractors from mutual information. Phys. Rev. A **33**(2), 1134–1140 (1986)
10. Gama, J.: Knowledge Discovery from Data Streams. Chapman and Hall/CRC, Atlanta (2010)
11. Garland, J., Bradley, E., Meiss, J.: Exploring the topology of dynamical reconstructions. Physica D **334**, 49–59 (2014)
12. Ghrist, R.: Barcodes: the persistent topology of data. Bull. Amer. Math. Soc. **45**(1), 61–75 (2008)
13. Kaczynski, T., Mischaikow, K., Mrozek, M.: Computational Homology. AMS, vol. 157. Springer, New York (2004)
14. Kennel, M.B., Brown, R., Abarbanel, H.D.I.: Determining minimum embedding dimension using a geometrical construction. Phys. Rev. A **45**, 3403–3411 (1992)
15. Muldoon, M., MacKay, R., Huke, J., Broomhead, D.: Topology from a time series. Physica D **65**, 1–16 (1993)
16. Munkres, J.: Elements of Algebraic Topology. Benjamin/Cummings, Menlo Park (1984)
17. Rakthanmanon, T., Keogh, E., Lonardi, S., Evans, S.: Time series epenthesis: clustering time series streams requires ignoring some data. In: Proceedings ICDM, pp. 547–556. IEEE, December 2011

18. Robins, V.: Computational topology for point data: Betti numbers of α-shapes. In: Mecke, K., Stoyan, D. (eds.) Morphology of Condensed Matter. Lecture Notes in Physics, vol. 600, pp. 261–274. Springer, Heidelberg (2002)
19. Robins, V., Abernethy, J., Rooney, N., Bradley, E.: Topology and intelligent data analysis. Intell. Data Anal. **8**, 505–515 (2004)
20. Robins, V., Turner, K.: Principal component analysis of persistent homology rank functions with case studies of spatial point patterns, sphere packing and colloids. Physica D **334**, 99–117 (2015)
21. de Silva, V., Carlsson, E.: Topological estimation using witness complexes. In: Alexa, M., Rusinkiewicz, S. (eds.) Eurographics Symposium on Point-Based Graphics, pp. 157–166. The Eurographics Association (2004)
22. Song, M., Wang, H.: Highly efficient incremental estimation of Gaussian mixture models for online data stream clustering. In: Priddy, K. (ed.) Proceedings of SPIE, vol. 5803, pp. 174–183. International Society for Optical Engineering (2005)
23. Xia, K., Feng, X., Tong, Y., Wei, G.: Persistent homology for the quantitative prediction of fullerene stability. J. Comput. Chem. **36**(6), 408–422 (2015)
24. Zomorodian, A.: Topological data analysis. In: Advances in Applied and Computational Topology, vol 70. American Mathematical Society (2012)

# Improving Cold-Start Recommendations with Social-Media Trends and Reputations

João Santos, Filipa Peleja, Flávio Martins, and João Magalhães$^{(\boxtimes)}$

NOVA Laboratory for Computer Science and Informatics, FCT NOVA,
Campus Caparica, Almada, Portugal
{jmes.espada.santos,filipapeleja}@gmail.com,
{fnm,jmag}@fct.unl.pt

**Abstract.** In recommender systems, the cold-start problem is a common challenge. When a new item has no ratings, it becomes difficult to relate it to other items or users. In this paper, we address the cold-start problem and propose to leverage on social-media trends and reputations to improve the recommendation of new items. The proposed framework models the long-term reputation of actors and directors, to better characterize new movies. In addition, movies popularity are deduced from social-media trends that are related to the corresponding new movie. A principled method is then applied to infer cold-start recommendations from these social-media signals. Experiments on a realistic time-frame, covering several movie-awards events between January 2014 and March 2014, showed significant improvements over ratings-only and metadata-only based recommendations.

**Keywords:** Cold-start · Recommendation · Online reputation · Sentiment analysis · Social-media

## 1 Introduction

Social-media services are a preferred way for users to obtain insight into various subjects, such as restaurants, events, books and movies. The information shared through social-media services is highly relevant to assert what everyone is consuming or looking at. Its importance is clear for recommender systems, whose goal is to compute suggestions concerning products of interest. Thus, automatic systems that mine these trends [1] and reputations across multiple social-media services are of great value in many recommendation scenarios.

In the most common scenario, user-item ratings are predicted for old items, i.e., movies that have already been rated by users trying to improve error or diversity [2]. A more challenging scenario for a recommender system is the cold-start scenario: given a new item that has not been rated or commented by any user, *how can we relate this item to other items or potential consumers?* In this paper, we argue that information shared in social-media services, can be useful to tackle the cold-start problem in recommender systems. Our main contribution concerns the decomposition of a new item (a movie in our case), into key components, whose reputation can be tracked over time (actors and directors). More specifically, we propose that the reputation of directors and

N. Adams et al. (Eds.): IDA 2017, LNCS 10584, pp. 297–309, 2017.
DOI: 10.1007/978-3-319-68765-0_25

actors obtained from textual feedback can be a quality measure for new movies. Although these movies have not yet been rated, their directors and actors have a track record. On top of this historical data, we also leverage on more immediate data, e.g. Twitter trends, to improve the recommendations. Experiments on data captured between January 2014 and March 2014, leveraged on the reputation record of actors and directors to improve recommendations for the cold-start scenario in the movies domain.

## 2  Related Work

One of the main applications of media monitoring is to capture popular topics in real-time. Twitter is one of the most popular micro-blogging services, favoring real-time user communication. Many authors have proposed methods to follow trends in Twitter [3–5]. TwitterMonitor [3] is a framework that exploits co-occurrences of bursty keywords to identify real-time trends on Twitter streams. Twevent [4] is a similar framework, which incorporates Wikipedia to identify realistic events and associate captured tweets to identified events. In turn, Ozdikis et al. [5] showed that the usage of *hashtags* improves the accuracy of event detection when compared to standard word-based vector generation methods.

A similar objective which also links to media monitoring is the reputation analysis of specific entities, such as movies, celebrities, products, etc. This process enables various applications, one of which is aiding in prediction tasks. Krauss et al. [6] used sentiment analysis on the IMDb discussion boards to predict Oscar nominations by obtaining the movies with best reputation. Joshi et al. [7] explored the popularity of old movies among online critic reviews to predict opening weekend revenues for new movies, by comparing the similarity in metadata of old highly rated movies with new ones. In a similar approach, Asur et al. [8] exploited bursty keywords on Twitter streams to predict box-office revenue for new movies. Oghina et al. [9] predicted the IMDb movie ratings by analyzing their popularity on social-media, namely YouTube and Twitter. In contrast, we use the reputation of actors and directors combined with data from Twitter streams to compute cold-start recommendations.

In some media review services such as IMDb or Metacritic, in addition to the traditional numeric ratings, users can also input feedback using textual reviews. Valuable information that is hard to represent by numeric ratings alone can be extracted from these textual reviews. To improve recommendation in the cold-start scenario, Moshfeghi et al. [10] explored the incorporation of emotions and semantic spaces. In their collaborative recommender system, information from user reviews and plot summaries is extracted in order to better describe items and users. Similarly, Jakob et al. [11] identified and clustered movie topics using movie reviews and a ratings matrix to recommend movies based on the prediction of user-topic similarity. Zhang et al. [12], obtained good results in online video recommendation with a framework that extracts a like/dislike rating from textual reviews and comments to minimize data sparsity. A different approach was proposed by Amatriain et al. [13], where "expert" opinions were extracted from a website widely considered as a trust-worthy source. Amatriain et al. applied *k-nn* to compute the similarity between users and experts, and recommend movies according to the most similar experts.

## 3  Cold-Start Recommendations with Social-Media Signals

To handle the cold-start problem, we propose a framework that computes personalized recommendations by exploring the reputation of new movies, directors and actors in social-media services, namely on Twitter and IMDb. Moshfeghi et al. [10] showed that recommendation of cold-start movies performs best when considering a movie meta-data and the sentiment expressed in movie reviews. Building on this idea, we represent a movie as the vector

$$m_j = (D_j, A_j, G_j, R_j, S_j), \tag{1}$$

where $D_j$ is the set of directors, $A_j$ is the set of participating actors, $G_j$ is the set of corresponding genres, $R_j$ is the set of associated user ratings and $S_j$ is the social-media feedback inferred by a monitoring process described in Sect. 3.2. The $S_j$ variable is composed of the Twitter posts (or tweets) about the movie $m_j$ as well as the reputation of its directors and actors, obtained from IMDb. As we will see, $S_j$ will be fundamental to improve cases of cold-start recommendations where $R_j = \emptyset$.

In this scenario, users rate the movies they have watched and from this data we compute their profiles in terms of personal preferences towards directors, actors and genres. Formally, a user $u_i$ is then represented as the vector

$$u_i = (D^i, A^i, G^i), \tag{2}$$

where $D^i$ is the set of directors, $A^i$ is the set of actors and $G^i$ is the set of genres. These three sets follow the same structure and are represented as

$$D^i = \left\{ (d_i^1, dr_i^1, df_i^1), \ldots, (d_i^n, dr_i^n, df_i^n), \ldots \right\}, \tag{3}$$

$$A^i = \left\{ (a_i^1, ar_i^1, af_i^1), \ldots, (a_i^n, ar_i^n, af_i^n), \ldots \right\}, \tag{4}$$

$$G^i = \left\{ (g_i^1, gr_i^1, gf_i^1), \ldots, (g_i^n, gr_i^n, gf_i^n), \ldots \right\}, \tag{5}$$

where the first element $d_i^n$ identifies the director, $dr_i^n$ is the average rating given by the user to the movies directed by that director and $df_i^n$ is the number of movies directed by $d_i^n$ that are rated by the user. The same rationale applies to $A^i$ and $G^i$.

### 3.1  Formal Model

We start by exploring the similarity of the movie profile Eq. (1) and the user profile Eq. (2). This similarity is obtained by quantifying how much a user likes each aspect of the movie separately, i.e., the values $\hat{d}_{ij}$, $\hat{a}_{ij}$ and $\hat{g}_{ij}$, and later combining them into a final similarity score.

To infer the user $u_i$ preference towards the directors of the movie $m_j$, we compute the weighted average of how much the user likes each director of the movie, i.e., the weighted average of the values $dr_i$ for each director on $D_j$. The weight, representing the contribution of each director rating to the average, is pondered according to the number of movies that the user rated where the director participated, i.e., each director's corresponding value $df_i$ on the user profile $D^i$. The reasoning, is that a user formulates a more refined and accurate opinion about a director if he/she watches more movies from that director. Hence, we consider that directors that have been watched more times by the user should have a stronger weight on the prediction. Let $D_{ij} = D^i \cap D_j$ be the set of the directors of movie $m_j$ that are on the user profile $D^i$. The weight $w_{d_{ij}^n}$ of the $n^{th}$ director $d_{ij} \in D_{ij}$ is then obtained by the expression

$$w_{d_{ij}^n} = \frac{df_i}{\sum_{p \in D_{ij}} df_p},$$

(6)

such that $\sum_n w_{d_{ij}^n} = 1$. Considering this, the preference of user $u_i$ towards the team of directors of the movie $m_j$ is obtained by the expression

$$\hat{d}_{ij} = \frac{\sum_{n \in D_{ij}} dr_i^n \cdot w_{d_{ij}^n}}{|D_{ij}|},$$

(7)

where $|D_{ij}|$ is the number of directors on $D_{ij}$. Since all director ratings $dr_{ij}$ are values between 1 and 10, the resulting average $\hat{d}_{ij}$ will also be a value between 1 and 10. Note that when none of the directors of movie $m_j$ are on the user's directors set $D^i$, $\hat{d}_{ij} = 0$.

How much the user $u_i$ likes the actors of the movie $m_j$ can be obtained similarly to how $\hat{d}_{ij}$ is obtained. Let $A_{ij} = A^i \cap A_j$ be the set of actors of movie $m_j$ that are on the user profile $A^i$. Thus, the user $u_i$ preference towards likes the actors of the movie $m_j$ is obtained by the expression

$$\hat{a}_{ij} = \frac{\sum_{n \in A_{ij}} ar_i^n \cdot w_{a_{ij}^n}}{|A_{ij}|},$$

(8)

where $|A_{ij}|$ is the number of actors on $A_{ij}$ and $w_a$ is the weight of the actor $a$. Similarly to $\hat{d}_{ij}$, when none of the actors of movie $m_j$ are on the user actors $A^i$, $\hat{a}_{ij} = 0$. In turn, let $G_{ij} = G^i \cap G_j$ be the set of the genres of movie $m_j$ that are on the user profile $G_i$. How much the user $u_i$ likes the genres of the movie $m_j$ is obtained by the expression

$$\hat{g}_{ij} = \frac{\sum_{n \in G_{ij}} gr_i^n \cdot w_{g_{ij}^n}}{|G_{ij}|},$$

(9)

where $|G_{ij}|$ is the number of genres on $G_{ij}$ and $w_g$ is the weight of the genre $g$. Like $\hat{d}_{ij}$, and $\hat{a}_{ij}$, $0 \le \hat{g}_{ij} \le 10$, with 0 occurring when none of movie genres are on the user genres $G^i$. The predicted rating $\hat{pr}_{ij}$ for user $u_i$ and the *cold-start* movie $m_j$ is obtained by the expression:

$$\widehat{pr}_{ij} = \frac{1}{T}\left(\theta_a \cdot \hat{a}_{ij} + \theta_d \cdot \hat{d}_{ij} + \theta_g \cdot \hat{g}_{ij}\right). \tag{10}$$

where $\theta_a$, $\theta_d$ and $\theta_g$ are constants controlling the contributions of directors, actors and genres to the rating predictions. Their values are estimated from a set of training data by finding the values that minimize Mean Average Error. Where $T$ is the number of feature set ratings $\hat{d}_{ij}$, $\hat{a}_{ij}$ and $\hat{g}_{ij}$ that are different from 0. In the following sections, we will extend the $\widehat{pr}_{ij}$ computation to include social-media feedback.

### 3.2   Social-Media Trends and Reputations

In this section, we formalize the social-media feedback,

$$S_j = \left\{ T(m_j), reps(m_j) \right\}, \tag{11}$$

as the set of tweets $T(m_j)$ where the movie $m_j$ is mentioned, and the reputation of all actors and directors participating on movie $m_j$.

**New-movies popularity on Twitter**
The social-media feedback about new movies is obtained from Twitter: tweets where the movie title is identified are stored and labelled according to the movie name. The captured tweets are then classified by a sentiment classifier such that, for each tweet, it is inferred if it is positive or negative reference to the movie. A tweets index is then constructed to allow fast look-ups for the cold-start recommendation. Formally, the resulting tweets for a certain movie $m_j$ are represented as the set

$$T(m_j) = \left\{ (t_{j1}, s_{j1}), \ldots, (t_{jl}, s_{jl}), \ldots, (t_{jM}, s_{jM}) \right\}, \tag{12}$$

where $t_{jl}$ is the tweet (talking about $m_j$) and $s_{jl}$ is the sentiment of the tweet such that $s_{jl} \in \{pos, neg\}$. We used a *k-nn* classifier and a domain specific lexicon, see [12].

**Actors and directors reputation on IMDb**
The social-media feedback on directors and actors is obtained from IMDb: movie reviews are crawled and used to build a sentiment graph linking named-entities, from which the reputation of directors and actors is computed (see [14] for details). This step allows us to obtain the reputation of the directors and actors of the new movies we want to recommend. Formally, the reputation of all the directors and actors participating on movie $m_j$ is represented by the expression

$$reps(m_j) = \{rep(e_1), \ldots, rep(e_k), \ldots\}, \tag{13}$$

where the reputation of each entity $e_k$ is $rep(e_k) \in [0.0, 1.0]$, with 0.0 being the worst reputation and 1.0 being the best reputation.

### 3.3    Recommendations with Social-Media Signals

Moshfeghi et al. [10] and Krauss et al. [6] obtained hidden latent factors, to correlate movies through sentiment analysis. Here, however, new movies do not have reviews and tweets about new movies are too scarce to infer relevant latent topics. Therefore, we explore emotion as a qualitative measure, in which we obtain and consider the inherent quality of new movies, directors and actors.

The rating prediction $\hat{r}_{ij}$ is obtained by considering both, how popular the movie is, $\text{pop}(m_j)$, and how much a user might enjoy the movie $m_j$, given the reputation $\text{reps}(m_j)$ of its participants. The proposed approach is formalized as

$$\hat{r}_{ij} = \alpha_t \cdot \left(\text{pop}(m_j) + bias_i\right) + (1 - \alpha_t) \cdot \widehat{pr}_{ij|\text{reps}(m_j)}, \tag{14}$$

where $\alpha_t$ is a constant reflecting the importance of the movies popularity to the final user-movie rating. Formally, the user $u_i$ bias accounts for the deviation of the user ratings from the general average rating:

$$bias_i = \frac{\sum_{r_i^k \in ur_i}(r_i^k - avg_{<k>})}{|ur_i|}. \tag{15}$$

Let $ur_i = \{r_i^1, \ldots, r_i^k, \ldots, r_i^K\}$ be the user $u_i$ ratings, $avg_{<k>}$ be the average rating of the movie $m_{<k>}$, and $|ur_i|$ is the number of ratings given by user $u_i$. Rewriting Eq. (10) with the new reputations information we have:

$$\widehat{pr}_{ij|\text{reps}(m_j)} = \frac{1}{T}\left(\theta_a \cdot \hat{a}_{ij|\text{reps}(m_j)} + \theta_d \cdot \hat{d}_{ij|\text{reps}(m_j)} + \theta_g \cdot \hat{g}_{ij}\right). \tag{16}$$

**Modeling user preferences $\hat{a}_{ij|\text{reps}(m_j)}$ and $\hat{d}_{ij|\text{reps}(m_j)}$ with social-media signals**

Up until this point, when predicting the values $\hat{d}_{ij}$ and $\hat{a}_{ij}$ (i.e., how much a user likes or dislikes the directors and actors of a movie), the entities that the user does not know were not considered. In this section, we propose to enhance the calculation of $\hat{d}_{ij}$ and $\hat{a}_{ij}$, given the reputations of directors and actors available in $\text{reps}(m_j)$. For this purpose, two new variables, $\widehat{ud}_{ij}$ and $\widehat{ua}_{ij}$, are introduced to express the reputation of the unknown directors and actors:

$$\widehat{ud}_{ij} = \frac{\sum_{d \in D_j}\text{rep}(d)}{|D_j|}, \quad \widehat{ua}_{ij} = \frac{\sum_{a \in A_j}\text{rep}(a)}{|A_j|}, \tag{17}$$

where $D_j - D^i$ and $A_j - A^i$ are the sets of directors and actors on movie $m_j$ that the user does not know.

To consider $\widehat{ud}_{ij}$ and $\widehat{ua}_{ij}$, in the calculation of $\widehat{pr}_{ij|\text{reps}(m_j)}$, one ought to note that $\hat{d}_{ij}$ and $\hat{a}_{ij}$ represent user preferences towards their known directors and actors. Thus, $\hat{d}_{ij|\text{reps}(m_j)} = \hat{d}_{ij}$ and $\hat{a}_{ij|\text{reps}(m_j)} = \hat{a}_{ij}$, when all the directors or actors of $m_j$ are known by the user, and $\hat{d}_{ij|\text{reps}(m_j)} = \widehat{ud}_{ij}$ and $\hat{a}_{ij|\text{reps}(m_j)} = \widehat{ua}_{ij}$, when the user does not know any directors or actors of the movie. The general case is when the user knows some of the directors and actors of the movie. Formally, the final directors and actors scores $\hat{d}_{ij|\text{reps}(m_j)}$ and $\hat{a}_{ij|\text{reps}(m_j)}$ are calculated by considering both the user preferences and the public opinion, i.e., a weighted average between the scores of the known entities and the unknown entities:

$$\hat{d}_{ij|\text{reps}(m_j)} = \delta_{ud} \cdot (\widehat{ud}_{ij} + bias_i) + (1 - \delta_{ud}) \cdot \hat{d}_{ij}, \tag{18}$$

$$\hat{a}_{ij|\text{reps}(m_j)} = \delta_{ua} \cdot (\widehat{ua}_{ij} + bias_i) + (1 - \delta_{ua}) \cdot \hat{a}_{ij}, \tag{19}$$

where the constants $\delta_{ud}$ and $\delta_{ua}$, represent the contribution of the unknown directors and actors to the computation of $\hat{d}_{ij}$ and $\hat{a}_{ij}$ respectively. They are computed as:

$$\delta_{ud} = \frac{|D_j - D^i|}{|D_j|}, \quad \delta_{ua} = \frac{|A_j - A^i|}{|A_j|}, \tag{20}$$

where $|D_j - D^i|$ is the number of directors on movie $m_j$ that the user $u_i$ does not know and $|A_j - A^i|$ is the number of actors on movie $m_j$ that the user does not know.

**Modeling a movie popularity pop$(m_j)$ with social-media trends**
So far, the predicted rating $\hat{pr}_{ij}$ captures an incomplete set of indicators about the movie, missing a key indicator which is the trendiness of that movie. Krauss et al. [6] has indeed showed that movie trendiness is projected in Oscar nominations, which are generally associated with highly rated movies. The set $T(m_j)$, containing tweets targeting movie $m_j$, can be used to predict its reputation. Oghina et al. [9] have shown that the fraction of likes/dislikes is the strongest feature for predicting IMDb movie ratings from social-media. Following this remarks, we consider the popularity of a movie $m_j$, to be measured as

$$\text{pop}(m_j) = \frac{|pos_j|}{|tweets_j|}, \tag{21}$$

where $|pos_j|$ is the number of positive tweets referring the movie $m_j$ and $|tweets_j|$ is the total number of tweets referring $m_j$.

## 4 Evaluation

### 4.1 Datasets

To set up a realistic evaluation scenario, we obtained our dataset by crawling IMDb user-movie ratings. We focused the extraction process on users who have rated at least one of a selection of 60 new movies, finalists on 5 popular movie awards ceremonies:

the 2014 editions of *The Golden Globes*, *The Critic's Choice Awards*, *The BAFTA Film Awards*, *The Independent Spirit Awards* and *The Oscars*. We selected such movies so as to capture a large number of tweets in a small period. We also crawled 52,236 tweets, between January 2014 and March 2014, regarding the new movies.

In total, we obtained a dataset with 1,064,766 ratings, given by 2,909 users to 60 new movies and 46,843 old movies. The computation of the actors and directors reputation [14] for the new movies used a total of 124,236 IMDb reviews. In total we considered 225 actors and 169 directors corresponding to the 60 new movies. Finally, the movies metadata for all rated movies were obtained through the publicly available OMDb API (www.omdbapi.com).

## 4.2   Methods

To evaluate the proposed approach we leveraged on data concerning 60 new movies collected from different sources. This created a realistic setting to assess the different elements of the model formalized in Eq. (14). The first method (MRep), assesses the contribution of the movie reputation $\text{pop}(m_j)$, which is inferred from tweets available through Twitter API. The second method (ERep), assesses the contribution of entities reputations $\hat{a}_{ij|\text{reps}(m_j)}$ and $\hat{d}_{ij|\text{reps}(m_j)}$, which were computed from IMDB reviews. The third method (FRep), uses the full spectrum of social-media reputation where both movies popularity and entities reputation are considered. We evaluate each variant by comparing the predicted ratings to the real user-movie ratings for the new movies. We use the Mean Average Error (MAE) to evaluate rating predictions and the F-Measure to evaluate recommendations.

The proposed approach, was also compared to baseline methods where recommendations are predicted from the movies metadata and past ratings. The first baseline is the *k-nn* algorithm which is widely successful for hybrid recommendations [13]. In the *k-nn* algorithm a movie is represented by its average rating and a binary vector representing its metadata (a user is represented by its rated movies). The second baseline, formalized in Eq. (10), does not include any social-media feedback: it is formal model where user ratings directly affect the importance of each aspect of a movie metadata. We distinguish the case where $\theta_d$, $\theta_a$ and $\theta_g$ are all equal to 1.0 and where these weights were estimated with 10-fold cross validation, resulting in $\theta_d = 0.35$, $\theta_a = 0.2$ and $\theta_g = 0.45$. We refer to these models as FM1 and FM2, respectively.

Before discussing the main results, we evaluate Twitter as a source of reliable movie feedback. We also estimate the best $\delta_t$ parameter for Eq. (21), relevant for Movie Reputation and Full Reputation, and discuss how the inclusion of user bias influences the results.

## 4.3   Results and Discussion

**Twitter for Estimating a Movie Popularity.** We use Twitter as a source of movie feedback to predict the reputation for new movies. We compare the predicted reputations with the average IMDb ratings of the target movies, captured several months

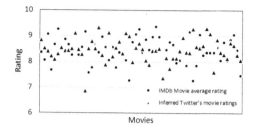

**Fig. 1.** Twitter-based Movie Ratings vs IMDb Movie Ratings.

after the movies' release data. Figure 1 shows the predicted ratings and the IMDb average ratings. From it, we can observe the overall deviation of the predicted ratings.

Overall, the MAE is 0.59, which translates into very accurate results, considering that the rating scale is very ample (from 1 to 10). The prediction errors varied from 0.026 (*Blue is the Warmest Colour*) to 2.29 (*Her*). By analysing the overall error, we can observe that movies with lower IMDb ratings are more likely to have a higher prediction error: for instance, while *Blue is the Warmest Colour* has an average IMDb rating of 8.0, examples of high error such as *The Invisible Woman* (MAE = 2.01) and *Computer Chess* (MAE = 1.73), have an average IMDb rating of 6.3. This leads us to believe that Twitter users are more likely to share positive tweets about movies than negative tweets, which makes our method more precise for highly rated movies.

**Estimation of the $\alpha_t$ parameter.** In Eq. (14), the parameter $\alpha_t$ controls the influence of the movie reputation in the user-movie rating predictions. In order to find the best value for $\alpha_t$, we computed rating predictions for various $\alpha_t$ values on a validation dataset. Figure 2 plots the MAE and F-Measure curves for a range of value. Both MRep and FRep present the best results for $\delta_t$ values below or equal to 0.40 – after this point both MAE and F-Measure start to deteriorate. For movie reputation, both the best MAE and F-Measure values are obtained at $\alpha_t = 0.35$ (MAE of 1.2266 and F-Measure of 87.2%). For full reputation, the best F-Measure is also obtained at 0.35 (87.7%), while the best MAE is obtained at 0.20. These results suggest that the general opinion about the movies, has a significant influence when predicting user-movie *cold-start* ratings. However, if the general opinion ($\alpha_t$) is too contrary to the personal preferences, the predicted user-movie rating looses the personalization component, leading to less accurate predictions. For subsequent experiments, we set $\alpha_t = 0.35$.

**User-bias.** User bias is considered in our methods to adjust the general opinions about movies, directors and actors to each user preferences. We predict user-movie ratings with the three main methods (MRep, ERep and FRep), both considering and not considering user bias ($bias_i = 0$). Table 2 presents the obtained MAE and F-Measure results for each method, both including and excluding user bias. As we can observe, the inclusion of user bias improves the results in all three approaches: the MAE decreases in all methods and the F-Measure increases. In terms of rating prediction, MRep presents the best results (1.227) and ERep presents the worst results (1.254), while FRep presents neither the best nor the worst result (1.245). These values suggest that

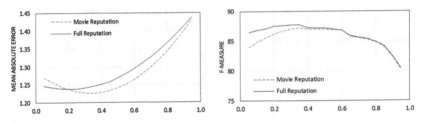

**Fig. 2.** Estimation of $\delta_t$ as a function of Mean Absolute Error and F-Measure.

the reputation of the movie itself is more useful for predicting the user-movie rating, when compared to the reputation of directors and actors.

When computing recommendations, Table 1, assessed by F-Measure, FRep presents the best results (87.7%), when compared to MRep (87.2%) and ERep (85.9%). Unlike when predicting the user-movie rating, considering both the reputation of movies and the reputation of directors and actors is the best approach to distinguishing relevant movies from irrelevant movies. For subsequent experiments, we will consider user bias.

**Table 1.** The influence of user bias.

| Methods | With bias | | Without bias | |
|---|---|---|---|---|
| | MAE | F-M | MAE | F-M |
| MRep | 1.227 | 87.2% | 1.241 | 86.4% |
| ERep | 1.254 | 85.9% | 1.261 | 85.3% |
| FRep | 1.245 | 87.7% | 1.248 | 86.8% |

**Methods Comparison.** Figure 3 shows the MAE and F-Measure results for users with different numbers of rated old movies. This enables us to compare how each method handles different levels of user sparsity. From all the methods, *k-nn* presents the worst results: it has the worst MAE for all levels of users; in recommendation, it only matches other methods for users with more than 40 rated movies. From the other baseline methods, FM2 presents better results than FM1 in both MAE and F-Measure for all values, suggesting that directors, actors and genres should weight differently when predicting user-movie ratings. The three main methods perform better than all baselines, both in rating prediction and recommendation. ERep performs better than MRep when recommending to users with less than 70 rated movies, while MRep performs better than ERep for users with a high number of rated movies. This shows that the reputation of a movie is only better to predict its ratings when accompanied with well-defined user preferences. FRep presents the best recommendation results overall: it performs very closely to ERep when recommending to users with less or equal to 70 rated movies and very closely to MRep when recommending for users with a lot of rated movies. Table 2 summarizes the overall results on all measures for all the methods, for all the dataset.

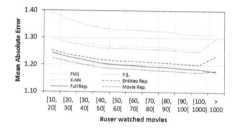
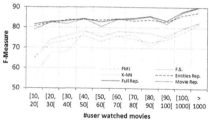

**Fig. 3.** Mean average error comparison.

**Table 2.** Overall comparitive results.

|       | MAE    | Prec. (%) | Rec. (%) | F-M (%) |
|-------|--------|-----------|----------|---------|
| k-nn  | 1.3933 | 70.1      | 86.5     | 78.3    |
| FM1   | 1.3058 | 70.3      | 85.2     | 77.8    |
| FM2   | 1.2962 | 71.4      | 87.7     | 79.6    |
| MRep  | 1.2266 | 76.0      | 98.5     | 87.2    |
| ERep  | 1.2536 | 75.6      | 96.1     | 85.9    |
| FRep  | 1.2450 | 76.0      | 99.4     | 87.7    |

Overall, all the methods that consider social-media information outperform the baselines. In terms of rating prediction, MRep presents the best MAE results. FRep, for instance, presents the best results in all recommendation measures, with a total F-Measure of 87.7%. The improvement in Recall values for the social-media methods relatively to the baselines shows that the reputation of movies, directors and actors help especially in identifying great movies, which are more usually relevant for users.

**Cases of extreme user cold-start.** When users have not rated many movies, their preferences cannot be well modelled: these users suffer from the *cold-start* problem. This happens mostly, but not exclusively, for users who are new to the system. We simulate a scenario where all our 2,909 test users suffer from the *cold-start* problem by not considering their previous ratings and perform experiments with MRep, ERep and FRep. Table 3 show both the obtained MAE and F-Measure results for each method. Note that users bias cannot be considered in this scenario since there are no previously given ratings from any user.

ERep obtains the worst results (MAE of 1.7741 and F-Measure of 63.5%) while MRep obtains the best results (MAE of 1.6236 and F-Measure of 74.75%). FRep presents intermediate results. These results show that the reputation of movies is a good

**Table 3.** Extreme user-side cold-start.

|       | MAE  | F-Measure |
|-------|------|-----------|
| MRep  | 1.62 | 74.75%    |
| ERep  | 1.77 | 63.50%    |
| FRep  | 1.64 | 69.38%    |

baseline predictor of the quality of a movie and is useful for recommending movies when the user preferences are not known. While the reputation of the movie directors and actors present much worse results, these also prove to be an average predictor of a movie quality, as a 63.5% recommendation accuracy is very good for a scenario where there is no information on users.

## 5    Conclusion

This paper addressed the problem of recommending new movies, affected by the cold-start problem, by monitoring information from social-media services. More specifically, it focused on exploring how the reputation of movies, directors and actors can be used to tackle this issue. Our experiments have shown that Twitter is a reliable source for predicting reputation of new movies: we were able to predict the actual IMDb average rating with a Mean Average Error (MAE) value of 0.59, in a rating scale from 1 to 10. In turn, the reputations of new movies, directors and actors have proven to be useful in cases of movie-side cold-start: we were able to improve recommendation by 8.1% when compared to our baseline method, reaching an F-Measure value of 87.7%.

Finally, the proposed framework has also shown to be useful in severe cases of both movie-side and user-side cold-start: we were able to recommend movies with an F-Measure of 74.75%. This is a major improvement over the baseline methods, where it would be not possible to recommend movies in this scenario.

**Acknowledgements.** This work has been partially funded by the CMU Portugal research project GoLocal Ref. CMUP-ERI/TIC/0033/2014, by the H2020 ICT project COGNITUS with the grant agreement No 687605 and by the project NOVA LINCS Ref. UID/CEC/04516/2013.

## References

1. Peleja, F., Dias, P., Martins, F., Magalhães, J.: A recommender system for the TV on the web: integrating unrated reviews and movie ratings. Multimedia Syst. **19**, 543–558 (2013)
2. Dias, P., Magalhaes, J.: Multi-user diverse recommendations through greedy vertex-angle maximization. In: International Symposium on Intelligent Data Analysis, pp. 96–107 (2014)
3. Mathioudakis, M., Koudas, N.: Twittermonitor: trend detection over the twitter stream. In: Proceedings of the 2010 ACM SIGMOD International Conference on Management of data, pp. 1155–1158 (2010)
4. Li, C., Sun, A., Datta, A.: Twevent: segment-based event detection from tweets. In: Proceedings of the 21st ACM international conference on Information and knowledge management, pp. 155–164 (2012)
5. Ozdikis, O., Senkul, P., Oguztuzun, H.: Semantic expansion of hashtags for enhanced event detection in twitter. In: Proceedings of the 1st International Workshop on Online Social Systems (2012)
6. Krauss, J., Nann, S., Simon, D., Gloor, P.A., Fischbach, K.: Predicting movie success and academy awards through sentiment and social network analysis. In: ECIS, pp. 2026–2037 (2008)

7. Joshi, M., Das, D., Gimpel, K., Smith, N.A.: Movie reviews and revenues: an experiment in text regression. In: Human Language Technologies: The 2010 Annual Conference of the North American Chapter of the Association for Computational Linguistics, pp. 293–296. Association for Computational Linguistics, Stroudsburg, PA, USA (2010)
8. Asur, S., Huberman, B.A.: Predicting the future with social media. In: Web Intelligence and Intelligent Agent Technology (WI-IAT), 2010 IEEE/WIC/ACM International Conference on, pp. 492–499 (2010)
9. Oghina, A., Breuss, M., Tsagkias, M., de Rijke, M.: Predicting IMDB Movie Ratings Using Social Media. In: Baeza-Yates, R., de Vries, Arjen P., Zaragoza, H., Cambazoglu, B.Barla, Murdock, V., Lempel, R., Silvestri, F. (eds.) ECIR 2012. LNCS, vol. 7224, pp. 503–507. Springer, Heidelberg (2012). doi:10.1007/978-3-642-28997-2_51
10. Moshfeghi, Y., Piwowarski, B., Jose, J.M.: Handling data sparsity in collaborative filtering using emotion and semantic based features. In: Proceedings of the 34th international ACM SIGIR conference on Research and development in Information - SIGIR 2011, p. 625. ACM Press, New York (2011)
11. Jakob, N., Weber, S.-H., Müller, M.-C., Gurevych, I.: Beyond the stars: exploiting free-text user reviews for improving the accuracy of movie recommendations. In: Proceedings of the 1st International CIKM Workshop on Topic-Sentiment Analysis for Mass Opinion Measurement, Hong Kong, pp. 57–64 (2009)
12. Zhang, W., Ding, G., Chen, L., Li, C.: Augmenting online video recommendations by fusing review sentiment classification. In: Workshop on Recommender Systems and the Social Web (2010)
13. Amatriain, X., Lathia, N., Pujol, J.M., Kwak, H., Oliver, N.: The wisdom of the few: a collaborative filtering approach based on expert opinions from the web. In: Proceedings of the 32nd international ACM SIGIR conference on Research and development in information retrieval, pp. 532–539 (2009)
14. Peleja, F., Santos, J., Magalhães, J.: Ranking linked-entities in a sentiment graph. In: The 2014 IEEE/WIC/ACM International Conference on Web Intelligence (2014)

# Hierarchical Novelty Detection

Paolo Simeone[1]([✉]), Raúl Santos-Rodríguez[2], Matt McVicar[3],
Jefrey Lijffijt[1], and Tijl De Bie[1]

[1] IDLab, Department of Electronics and Information Systems, Ghent University,
Ghent, Belgium
{paolo.simeone,jefrey.lijffijt,tijl.debie}@ugent.be
[2] Department of Engineering Mathematics, University of Bristol, Bristol, UK
enrsr@bristol.ac.uk
[3] Jukedeck Tech Hub, 20 Ropemaker Street, London EC2Y 9AR, UK
mattjamesmcvicar@gmail.com

**Abstract.** Hierarchical classification is commonly defined as multi-class classification where the classes are hierarchically nested. Many practical hierarchical classification problems also share features with multi-label classification (i.e., each data point can have any number of labels, even non-hierarchically related) and novelty detection (i.e., some data points are novelties at some level of the hierarchy). A further complication is that it is common for training data to be incompletely labelled, e.g. the most specific labels are not always provided. In music genre classification for example, there are numerous music genres (multi-class) which are hierarchically related. Songs can belong to different (even non-nested) genres (multi-label), and a song labelled as Rock may not belong to any of its sub-genres, such that it is a novelty *within* this genre (novelty-detection). Finally, the training data may label a song as Rock whereas it really could be labelled correctly as the more specific genre Blues Rock. In this paper we develop a new method for hierarchical classification that naturally accommodates every one of these properties. To achieve this we develop a novel approach, modelling it as a Hierarchical Novelty Detection problem that can be trained through a single convex second-order cone programming problem. This contrasts with most existing approaches that typically require a model to be trained for each layer or internal node in the label hierarchy. Empirical results on a music genre classification problem are reported, comparing with a state-of-the-art method as well as simple benchmarks.

**Keywords:** Hierarchical classification · Novelty detection · Optimization · Music genre classification · Music information retrieval

## 1 Introduction

Multi-Class Classification (MCC) is defined as the task of assigning one of three or more labels to a training instance. Most approaches to MCC break this problem down in a set of simpler problems, typically two-class classification problems.

© Springer International Publishing AG 2017
N. Adams et al. (Eds.): IDA 2017, LNCS 10584, pp. 310–321, 2017.
DOI: 10.1007/978-3-319-68765-0_26

Examples include One vs All, One vs One and Error Correcting Output Code (ECOC) approaches [1]. These approaches, however, fail to take into account particular structure amongst the labels. A particular case of interest is Hierarchical Classification (HC), where labels are hierarchically nested. For example in Music Genre Classification (MSC), if a song is labelled with Blues Rock the hierarchical relation among genre labels implies that it can be labelled with Rock as well. Also in domains such as image [2], text [17] and phoneme [5] classification, assuming hierarchical label relations provides both enhanced accuracy and more straightforward interpretation.

As a motivation for the present paper, we use the Music Genre Classification (MGC) task [4,10,11,13,15]. MGC is confronted with a number of issues:

1. Music genres are hierarchically organised (see Sect. 2.1 for a discussion). This means that if a label applies (e.g. Blues Rock), then all more generic labels must automatically apply as well (e.g. Rock).
2. Some songs are influenced by several non-hierarchically related genres. Thus, several labels can apply, even if not directly related in the genre hierarchy.
3. Some songs can only sensibly be categorised into high-level genres (commonly cited examples: Rock, Classical) and not easily into any subgenre, whereas others might be highly specific (Melodic Death Metal).
4. However, the labels provided for some data points in commonly available training sets are not always maximally specific, i.e., a Blues Rock song may be labelled as Rock but not as Blues Rock. Some genre labels may also be altogether missing, e.g. because the genre label was unknown by the data annotator. In other words, while the *presence* of a label in the training implies that this label applies to that song, the *absence* of a label does not imply that it does not apply.

The first issue makes MGC a HC problem. Due to the second issue, MGC is also inherently a Multi-Label Classification problem. The third issue means that for each class there are really only positive examples available in the training set, as the absence of a label for a data point does not imply that data point is a negative example for that class. MGC is thus reminiscent of Novelty Detection (ND) [14]. Finally, to account for the fourth issue, a suitable method must be capable of predicting a (set of) label(s) that are not necessarily maximally specific. We believe that these four issues are relevant in many practical problems beyond MGC.

A naive approach to tackle issues 2–4 would be train a ND method for each of the different classes. Given a song, it can then be labelled with the genres for which the song is not a novelty. Unfortunately, this approach neglects the hierarchical nature of the set of genre labels (issue 1).

In this paper we are interested in extending a well-known ND method to the hierarchical setting, leading to a method that we call Hierarchical Novelty Detection (HND). To the best of our knowledge, HND is the first method that accounts for *all four issues* highlighted above. A further benefit of HND is that it can be trained by solving a single convex second-order cone programming problem, making it a global or 'big bang' approach—such approaches are considered preferable over the more common approaches that train a collection of models for different levels in the hierarchy of labels [13]. Additionally the method allows for labels to belong to various branches. Whether a novelty or not, we are able to enrich the information about data points and this is relevant in MGC at least for automatic identification of erroneously-annotated genre labels and for identification of emerging music styles. To validate the method, we present experiments on a set of samples from a publicly available dataset tagged by `Last.fm` users. We defined a hierarchy over the genres in these data, but intentionally left a subset out, to assess the extent to which our algorithm would identify songs which belonged to this missing node in the graph as being novel.

The remainder of this paper is structured as follows. In Sect. 2 we review existing approaches before introducing our proposed method in Sect. 3. Experiments and conclusions are presented in Sect. 4.

## 2    Background

In this Section we delve deeper into how hierarchical classification and novelty detection operate and outline the terminology and notation using MGC as domain, so that we can introduce our method concisely in Sect. 3. We begin with a discussion on how relationships between genres can be represented mathematically.

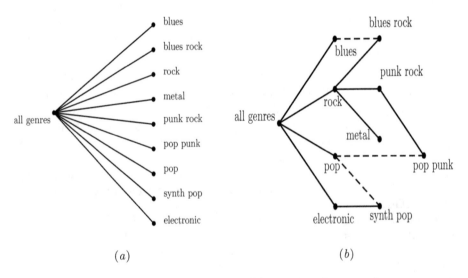

**Fig. 1.** Example of music genre relationships. (*a*): flat classification, where every genre is a subgenre of the root. (*b*) tree structure, where the hierarchical nature of genres is explicit. If links such as `Blues → Blues Rock` are included, the tree becomes a DAG.

## 2.1   Genres Relationships: Graphs, DAGs or Trees?

For many problems, as in MGC, the relationship between classes can be formalised mathematically as a graph, with nodes equal to genres and edges between nodes representing relationships between genres. The exact nature of the graph is the subject of current scholarly work. The model could be a directed or undirected graph, with or without weights on the edges, with or without cycles (in the directed case). Commonly the choice is between trees or Direct Acyclic Graphs (DAG), but any election of a graph structure is only an approximation of reality, a compromise reached for pure modelling purposes. Here, the set of music genres is modelled as a Directed Acyclic Graph (DAG); see Fig. 1(b). The *root* in this DAG will correspond to *all genres*, while the *leaf nodes* correspond to the *most specific genres*. An edge connecting genre A to genre B means that genre B is a subgenre of genre A. The flexibility of a DAG structure means that, for example, Blues Rock may be a subgenre of both Blues and Rock.

## 2.2   Hierarchical Classification

A very naive approach is a single multi-class classifier trained to discriminate between each of the classes. This model ignores the hierarchical structure of the labels given by the nature of music genres and therefore often performs poorly when compared to models that explicitly exploit this information. The three main hierarchical approaches are listed below. For a comprehensive review of HC we refer the reader to [13] and the references therein.

**Flat Classifier Approach.** A flat classification approach ignores the hierarchy being able to predict only the leaf nodes, i.e., the most specific subclasses, and then considers the *IS-A* relationship in the hierarchy for a multi-label classification. Such an approach is unlikely to be beneficial for the performance of the classifier.

**Local Classifier Approach.** These methods are designed as a sequence of flat classifiers. They are often top-down in nature, first classifying each test point into one of the children of the root, and then iterating over the children. Either the process continues until the test point reaches a leaf (*mandatory leaf node prediction* in the language of [13]), or some stopping criteria are applied (*non-mandatory leaf node prediction*).

**Global Classifier Approach.** The most sophisticated approaches yield a single overall model, trained at once on the entirety of the data. These are known as 'big-bang' approaches and are the most principled. However, only a handful of methods exist that fall into this category [7,9,12]. Our method falls within this category.

## 2.3  Novelty Detection

The goal of Novelty Detection (ND) is to decide if new points belong to (one or more) classes present in the data. A natural way to approach this problem is to enclose the data within a decision surface, with any points that fall outside the surface classified as novel. The most common surface is the hypersphere [14].

**Hard Margin Novelty Detection.** Let $\mathbf{x} = \{x_1, \ldots, x_n\}$ be a set of $n$ data points with $x_i \in \mathbb{R}^d, i = 1, \ldots, n$. The simplest form of novelty detection fits a hypersphere $\mathcal{H} = (\mu, R)$ around the data, specified by a centre vector $\mu \in \mathbb{R}^d$ and radius $R \in \mathbb{R}$. Finding the $\mathcal{H}$ which most tightly fits the data $\mathbf{x}$ amounts to solving the following optimisation problem:

$$\min_{R, \mu} R^2, \quad \text{subject to} \tag{1}$$

$$R \geq 0 \tag{2}$$

$$||\mathbf{x}_i - \mu||^2 \leq R^2, \quad i = 1, \ldots, n, \tag{3}$$

where $|| \cdot ||$ represents Euclidean norm. The objective (3) is convex in $R^2$ and $\mu$, meaning the above problem can be solved efficiently using gradient descent or similar methods.

**Soft Margin Novelty Detection.** A common variant of the problem above is to allow some points to be slightly outside the enclosing hypersphere. This allows for extreme values to be ignored and is more robust in train/test settings. Introducing slack variables $\xi \in \mathbb{R}, i = 1, \ldots, n$ and a hyper-parameter $C \in \mathbb{R}$, the objective then becomes:

$$\min_{R, \mu, \xi} R^2 + C \sum_{i=1}^{n} \xi_i, \quad \text{subject to} \tag{4}$$

$$\xi_i \geq 0, \quad R \geq 0 \tag{5}$$

$$||\mathbf{x}_i - \mu||^2 \leq R^2 + \xi_i, \quad i = 1, \ldots, n, \tag{6}$$

This problem is still convex, but has an increase in computational complexity due to the increased number of parameters.

**Soft Margin Novelty Detection with Multiple Classes.** Similarly to the multiple one-class SVM-based method [8], it is possible to perform novelty detection with multiple known classes. One simply solves the problem (4) subject to (6) for each class. Suppose that for each of the $x_i$ we have a label $y_i \in \{1, \ldots, K\}$. Then it is required to find $K$ hyperspheres $\{\mathcal{H}_1, \ldots, \mathcal{H}_k\} = \{(\mu_1, R_1), \ldots, (\mu_K, R_K)\}$, which may be done via:

$$\min_{R,\mu,\xi} \sum_{i=1}^{K} R_k^2 + C \sum_{i=1}^{n} \xi_i, \quad \text{subject to} \tag{7}$$

$$\xi_i \geq 0, \quad R_k \geq 0 \tag{8}$$

$$||\mathbf{x}_i - \mu_{y_i}||^2 \leq R_{y_i}^2 + \xi_i, \quad i = 1, \dots, n. \tag{9}$$

Note that since there is no interaction between any hypersphere $\mu_k$ or $R_k$, this problem can be solved independently for each $k$. However, using the above formulation is faster in practice due to the reduced overhead.

Using 2-dimensional toy data, Fig. 2 illustrates the differences between flat classification, hierarchical classification, novelty detection and our proposed hierarchical novelty detection, that we introduce below.

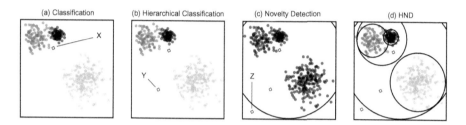

**Fig. 2.** Flat classification, HC, ND and the proposed HND within a three classes example: black, grey (upper left corner) and light grey (lower right corner). In (a), the point $x$ is classified as belonging to the grey class. In (b) and subsequent plots, grey and black are assumed to share the same parent. Here, $x$ is classified as grey_or_black and then possibly into one of grey or black, and $y$ is classified as light grey. In (c), $x$ and $y$ are classified as normal whilst $z$ is classified as novel. Finally in (d), $x$ is classified as grey_or_black, but novel to the classes grey, black and light grey, whilst $y$ is classified to be belonging to the dataset—see (c)—but none of its subclasses. $z$ is considered novel as before. Note that for (c) and (d) the slack variables allow some points to fall slightly outside the spheres.

## 3  Hierarchical Novelty Detection and Classification

As previously illustrated in Sect. 1, hierarchical classification methods must satisfy a number of design criteria. They must be able to operate in the *multi-label setting* (both for the training set and for prediction). They must be able to work with *hierarchies that can be DAGs*, not just trees. They must predict label sets that are *upper sets* of the DAG: for each predicted genre, all of its more generic genres (higher in the DAG) must be predicted as well. We will refer to this criterion as the *consistency criterion* of the label set. And finally, they must be able to deal with training examples for which the *label sets are incomplete*. To the best of our knowledge, no existing methods satisfy all of the design criteria listed above and, importantly, none of them naturally deals with the fact that the training label sets may be incomplete.

### 3.1 Hierarchical Novelty Detection

We take inspiration from the novelty detection methods listed in Sect. 2.3, with the only addition that we enforce relationships between the hyperspheres $(\mathcal{H}_1, \ldots, \mathcal{H}_K)$ by encoding the dependencies present in the hierarchy. Suppose that the labels $y_i$ are arranged in a *hierarchy*, which may be realised as a rooted DAG with nodes labelled $\{1, \ldots, K\}$. Assuming w.l.o.g. that the label 1 is the *root* of the DAG, each label $k \in \{2, \ldots, K\}$ has a *parent* label set, which we denote by $\mathrm{Pa}(k)$. We then represent the subclass constraint by insisting that the hyperspheres for the subclasses are nested:

$$\min_{R, \mu, \xi} \sum_{k=1}^{K} R_k + C \sum_{i=1}^{n} \xi_i, \quad \text{subject to} \tag{10}$$

$$\xi_i \geq 0, \quad R_k \geq 0 \tag{11}$$

$$||\mathbf{x}_i - \mu_{y_i}|| \leq R_{y_i} + \xi_i, \quad i = 1, \ldots, n. \tag{12}$$

$$||\mu_{\mathrm{Pa}(k)} - \mu_k|| \leq R_{\mathrm{Pa}(k)} - R_k, \quad k = 2, \ldots, K, \tag{13}$$

This is a convex problem and is always feasible as we can trivially set all $\mu_k$ to be the centre of the data, $R_k$ equal to each other and large enough to encapsulate all the data, and $\xi_i$ equal to 0. It is clear that in this case Eqs. (11–13) are satisfied, and since the problem is convex we may use this as an initial solution and improve our objective to a global minimum. It also admits an interpretation of the $C$ parameter, as we demonstrate below.

### 3.2 Interpretation of the Hyperparameter $C$

Along with convexity, Eqs. (10–13) present another interesting property regarding the interpretation of the hyperparameter $C$. Suppose that (10) is solved optimally for a given set of data points, and that the optimal objective is found to be $L^* = \sum_{k=1}^{k} R_k^* + C \sum_{i=1}^{n} \xi_i^*$. Now suppose that there is an increase in all of the $R_k^*$, so that $R_k^* \to R_k^* + \Delta$, where $\Delta \in \mathbb{R}^+$. Adding $\Delta$ to each of the $K$ radii corresponds to adding $K\Delta$ to the objective and, since $L^*$ is optimal, this means there must be a reduction in $C \sum_{i=1}^{n} \xi_i^*$ of $K\Delta$. Therefore we want to determine how the slacks $\xi_i^*$ adjust in order to meet this arbitrage. With the original values of $R^*, \xi_i^*$ satisfied,

$$\xi_i^* \geq ||x_i - \mu_{y_i}^*|| - R_{y_i}^* \tag{14}$$

by simply re-arranging constraint in (12). Denoting the adjusted slack variables as $\hat{\xi}_i$, for values of the new $R^*$, they must satisfy,

$$\hat{\xi}_i \geq ||x_i - \mu_{y_i}^*|| - R_{y_i}^* - \Delta. \tag{15}$$

Taking the difference of (14) and (15), summing over $i$, and multiplying by $C$ we arrive at,

$$C \sum_{i=1}^{n} \xi_i^* - \hat{\xi}_i \geq Cn\Delta. \tag{16}$$

However, note that for some of the $x_i$, the slacks will not need to be adjusted, as they will remain within $R_{y_i}^*$ of $\mu_{y_i}$. Let $\mathcal{J} \subseteq \{1,\ldots,n\}$ be the set of points for which the slacks need to be adjusted, with $|\mathcal{J}| = J \leq n$. $\mathcal{J}$ can be understood as the number of *outliers* in the optimal solution, as it corresponds to the number of points which are outside their corresponding hypersphere. For all other $i \notin \mathcal{J}$ we may set $\hat{\xi}_i = \xi_i^*$ and so (16) becomes,

$$C \sum_{j \in \mathcal{J}} \xi_j^* - \hat{\xi}_j \geq CJ\Delta. \tag{17}$$

Recall that we require this change to be $K\Delta$ at the optimum, so we conclude that $CJ\Delta \geq K\Delta$ or equivalently $C \geq \frac{K}{J}$. Remarkably, this means that we may adjust $C$ to directly control how many outliers we are willing to tolerate in our problem. For example, if we wish to tolerate exactly one outlier, we should set $C = K$.

## 4    Experimental Results

To validate our method we considered a comparison, using Hierarchical Precision and Recall as defined in [13], against our implementations of the methods depicted in Sect. 2.2 as well as the hierarchical clustering algorithm from CLUS library [16]. The former were implemented using one vs all approach with linear Support Vector Machines as base classifier: a general one discriminating between each class, a flat and a local hierarchical classifier. For the local classifier we used a 'siblings policy'[6] for every node, i.e., considering samples for a parent node and the child siblings as labels. Experiments were run over a subset of the samples provided by the Million Song Dataset [3] selected according to a user defined taxonomy meant to allow for the possibility of discovering novelty genres in the dataset. We will first proceed with the definition of the performance measures and then with a description of the dataset before the final discussion.

### 4.1    Hierarchical Precision and Recall

Let us consider $\hat{P}_i$ as one element in the set of predicted values for a sample by any of the classifiers and $\hat{T}_i$ as the set of real labels for the same sample. The definitions for hierarchical precision $(hP)$ and recall $(hR)$ are:

$$hP = \frac{\sum_i |\hat{P}_i \cap \hat{T}_i|}{\sum_i |\hat{P}_i|}, \quad hR = \frac{\sum_i |\hat{P}_i \cap \hat{T}_i|}{\sum_i |\hat{T}_i|} \tag{18}$$

It is worth noting how this definition is the most inclusive possible as it considers the possibility of multiple outputs for the classifier, although flat and local solutions will mostly output a value in the hierarchy while considering the superclasses labels. This is a well known problem affecting different methods thoroughly discussed in [13]. In this sense, CLUS is among all methods the most similar to HND for its ability to produce multiple labels along different branches.

## 4.2    Million Song Dataset

The Million Song Dataset [3] is a publicly available dataset constituted by features collected by the Echo Nest, which has been additionally enhanced by a collection of annotations retrieved from the Last.fm website by means of tags (labels) for every song in the set. These tags were used to select a subset of samples according to a user defined hierarchy of 13 different music genres. However only 9 of them are featured in the taxonomy represented by the thick lines in Fig. 1(b). For the purpose of testing, the algorithm was required to discover the remaining 4 genres as novelties by labelling them as all genres samples.

The selection considers all the available tags for a single song. For example a sample tagged as both rock and pop rock by the users was considered as belonging to both classes. No limit over the number of labels was imposed except for the fact that only labels belonging to the taxonomy were considered. The choices resulted in an expansion of both training and test set: the previously described example was used as training sample for both the classes in such a context. For this reason the set-up affected and in some way penalized the performance according to the given definition of $hP$ and $hR$ because a prediction over the same example was considered partially or fully correct depending on the number of correct tags in the output. Table 1 gives an overview of the sample distribution. Training and test sets were built by splitting the collection in half.

**Table 1.** Summary of the dataset. Columns from left to right indicate respectively name of the main tag/class for a sample, number of samples included in the dataset, maximum number of tags per sample, average number of tags per sample and whether the current class was meant to be discovered as a novelty.

| Label | Samples | Max tags | Avg tags | Novelty |
|---|---|---|---|---|
| blues rock | 100 | 4 | 2.07 | No |
| blues | 100 | 5 | 1.64 | No |
| classical | 100 | 3 | 1.05 | **Yes** |
| country | 100 | 5 | 1.46 | **Yes** |
| electronic | 100 | 4 | 1.58 | No |
| folk | 100 | 6 | 1.69 | **Yes** |
| jazz | 100 | 5 | 1.37 | **Yes** |
| metal | 100 | 4 | 1.33 | No |
| pop punk | 100 | 6 | 1.89 | No |
| Pop | 100 | 4 | 1.58 | No |
| punk rock | 100 | 5 | 1.69 | No |
| rock | 100 | 4 | 1.56 | No |
| synth pop | 100 | 4 | 1.67 | No |
| all genres | 1300 | 6 | 1.58 | No |

## 4.3    Discussion and Conclusions

Experiments were performed to first study the effect of the variation of $C$ on the performance in terms of $hP$ and $hR$. The chosen range for the fraction of novelties was $[0.005, 0.7]$. As shown in Fig. 3($a$) HND scored better, as expected, than a traditional method as a general One vs All (OVA) multi-class classifier, similar to the FLAT classifier for a certain range of C and it was outperformed by other hierarchical strategies like CLUS and the local classifier (HOVA). We believe that in the case of CLUS, better performances are due to the specific clustering strategy which applies discriminant weights while descending down the hierarchy. Unfortunately, our formulation does not allow a straightforward integration (with relative interpretation) of a weighting strategy which might reduce the performance gap.

By definition HOVA and FLAT are able to output multi-class labels, but they are limited to the set of node labels that are on the path between the root and the lowest classified node, but still HOVA is comparable with CLUS. CLUS and HND, especially the latter, may *de facto* be penalized by their ability to descend through different branches of the hierarchy and defected by excessive specialization. The measures defined in (18) may not be adequate to actually evaluate our algorithm.

Hierarchies and their arbitrariness usually represent the weakness of hierarchical approaches as there is no unifying framework for the taxonomies. For instance, consider a sample which is simply labelled pop punk in the hierarchy of Fig. 1($b$). It is more likely that it belongs to pop even if it is just labelled

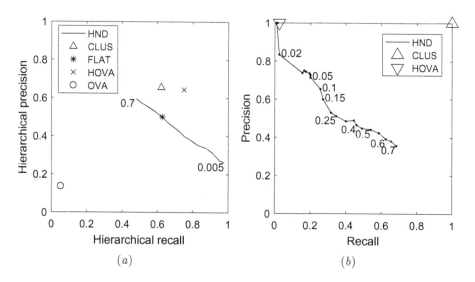

$(a)$                                    $(b)$

**Fig. 3.** Evaluation of precision and recall with significant values of $C$ specified along HND plot: ($a$) Hierarchical precision and recall for all evaluated methods. ($b$) Precision and recall for novelty detection.

as pop punk by many users. In Fig. 3($a$) we explicitly marked the values of $C$ close to our method's plot in order to show how increasing the 'discovery' factor affects our precision as the denominator in $hP$ is raised a lot by the tendency to predict too many labels. Conversely high values of recall are due to this tendency of hyperspheres from apparently far genre nodes in the hierarchy being instead very super-imposed. Such observations also suggest how a different model, e.g. ellipsoids, or a kernel version could lead to further improvements. This will be explored as future work together with the possibility for samples to be labelled as external to the set of nested hyperspheres, which may imply that a genre do not even belong to the range included in all genres.

The ability to discover novelties is presented in Fig. 3($b$). Both precision and recall were evaluated by considering novelty and non-novelty samples as positive and negative samples, respectively. As expected $C$ varies according to the fraction of novelties and therefore to recall values. However here CLUS shows how its precision may be very high on genres which are effectively far from those belonging to hierarchy. Being those the majority of the test samples, and given the effects of the weighting strategy on favouring higher levels of the hierarchy, the performance values were definitely raised.

**Acknowledgments.** The research leading to these results has received funding from the European Research Council under the European Union's Seventh Framework Programme (FP/2007-2013)/ERC Grant Agreement no. 615517, from the EPSRC (EP/M000060/1), from the FWO (project no. G091017N, G0F9816N), and from the European Union's Horizon 2020 research and innovation programme and the FWO under the Marie Sklodowska-Curie Grant Agreement no. 665501.

# References

1. Allwein, E.L., Schapire, R.E., Singer, Y.: Reducing multiclass to binary: a unifying approach for margin classifiers. J. Mach. Learn. Res. **1**, 113–141 (2000)
2. Barutcuoglu, Z., DeCoro, C.: Hierarchical shape classification using Bayesian aggregation. In: Proceedings of the IEEE International Conference on Shape Modeling (2006)
3. Bertin-Mahieux, T., Ellis, D.P., Whitman, B., Lamere, P.: The million song dataset. In: Proceedings of the 12th International Conference on Music Information Retrieval (2011)
4. DeCoro, C., Barutcuoglu, Z., Fiebrink, R.: Bayesian aggregation for hierarchical genre classification. In: ISMIR, pp. 77–80 (2007)
5. Dekel, O., Keshet, J., Singer, Y.: An online algorithm for hierarchical phoneme classification. In: Bengio, S., Bourlard, H. (eds.) MLMI 2004. LNCS, vol. 3361, pp. 146–158. Springer, Heidelberg (2005). doi:10.1007/978-3-540-30568-2_13
6. Fagni, T., Sebastiani, F.: On the selection of negative examples for hierarchical text categorization. In: Proceedings of the 3rd Language Technology Conference, pp. 24–28 (2007)
7. Garcia-Garcia, D., Santos-Rodriguez, R.: Sphere packing for clustering sets of vectors in feature space. In: Proceedings of the IEEE International Conference on Acoustics, Speech, and Signal Processing, pp. 2092–2095. IEEE (2011)

8. Jumutc, V., Suykens, J.A.K.: Multi-class supervised novelty detection. IEEE Trans. Pattern Anal. Mach. Intell. **36**(12), 2510–2523 (2014)

9. Kiritchenko, S., Matwin, S., Famili, A.F.: Functional annotation of genes using hierarchical text categorization. In: Proceedings of the BioLINK SIG: Linking Literature, Information and Knowledge for Biology (2005)

10. Li, T., Ogihara, M.: Music genre classification with taxonomy. In: IEEE International Conference on Acoustics, Speech, and Signal Processing, vol. 5, pp. v/197–v/200 (2005)

11. Pachet, F., Cazaly, D., et al.: A taxonomy of musical genres. In: Content-Based Multimedia Information Access, pp. 1238–1245 (2000)

12. Shawe-Taylor, J., Cristianini, N.: Kernel Methods for Pattern Analysis. Cambridge University Press, New York (2004)

13. Silla Jr., C., Freitas, A.: Novel top-down approaches for hierarchical classification and their application to automatic music genre classification. In: IEEE International Conference on Systems, Man and Cybernetics, pp. 3499–3504. IEEE (2009)

14. Tax, D.M., Duin, R.P.: Support vector data description. Mach. Learn. **54**(1), 45–66 (2004)

15. Tzanetakis, G., Cook, P.: Musical genre classification of audio signals. IEEE Trans. Speech Audio Process. **10**(5), 293–302 (2002)

16. Vens, C., Struyf, J., Schietgat, L., Dzeroski, S., Blockeel, H.: Decision trees for hierarchical multi-label classification. Mach. Learn. **73**(2), 185–214 (2008)

17. Xue, G.R., Xing, D., Yang, Q., Yu, Y.: Deep classification in large-scale text hierarchies. In: Proceedings of the International ACM SIGIR Conference on Research and Development in Information Retrieval, New York, NY, USA, pp. 619–626 (2008)

# Towards Automatic Evaluation of Asphalt Irregularity Using Smartphone's Sensors

Vinicius M.A. Souza[1]([✉]), Everton A. Cherman[1], Rafael G. Rossi[2],
and Rafael A. Souza[3]

[1] Onion Tecnologia, São Carlos, SP, Brazil
{vinicius,everton}@oniontecnologia.com.br
[2] Department of Information Systems, Federal University of Mato Grosso do Sul,
Três Lagoas, MS, Brazil
rafael.g.rossi@ufms.br
[3] Department of Civil Engineering, State University of Maringá,
Maringá, PR, Brazil
rsouza@uem.br

**Abstract.** The quality of the pavement of roads and streets has significant influence in the final price of goods and services, in the safety of pedestrians and also in the driver's comfort. Thus, the development of tools for continuous monitoring of the pavement, intending to obtain a more precise and adequate maintenance plan is essential. In order to reduce the manual effort of inspections made by experts, the use of high-cost equipment as laser profilometer and allowing evaluations in real-time, the use of motion sensor of smartphones to monitor the asphalt irregularity is proposed. In this paper, the present problem is modeled as a classification task that can be performed by supervised learning algorithms and aided by signal processing techniques for features extraction from the acceleration data. The proposed approach shows promising accuracies for the identification of asphalt irregularity (around 99%) and for identification of obstacles as speed bumps, raised crosswalk, pavement markers, and asphalt patches (around 87%).

**Keywords:** Accelerometer · Asphalt evaluation · Time series classification

## 1   Introduction

Streets and roads play a key role in the transportation of cargo, products, and passengers by means of vehicles. For example, roads are responsible for 61.2% of cargo movements and 95% of the transport of people in Brazil. However, out of a total of almost 2 million km of Brazilian roadways, only 12.3% are paved and half of them are considered inadequate [2]. In Europe, statistics findings of Eurostat[1]

---

[1] Eurostat is the statistical office of the European Union, based in Luxembourg. It publishes official, harmonized statistics on the European Union and the euro area.

© Springer International Publishing AG 2017
N. Adams et al. (Eds.): IDA 2017, LNCS 10584, pp. 322–333, 2017.
DOI: 10.1007/978-3-319-68765-0_27

in 2013 show that the inland freight that was transported by road (74.9%) was more than four times as high as the share transported by rail (18.2%). The quality of the pavement has significant influence in the final price of goods and services, the safety of drivers, pedestrians and passengers, and driver's comfort. Thus, it is essential the use of tools that allow the constant monitoring of pavement conditions by the Government authorities or private entities for more precise interventions in the maintenance planning with fewer expenses. This task is mainly important in developing countries where there is a lack of technology and reduced budget to maintenance.

In order to reduce the manual effort and time cost of inspections made by experts, the need of use of high-cost equipment as laser profilometer, smartphone sensors have been used to evaluate the irregularity of the asphalt pavement. Due to the popularity of smartphones, this tool can make possible that different users help to monitor the pavement quality in a ubiquitous way during driving periods without effort. Given the constant changes over time in the asphalt conditions by new repairs and degradations, this application is a concrete example of data stream problem where the learning process is continuous [8,16].

The problem of asphalt irregularity evaluation by the motion sensor is modeled in this paper as a classification task that can be performed by supervised learning algorithms. The proposed approach has five steps: ($i$) collect labeled accelerometer data to build a training set, ($ii$) data preprocessing to make the smartphone inclination/rotation invariant for any position, ($iii$) features extraction to discriminate different classes and to input machine learning algorithms, ($iv$) build a classification model through supervised learning, and ($v$) perform the automatic classification of new data.

Two scenarios of classification were considered in the experimental evaluation. The first scenario is more generic and the goal is to predict if a segment of asphalt show some irregularity or not. In the second scenario, we distinguish four different obstacles (speed bump, vertical patch, raised pavement markers, and raised crosswalk) that can be wrongly interpreted as an irregularity in the asphalt. In both scenarios, promising accuracies results was achieved, around 99% for the former and 87% for the second scenario.

The outline of the paper is as follows. In Sect. 2, some related work that uses data from smartphones to evaluate asphalt conditions are presented. In Sect. 3, the procedures of data collection and preprocessing are detailed. The features extracted from the acceleration data are shown in Sect. 4. In Sect. 5, the experimental setup is described and the results are reported and discussed. Conclusions and future works are presented in Sect. 6.

## 2    Related Work

Most part of works has been devoted only to detect and report potholes. Just a few researches combine more advanced techniques of signal processing with machine learning algorithms, as proposed in this paper. In this sense, we can consider that this work approaches the problem of pavement evaluation in a more comprehensive way with a wide experimental evaluation, never considered by other works. The works closest to our are: Pothole Patrol [5], Nericell [13] and the system proposed by Mednis et al. [12].

In Pothole Patrol [5], the three-axis acceleration data is gathered at high frequency (380 times per second) and the main goal of the system is to detect potholes. Signal processing filters are applied to reject one or more non-pothole event types as a manhole, expansion joint, and railroad crossing. The filters are based on the analysis of speed (to reject door slams and curb ramps), peaks of acceleration in the z-axis, and peaks of acceleration in the z-axis when the acceleration in the x-axis is below a threshold value. The authors report an accuracy of 92% for pothole identification.

Nericell [13] is a system to monitor road and traffic conditions. It detects potholes, braking, speed bumps, and horn using accelerometer, microphone, GSM radio and GPS sensors of smartphones. For potholes, the detection is based on the analysis of z-values of the accelerometer. It also provides two heuristics based upon the speed of the vehicle. If speed is greater than $25\,km/h$, it uses z-peak heuristic where a spike along z-value above a specific threshold is classified as a pothole. At low speed, the z-sus heuristic is used which detects a sustained dip in z-value for at least 20 ms. The false negative rate for both detectors is high (20–30%) for speed bump detection.

The system proposed by Mednis et al. [12] uses four simple algorithms to detect potholes: Z-THRESH, Z-DIFF, STDEV(Z) and G-ZERO. Z-THRESH identifies the type of pothole (small pothole, a cluster of potholes, large potholes) based on the values of acceleration observed in the z-axis above a specific threshold level. In Z-DIFF, is performed a search in the acceleration values of z-axis for two consecutive measurements with a difference between their values above a specific threshold level. STDEV(Z) algorithm calculates the standard deviation of accelerometer data in the vertical direction over a specified window size. G-ZERO searches for the case where all the three-axis data values are near to 0g. This measure indicates that the vehicle it is in a temporary free fall. The best result is achieved by Z-DIFF, with 92% of true positives to the identification of pothole type.

# 3   Asphalt Temporal Data

In this section, the procedures for data collection for evaluation of irregularities and obstacles recognition using smartphones are presented, as well the preprocessing and segmentation steps applied in the accelerometer data.

## 3.1   Data Collection

A smartphone was installed inside the vehicle using a flexible suction holder near the dashboard in order to collect the data. An expert is responsible for driving the vehicle while the device runs an Android application called *Asfault*, developed specifically to store the current asphalt condition continuously over time. This application is currently under development by our research group and it will be incorporated in a commercial tool in the future. *Asfault* stores the time-stamp, acceleration forces in $m/s^2$ along the three physical axes $(x, y, z)$, latitude, longitude, and velocity.

The acceleration forces are given by the accelerometer sensor. Latitude, longitude, and velocity are given by the GPS. Each one of this information is a different continuous time series. A time series $T$ can be defined as a sequence of $n$ real numbers obtained through repeated measurements over time, i.e., $T = \{t_1, t_2, \ldots, t_n\}$. A sampling rate of 100 Hz, which means 100 observations per second for each time series, was considered in this paper. It is important to note that this is a more reachable configuration for most of the devices. Some related works have considered a higher sampling rate, as 310 Hz [13].

In order to obtain labeled data, the *Asfault* application allows the expert to inform the asphalt condition before collecting the data. Initially, a problem with two classes based on the comfort felt by the driver according to the asphalt condition was considered:

- *Regular*: when the driver comfort is very little changed over time;
- *Irregular*: when is observed some irregularities and roughness on the asphalt that are responsible for transferring vibrations to the cabin of the vehicle, reducing the comfort of the driver.

This dataset has been named Asphalt-Regularity. Figure 1 shows their classes. In a first time, the classes *Regular* and *Irregular* also can be interpreted as "asphalt without problems" and "asphalt with problems", respectively. However, we observed that the presence of obstacles on the road is responsible for transferring vibrations to the interior of the vehicle without necessarily being a problem in the asphalt. With this, was built a second dataset considering four popular obstacles in the region of data collection. This dataset has been named as Asphalt-Obstacles and has the following classes: (*i*) speed bump, (*ii*) asphalt vertical patch, (*iii*) raised pavement markers, and (*iv*) raised crosswalk.

(a) Regular                    (b) Irregular

**Fig. 1.** Photos that illustrate the two classes of the Asphalt-Regularity dataset.

Examples of the classes of Asphalt-Obstacles are shown in Fig. 2. In this dataset is observed a considerable variability of the examples of the same class, given the lack of pattern in terms of width, height, size, and material.

(a) Speed bump                    (b) Vertical patch

(c) Raised pavement markers        (d) Raised crosswalk

**Fig. 2.** Photos that illustrate the four classes of the Asphalt-Obstacles dataset.

Both datasets were collected in the Brazilian cities of São Carlos and Maringá using a medium sized hatchback car (Hyundai i30) and two different devices (Samsung Galaxy A5 and Samsung S7) with different hardware specifications.

## 3.2   Data Preprocessing

A three-axis accelerometer detects the acceleration forces $A$ in three perpendicular directions $(x, y, z)$, resulting in the time series $A_x, A_y, A_z$. Thus, an accelerometer at rest on the surface of the Earth with an inclination of $90°$, will measure a continuous acceleration $A_z \approx 9.81 \ m/s^2$, given the Earth's gravity. If the inclination of the sensor is changed for any direction, this acceleration force is distributed along the axes. In order to make the accelerometer data invariant to the inclination of the device, the force of gravity is removed in a preprocessing step. To do this is applied a high-pass frequency filter ($4^{th}$ order Butterworth filter with a cutoff frequency $\omega = 0.25 \ Hz$) in each axis.

A more simple procedure for gravity removal was also considered. In this case, the input time series represented by the vectors $A_x, A_y, A_z$ are converted in a unique time series that represents the acceleration magnitude $A_m$. The entire Matlab code to calculate the acceleration magnitude without the gravity force is presented in Table 1.

**Table 1.** Matlab code to calculate the acceleration magnitude.

```
1   function Am = accelMagnitude(Ax, Ay, Az)
2   mag = sqrt(sum(Ax.^2 + Ay.^2 + Az.^2, 2));
3   Am = mag - mean(mag);
```

Figure 3 shows an example of acceleration data obtained after crossing a raised pavement markers. The example also illustrates the preprocessed data after the gravity force removal by the high-pass filter and the acceleration magnitude $A_m$.

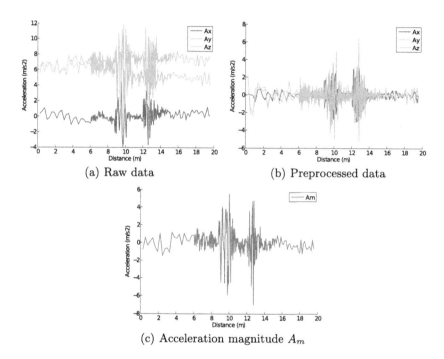

(a) Raw data     (b) Preprocessed data

(c) Acceleration magnitude $A_m$

**Fig. 3.** Example of acceleration data obtained crossing a raised pavement markers, before and after the removal of gravity force by a high-pass filter *(top)* and the acceleration magnitude $A_m$ *(bottom)*.

## 3.3   Data Segmentation

For Asphalt-Regularity dataset, was performed a procedure of data segmentation or windowing. In this dataset, the data was collected for a dozen of kilometers and the split of these data in minor segments is necessary to perform the evaluation procedures and to better indicate the location of asphalt problems on a map.

In order to better visualize the results in a map in our application in development, a length of approximately 30 m for each segment data was considered. To find the segments, the values of velocity (in $m/s^2$) of each observation divided by the current sampling rate are summed until this value reaches approximately 30 m. This procedure is because the sampling rate obtained by the smartphones used in the experiments varies from 95 to 105 Hz. Thus, each segment of data represented by a time series has a different number of observations according to the velocity and sampling rate. For example, if we collect data in a constant velocity of 40 km/h (or 11.11 m/s) and a constant sampling rate of 100 Hz, a time series with approximately 30 m will have 270 observations. If the sampling rate is reduced to 95 Hz, the number of observations will be 256. However, in practice, the velocity and sampling rate are not constant over time.

For the Asphalt-Regularity dataset 1,392 examples were collected and for the Asphalt-Obstacles dataset, 781 examples were collected. As each example has 30 m in average, our datasets constitute more than 65 Km of collected data. The distribution of the examples over the classes are presented in Table 2.

**Table 2.** Class distribution of Asphalt-Regularity and Asphalt-Obstacles datasets.

| Dataset | Class | Examples | Distribution (%) |
|---------|-------|----------|------------------|
| Asphalt-Regularity | Regular | 670 | 48.13 |
| | Irregular | 722 | 51.87 |
| Asphalt-Obstacles | Speed bump | 212 | 27.14 |
| | Vertical patch | 222 | 28.43 |
| | Raised markers | 187 | 23.94 |
| | Raised crosswalk | 160 | 20.49 |

## 4    Features Extraction

The time series of accelerometer data gathered by smartphones can be weakly representative as input features for supervised machine learning algorithms. For this reason, the strategy of extracting features from data using different signal processing methods was chosen. Briefly, these methods are responsible for changing the original data representation in a more representative feature set.

To facilitate the performance analysis of different signal processing methods for features extraction, the methods were divided into four major groups:

- **Temporal statistics.** 23 different statistics were extracted from the raw data. These statistics include short-time energy, mean, median, variance, standard deviation, centroid, zero-crossing rate, complexity estimate, skewness, kurtosis, entropy, maximum-to-minimum difference, mean number of peaks, average distance of peaks;
- **Spectral statistics.** The raw data were converted to the frequency domain by the fast Fourier transform (FFT) and 39 statistics were extracted from the signal's spectrum. The statistics include the same features extracted in the

temporal domain and other statistics as auto-correlation, roll-off, irregularity, flux, height and position of first 6 peaks, inter-quartile range;
- **LSF coefficients.** The Linear Spectral Frequencies (LSF) [9] is a signal representation derived from Linear Predictive Coding [11], in which a signal is represented as a linear combination of previously observed values. Thus, LSF can represent a large signal using a small number of coefficients. In the experiments, 20 coefficients were considered as feature set. This value was found empirically according to preliminary results;
- **CWT statistics.** The continuous wavelet transform (CWT) was applied to the raw data using the Haar function [10] and 36 different statistics were extracted from the coefficients as mean, median, maximum, standard deviation, entropy, and results of coefficients summed across scales. These features are extracted using the HCTSA Toolbox [6]. The use of CWT allows a better localization of the frequency components than FFT.

## 5  Experimental Evaluation

The experimental evaluation also aims to answer some practical questions that will be determinant in a real application currently in development in our research group, such as:

- Among different features extracted from the accelerometer data, which ones are adequate to discriminate our data?
- Do we need to extract features from each acceleration axis $(A_x, A_y, A_z)$ or are the features from the acceleration magnitude $(A_m)$ sufficient to discriminate our data?
- Among different supervised learning algorithms, which ones stand out?

In the experimental evaluation, the four groups of features sets extracted from the three acceleration axes $(A_x, A_y, A_z)$ and from the acceleration magnitude $(A_m)$ were considered. All the features set combined are also considered. Four supervised machine learning algorithms were chosen, each one with a different learning bias: Support Vector Machines (SVM) with the polynomial kernel $(c = 0.5$ and $e = 1)$, Random Forests (RF), Naive Bayes (NB), and One-Nearest Neighbor (1NN). Given all the combinations of features and classifiers, were generated 60 results for each dataset.

Two-fold cross-validation was used to partition the datasets into training and test sets. In order to reduce results variance by chance, we have repeated this process five times, randomizing the order of examples between executions, i.e., we performed 5×2-fold cross-validation as suggested by Dietterich [3].

In order to compare the results, the accuracies achieved by the One-Nearest Neighbor algorithm with Dynamic Time Warping as the distance measure (1NN-DTW) was considered as a baseline. This more simple approach only considers the similarity between the series without the need to extract features. The main advantage of Dynamic Time Warping (DTW) distance over other measures as Euclidean, is that DTW achieves the optimal non-linear alignment between the

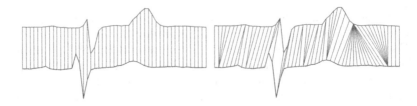

**Fig. 4.** Alignments achieved by Euclidean distance (*left*) and Dynamic Time Warping (*right*) [15]. In order to compare the results achieved by different classifiers after the features extraction phase, 1NN with Dynamic Time Warping is used as baseline. (Color figure online)

observations of two time series. Moreover, DTW allows the comparison of series with different length. Given two times series $S1$ and $S2$, Fig. 4 illustrates the linear alignment obtained by the Euclidean distance (*left*) and the non-linear alignment obtained by DTW (*right*) in the gray lines between the series.

1NN-DTW is known to be a very competitive classifier and has been widely used for time series classification [4]. For simplicity, our baseline based on the 1NN-DTW classifier was carried out only considering the time series of acceleration magnitude ($A_m$).

For the Asphalt-Regularity dataset, the baseline classifier 1NN-DTW showed 67.44% of average accuracy and 0.98 of standard deviation over the 5 runs. The results achieved by different feature sets and classifiers are shown in Table 3 and the result of the baseline is shown in the last row of the table. The best result achieved for each feature set is highlighted in bold.

**Table 3.** Mean accuracy and standard deviation for Asphalt-Regularity dataset.

| Feature set | Time series | # features | Avg. accuracy (std. deviation) | | | |
|---|---|---|---|---|---|---|
| | | | SVM | RF | NB | 1NN |
| Temporal stats | $A_x, A_y, A_z, A_m$ | 92 | 98.79 (0.48) | 98.82 (0.52) | 96.32 (0.89) | **98.95 (0.30)** |
| | $A_x, A_y, A_z$ | 69 | 98.43 (0.45) | 98.81 (0.46) | 95.40 (1.05) | 98.89 (0.33) |
| | $A_m$ | 23 | 98.12 (0.44) | 98.05 (0.27) | 94.77 (1.16) | 97.05 (0.62) |
| Spectral stats | $A_x, A_y, A_z, A_m$ | 156 | 97.89 (0.43) | **98.59 (0.48)** | 95.98 (0.50) | 96.01 (0.56) |
| | $A_x, A_y, A_z$ | 117 | 97.90 (0.60) | 98.00 (0.54) | 94.63 (0.81) | 94.51 (0.46) |
| | $A_m$ | 39 | 97.21 (0.58) | 97.41 (0.44) | 95.20 (0.73) | 93.61 (0.94) |
| LSF coeffs | $A_x, A_y, A_z, A_m$ | 80 | 93.20 (0.83) | 91.19 (0.97) | **93.20 (0.47)** | 91.62 (0.29) |
| | $A_x, A_y, A_z$ | 60 | 93.09 (1.02) | 91.47 (1.05) | 93.09 (0.43) | 91.87 (0.70) |
| | $A_m$ | 20 | 84.97 (1.03) | 85.53 (1.33) | 85.66 (0.80) | 79.78 (1.37) |
| CWT stats | $A_x, A_y, A_z, A_m$ | 144 | 98.29 (0.40) | **98.55 (0.50)** | 90.60 (0.43) | 97.53 (0.49) |
| | $A_x, A_y, A_z$ | 108 | 98.30 (0.61) | 98.18 (0.78) | 90.33 (0.43) | 97.36 (0.45) |
| | $A_m$ | 36 | 97.20 (0.49) | 97.47 (0.66) | 89.50 (0.86) | 96.81 (0.69) |
| All features | $A_x, A_y, A_z, A_m$ | 472 | **98.98 (0.27)** | 98.71 (0.42) | 98.05 (0.46) | 98.62 (0.39) |
| | $A_x, A_y, A_z$ | 354 | 98.66 (0.38) | 98.84 (0.49) | 97.80 (0.75) | 98.43 (0.45) |
| | $A_m$ | 118 | 98.05 (0.24) | 97.83 (0.32) | 97.07 (0.66) | 97.70 (0.65) |
| 1NN-DTW | $A_m$ | 67.44 (0.98) | | | | |

We can note in Table 3 that the best result (98.98%) was achieved by the SVM classifier using the four feature sets together extracted from the acceleration axes $A_x, A_y, A_z$, and the acceleration magnitude $A_m$. However, equivalent results were also achieved by different classifiers using Temporal, Spectral, and CWT statistics as feature set. In general, the results are substantially superior to those achieved by the baseline classifier.

For the Asphalt-Obstacle dataset, 1NN-DTW showed 79.95% of classification accuracy and 1.63 of standard deviation. The results achieved by different feature sets and classifiers are shown in Table 4.

**Table 4.** Mean accuracy and standard deviation for Asphalt-Obstacles dataset.

| Feature set | Time series | # features | Avg. accuracy (std. deviation) | | | |
|---|---|---|---|---|---|---|
| | | | SVM | RF | NB | 1NN |
| Temporal stats | $A_x, A_y, A_z, A_m$ | 92 | 82.43 (1.40) | 79.98 (2.09) | 73.65 (1.46) | 74.50 (2.35) |
| | $A_x, A_y, A_z$ | 69 | 77.88 (2.25) | 79.54 (1.79) | 71.73 (2.08) | 70.94 (2.20) |
| | $A_m$ | 23 | **85.07 (1.26)** | 77.34 (1.42) | 71.32 (2.06) | 72.91 (2.14) |
| Spectral stats | $A_x, A_y, A_z, A_m$ | 156 | 79.16 (2.97) | 76.78 (2.10) | 70.37 (1.82) | 67.05 (3.01) |
| | $A_x, A_y, A_z$ | 117 | 67.68 (1.83) | 66.20 (2.76) | 57.49 (2.52) | 55.75 (2.09) |
| | $A_m$ | 39 | **80.41 (1.52)** | 78.03 (1.82) | 71.60 (1.66) | 67.84 (2.32) |
| LSF coeffs | $A_x, A_y, A_z, A_m$ | 80 | 70.40 (2.39) | 67.58 (2.57) | **72.11 (1.88)** | 57.03 (2.81) |
| | $A_x, A_y, A_z$ | 60 | 56.82 (2.35) | 54.67 (2.72) | 64.64 (1.76) | 46.30 (1.23) |
| | $A_m$ | 20 | 64.05 (2.00) | 61.92 (2.31) | 65.48 (2.01) | 56.70 (2.64) |
| CWT stats | $A_x, A_y, A_z, A_m$ | 144 | 75.78 (1.28) | 78.18 (1.67) | 76.13 (2.45) | 72.83 (2.04) |
| | $A_x, A_y, A_z$ | 108 | 76.70 (1.28) | **78.23 (1.71)** | 77.36 (2.61) | 71.98 (1.69) |
| | $A_m$ | 36 | 77.21 (0.95) | 75.01 (2.54) | 68.55 (2.50) | 67.43 (1.73) |
| All features | $A_x, A_y, A_z, A_m$ | 472 | **86.97 (1.93)** | 81.33 (1.38) | 79.69 (1.45) | 78.41 (1.32) |
| | $A_x, A_y, A_z$ | 354 | 81.56 (1.64) | 79.95 (2.02) | 75.36 (1.34) | 75.06 (2.37) |
| | $A_m$ | 118 | 82.00 (2.08) | 81.74 (1.13) | 81.05 (1.22) | 77.08 (2.77) |
| 1NN-DTW | $A_m$ | | 79.95 (1.63) | | | |

For the Asphalt-Obstacles dataset, the best result (86.97%) was achieved by the SVM using all feature sets together extracted from the acceleration axes $A_x, A_y, A_z, A_m$ (see the confusion matrix and detailed results for each class in Table 5). However, an equivalent result (85.07%) was achieved by the SVM using the Temporal statistics extracted from the acceleration magnitude ($A_m$). Although this result is slightly lower, it was achieved using a reduced number of features (23 Temporal features instead 472 of all features combined), which turns the classification faster. LSF and CWT showed accuracies below the baseline classifier, even in their best configurations.

Friedman test with the Nemenyi post hoc with 95% as confidence level was performed to statistically compare the results. The critical difference diagrams that illustrate the results of the test considering the evaluation of different feature sets, acceleration axes, and classifiers, are shown in Fig. 5. In this diagram, the methods are sorted according to their average ranking and the methods connected by a line do not present significant differences among them.

From the analysis of the diagrams in Fig. 5, it is possible to answer the three questions raised at the beginning of this section. In short, the Temporal

**Table 5.** Confusion matrix and detailed results achieved by the best classifier for Asphalt-Obstacles dataset for one of 5×2-fold cross-validation run.

| Actual | Predicted | | | | Precision | Recall | F-Measure |
|---|---|---|---|---|---|---|---|
| | SB | VP | RM | RC | | | |
| Speed bump (SB) | **197** | 1 | 2 | 12 | 0.89 | 0.93 | 0.91 |
| Vertical patch (VP) | 4 | **185** | 33 | 0 | 0.84 | 0.83 | 0.84 |
| Raised markers (RM) | 5 | 34 | **147** | 1 | 0.80 | 0.79 | 0.79 |
| Raised crosswalk (RC) | 15 | 0 | 1 | **144** | 0.92 | 0.90 | 0.91 |

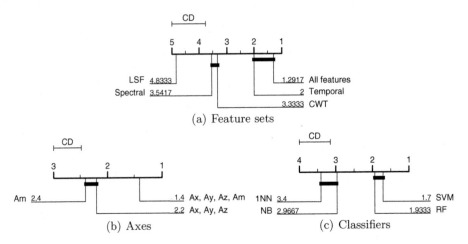

(a) Feature sets

(b) Axes

(c) Classifiers

**Fig. 5.** Critical difference diagrams considering the average results achieved by different feature sets, acceleration axes, and classifiers.

feature set is the most adequate to discriminate our data, although the use of all features together leads to slightly better results (but not statistically significant). Temporal features have a low cost for their extraction, allowing them to be calculated in real time by the device responsible to collect data. The features extracted from the acceleration magnitude $A_m$ together with the three axes $A_x, A_y, A_z$ provided better results with statistically significant difference than the obtained by only $A_m$. Finally, among the evaluated classifiers, SVM and RF showed the best results and both can be used in the practice.

## 6    Conclusions

In this paper, a real data stream problem of asphalt evaluation using acceleration data gathered by smartphones was considered. This problem is modeled as a classification task that can be performed by supervised learning algorithms. In our experiments on two datasets, four groups of features sets and four supervised classification algorithms were evaluated. The proposed approach shows promising accuracies results for the identification of asphalt irregularity (around 99%) and for obstacles recognition (around 87%). To the best of our knowledge, our evaluation is the most comprehensive nowadays and shows the best results.

In future work, we intend to collect more data considering different vehicles to evaluate the effectiveness of the features sets on diversified conditions. We also want to include new classes to the datasets and explore similarity search methods in time series as CID [1], DDTW [7], RPCD [14], and others [4].

**Acknowledgments.** This work was supported by Fundação de Amparo à Pesquisa do Estado de São Paulo (FAPESP), Grant Number #2016/07767-3.

# References

1. Batista, G., Keogh, E., Tataw, O.M., Souza, V.M.A.: CID: an efficient complexity-invariant distance for time series. Data Min. Knowl. Discov. **28**(3), 634–669 (2014)
2. C.N.T.: Brazillian confederation of transport survey of highways 2016: management report. Technical report 20, CNT:SEST:SENAI (2016)
3. Dietterich, T.G.: Approximate statistical tests for comparing supervised classification learning algorithms. Neural Comput. **10**(7), 1895–1923 (1998)
4. Ding, H., Trajcevski, G., Scheuermann, P., Wang, X., Keogh, E.: Querying and mining of time series data: experimental comparison of representations and distance measures. VLDB Endow. **1**(2), 1542–1552 (2008)
5. Eriksson, J., Girod, L., Hull, B., Newton, R., Madden, S., Balakrishnan, H.: The pothole patrol: using a mobile sensor network for road surface monitoring. In: MobiSys, pp. 29–39 (2008)
6. Fulcher, B.D., Little, M.A., Jones, N.S.: Highly comparative time-series analysis: the empirical structure of time series and their methods. J. R. Soc. Interface **10**(83), 20130048 (2013)
7. Górecki, T., Łuczak, M.: Using derivatives in time series classification. Data Min. Knowl. Disc. **26**, 310–331 (2013)
8. Hulten, G., Spencer, L., Domingos, P.: Mining time-changing data streams. In: ACM SIGKDD, pp. 97–106 (2001)
9. Itakura, F.: Line spectrum representation of linear predictor coefficients of speech signals. J. Acoust. Soc. Am. **57**(S1), S35–S35 (1975)
10. Lilly, J.M., Olhede, S.C.: Higher-order properties of analytic wavelets. IEEE Trans. Signal Process. **57**(1), 146–160 (2009)
11. Makhoul, J.: Linear prediction: a tutorial review. Proc. IEEE **63**(4), 561–580 (1975)
12. Mednis, A., Strazdins, G., Zviedris, R., Kanonirs, G., Selavo, L.: Real time pothole detection using android smartphones with accelerometers. In: DCOSS, pp. 1–6 (2011)
13. Mohan, P., Padmanabhan, V.N., Ramjee, R.: Nericell: rich monitoring of road and traffic conditions using mobile smartphones. In: ACM SenSys, pp. 323–336 (2008)
14. Silva, D.F., Souza, V.M.A., Batista, G.: Time series classification using compression distance of recurrence plots. In: IEEE ICDM, pp. 687–696 (2013)
15. Souza, V.M.A., Silva, D.F., Batista, G.: Extracting texture features for time series classification. In: ICPR, pp. 1425–1430 (2014)
16. Souza, V.M.A., Silva, D.F., Gama, J., Batista, G.: Data stream classification guided by clustering on nonstationary environments and extreme verification latency. In: SIAM SDM, pp. 873–881 (2015)

# A Structural Benchmark for Logical Argumentation Frameworks

Bruno Yun[1]([✉]), Srdjan Vesic[2], Madalina Croitoru[1], Pierre Bisquert[3], and Rallou Thomopoulos[3]

[1] Inria GraphIK, LIRMM, Université de Montpellier, Montpellier, France
yun@lirmm.fr
[2] CRIL - CNRS, Université d'Artois, Lens, France
[3] Inria GraphIK, INRA, Montpellier, France

**Abstract.** This paper proposes a practically-oriented benchmark suite for computational argumentation. We instantiate abstract argumentation frameworks with existential rules, a language widely used in Semantic Web applications and provide a generator of such instantiated graphs. We analyse performance of argumentation solvers on these benchmarks.

## 1 Introduction

Amongst the plethora of tools for reasoning in presence of inconsistency, argumentation has always held a particular place because of its proximity with real world interaction (Leite et al. 2015). In this paper, we focus on *logic based argumentation* where abstract argumentation frameworks (Dung 1995) are instantiated by constructing arguments and attacks from inconsistent knowledge bases. Logic-based argumentation has been studied with *many frameworks proposed*: assumption-based argumentation frameworks (Bondarenko et al. 1993), DeLP (García and Simari 2004), deductive argumentation (Besnard and Hunter 2008) or ASPIC/ASPIC+ (Amgoud et al. 2006; Modgil and Prakken 2014).

Despite argumentation being a mature field, *practically inspired benchmarks are currently missing*. As a rare example of a practical argumentation benchmark consider NoDE[1], which contains graphs that model debates from Debatepedia[2], the drama "Twelve Angry Men" by Reginald Rose and Wikipedia revision history. However, the graphs from this benchmark are small (many of them have less than 10 arguments) and their structure is simplistic. The lack of benchmark was acknowledged by the community long time ago, but became obvious with the appearance of the International Competition on Computational Models of Argumentation (ICCMA)[3]. This is why new algorithms are always tested on randomly generated graphs, e.g. Nofal et al. (2014) and Cerutti et al. (2014).

The goal of this paper is to address this drawback by *generating argumentation graphs from knowledge bases and studying their properties empirically*

---

[1] http://www-sop.inria.fr/NoDE/.
[2] http://debatepedia.org/.
[3] http://argumentationcompetition.org/.

© Springer International Publishing AG 2017
N. Adams et al. (Eds.): IDA 2017, LNCS 10584, pp. 334–346, 2017.
DOI: 10.1007/978-3-319-68765-0_28

(by running the argumentation solvers). We use an existing logic-based argumentation framework (Croitoru and Vesic 2013; Croitoru et al. 2015) instantiated with existential rules. This language was chosen because of its practical interest on the Semantic Web (Thomazo and Rudolph 2014; Thomazo 2013; Zhang et al. 2016). Existential rules generalise Description Logics fragments (such as DL-Lite, etc.) that are underlying OWL profiles. Therefore, the choice of this language is significant for Semantic Web applications, notably Ontology-Based Data Access (OBDA) applications. Given the amount of ontologies and data sources made available by such applications, the paper positioning within this language demonstrates its practical interest and relevance for benchmarking argumentation frameworks. Existential rules possess particular features of interest for logic-based argumentation frameworks such as n-ary (as opposed to binary only) negative constraints or existential variables in the head of rules.

The contribution of the paper is the first benchmark in the literature that uses graphs generated from knowledge bases expressed with existential rules instead of random graphs. Using a suite of parametrised existential rule knowledge bases, *we produced the first large scale practically-oriented benchmark in the literature.* Furthermore, we run the top six solvers from ICCMA 2015 *on the generated benchmark and show that the ranking is considerably different* from the one obtained during the competition on randomly generated graphs.

This paper is of interest to both argumentation community and data analysis community. Indeed, for data analysis, the existence of real benchmarks of arguments could be of interest because it can pave the way for intelligent analysis of such instances. These results could then further our comprehension of argumentation graphs structural properties.

## 2  Background Notions

In this paper we use the existential rule instantiation of argumentation frameworks of Croitoru and Vesic (2013). The existential rules language (Calì et al. 2009) extends plain Datalog with *existential variables* in the rule head and is composed of formulae built with the usual quantifiers ($\exists, \forall$) and *only* two connectors: implication ($\rightarrow$) and conjunction ($\wedge$). A subset of this language, also known as $Datalog^{\pm}$, refers to identified decidable existential rule fragments (Gottlob et al. 2014; Baget et al. 2011). The language has attracted much interest recently in the Semantic Web and Knowledge Representation community for its suitability for representing knowledge in a distributed context such as Ontology Based Data Access applications (Baget et al. 2011; Thomazo and Rudolph 2014; Thomazo 2013; Magka et al. 2013; Zhang et al. 2016. The language is composed of the following elements. A *fact* is a ground atom of the form $p(t_1, \ldots, t_k)$ where $p$ is a predicate of arity $k$ and $t_i, i \in [1, \ldots, k]$ constants. An existential *rule* is of the form $\forall \overrightarrow{X}, \overrightarrow{Y} \ H[\overrightarrow{X}, \overrightarrow{Y}] \rightarrow \exists \overrightarrow{Z} C[\overrightarrow{Z}, \overrightarrow{X}]$ where $H$ (called the hypothesis) and $C$ (called the conclusion) are existentially closed atoms or conjunctions of existentially closed atoms and $\overrightarrow{X}, \overrightarrow{Y}, \overrightarrow{Z}$ their respective vectors of variables. A *rule is applicable* on a set of facts $\mathcal{F}$ iff there exists a homomorphism from the hypothesis of the rule to $\mathcal{F}$. Applying a rule to a set of facts (also called *chase*) consists of adding the set of atoms of the conclusion of the rule to the facts according to the application homomorphism. A *negative constraint* (NC) is a particular kind of rule where $C$ is $\perp$ (*absurdum*). It implements weak negation. A *knowledge base* $\mathcal{K} = (\mathcal{F}, \mathcal{R}, \mathcal{N})$ is composed of a finite set of facts $\mathcal{F}$, a set of rules $\mathcal{R}$ and a set of negative constraints $\mathcal{N}$. We denote by $\mathcal{Cl}_{\mathcal{R}}^{*}(\mathcal{F})$ the *closure* of $\mathcal{F}$ by $\mathcal{R}$ (computed by all possible rule $\mathcal{R}$ applications over $\mathcal{F}$ until a fixed point). $\mathcal{Cl}_{\mathcal{R}}^{*}(\mathcal{F})$ is said to be $\mathcal{R}$-*consistent* if no negative constraint hypothesis can be deduced. Otherwise $\mathcal{Cl}_{\mathcal{R}}^{*}(\mathcal{F})$ is $\mathcal{R}$-*inconsistent*. Note that different *chase* mechanisms use different simplifications that prevent infinite redundancies (Baget et al. 2011). In fact, $\mathcal{Cl}_{\mathcal{R}}^{*}(\mathcal{F})$ is a finite set when we restrict ourselves to recognisable *finite extension set* classes (Baget et al. 2011) of existential rules (i.e. those sets of rules that when applied over a set of facts guarantee a finite closure) and use a skolem chase (i.e. the rule application operator that replaces every existential variable with a function depending on the hypothesis' variables) for saturation (Marnette 2009).

*Example 1.* Consider the following simple knowledge base $\mathcal{K}$: James is a cat. James is affectionate. James is handsome. James is intelligent. All cats are mammals. One cannot be affectionate, handsome and intelligent at the same time[4].
    Formally, $\mathcal{K} = (\mathcal{F}, \mathcal{R}, \mathcal{N})$, where:

$$\mathcal{F} = \{cat(James), affectionate(James),$$
$$handsome(James), intelligent(James)\}.$$
$$\mathcal{R} = \{\forall x \; cat(x) \rightarrow mammal(x)\}.$$
$$\mathcal{N} = \{\forall x \; (affectionate(x) \wedge handsome(x)$$
$$\wedge intelligent(x) \rightarrow \perp)\}.$$

We can see that the set of facts is $\mathcal{R}$-inconsistent. Indeed, by using solely $\mathcal{F}$ we are able to deduce the hypothesis of the negative constraint in $\mathcal{N}$.

An argument (Croitoru and Vesic 2013) in $Datalog^{\pm}$ is composed of a minimal (with respect to set inclusion) set of facts called *support* and a *conclusion* entailed from the support. The Skolem chase coupled with the use of decidable classes of $Datalog^{\pm}$ ensures the finiteness of the proposed argumentation framework (Baget et al. 2011).

**Definition 1.** *Let* $\mathcal{K} = (\mathcal{F}, \mathcal{R}, \mathcal{N})$ *be a knowledge base. An argument* $a$ *is a tuple* $(H, C)$ *with* $H$ *a non-empty* $\mathcal{R}$-*consistent subset of* $\mathcal{F}$ *and* $C$ *a set of facts such that:*

- $H \subseteq \mathcal{F}$ *and* $\mathcal{Cl}_{\mathcal{R}}^{*}(H) \not\models \perp$ *(consistency)*
- $C \subseteq \mathcal{Cl}_{\mathcal{R}}^{*}(H)$ *(entailment)*
- $\nexists H' \subset H$ *s.t.* $C \subseteq \mathcal{Cl}_{\mathcal{R}}^{*}(H')$ *(minimality)*

---

[4] The example is obviously fictitious.

The support $H$ of an argument $a$ is denoted by $Supp(a)$ and the conclusion $C$ by $Conc(a)$. If $X$ is a set of arguments, we denote by $Base(X) = \bigcup_{a \in X} Supp(a)$.

*Example 1 (cont.).* An example of an argument is $a_1 = (\{affectionate(James)\}, \{affectionate(James)\})$ which states that James is affectionate. Moreover, the minimality implies that arguments that possess excess information in their supports like $(\{affectionate(James), cat(James)\}, \{affectionate(James)\})$ are not considered. Another example of argument is $a_2 = (\{intelligent(James), handsome(James)\}, \{intelligent(James), handsome(James)\})$. The argument $a_3 = (\{cat(James)\}, \{mammal(James)\})$ is another example of an argument.

To capture inconsistencies between arguments, we consider the undermining attack (Croitoru and Vesic 2013): $a$ attacks $b$ iff the union of the conclusion of $a$ and an element of the support of $b$ is $\mathcal{R}$-inconsistent.

**Definition 2.** *An argument $a$ attacks an argument $b$ denoted by $(a, b) \in \mathcal{C}$ (or $a\mathcal{C}b$) iff $\exists \phi \in Supp(b)$ s.t. $Conc(a) \cup \{\phi\}$ is $\mathcal{R}$-inconsistent.*

Now that we defined the structure of arguments and attacks, the argumentation graph corresponding to a knowledge base simply consists of all arguments and attacks that can be generated.

**Definition 3.** *An argumentation graph $\mathcal{AS}$ is a tuple $(\mathcal{A}, \mathcal{C})$ where $\mathcal{A}$ is a set of arguments and $\mathcal{C} \subseteq \mathcal{A} \times \mathcal{A}$ is a binary relation between arguments called attacks. The argumentation graph instantiated over a knowledge base $\mathcal{K}$ is denoted by $\mathcal{AS}_{\mathcal{K}}$, where the set of arguments and attacks follow Definitions 1 and 2 respectively.*

*Example 1 (cont.).* We have that $a_2$ attacks $a_1$ but $a_1$ does not attack $a_2$ (as we consider a subset of the support of $a_2$ we cannot entail a negative constraint). This is an example that shows that the graph is not symmetric, which is due to the presence of $n$-ary constraints.

The complete graph for the knowledge base of Example 1 is composed of 27 arguments and 144 attacks and is represented in Fig. 1. For example, the argument $a_{7\_0} = (\{intelligent(James)\}, \{intelligent(James)\})$ is attacked by the argument $a_{5\_2} = (\{affectionate(James), handsome(James)\}, \{affectionate(James), handsome(James)\})$.

When considering an argumentation graph $\mathcal{AS} = (\mathcal{A}, \mathcal{C})$, one is often interested in the several consistent viewpoints (or subsets of arguments) that can be inferred. Let $E \subseteq \mathcal{A}$ and $a \in \mathcal{A}$. We say that $E$ is *conflict-free* iff there exists no arguments $a, b \in E$ such that $(a, b) \in \mathcal{C}$. $E$ *defends* $a$ iff for every argument $b \in \mathcal{A}$, if we have $(b, a) \in \mathcal{C}$ then there exists $c \in E$ such that $(c, b) \in \mathcal{C}$. $E$ is *admissible* iff it is conflict-free and defends all its arguments. $E$ is a *complete* (CO) extension iff $E$ is an admissible set which contains all the arguments it defends. $E$ is a *preferred extension* (PR) iff it is maximal (with respect to set inclusion) admissible set. $E$ is a *stable extension* (ST) iff it is conflict-free and for all $a \in \mathcal{A} \backslash E$, there exists an argument $b \in E$ such that $(b, a) \in \mathcal{C}$. $E$ is a *grounded extension* (GR) iff $E$ is a minimal (for set inclusion) complete extension.

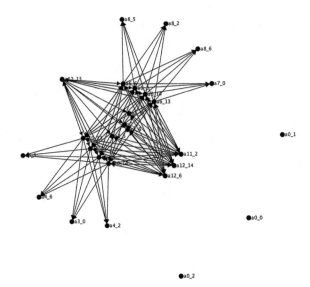

**Fig. 1.** Graph representation of the instantiated argumentation framework constructed on the knowledge base of Example 1

## 3   The Benchmark

The aim of this section is to detail the generation of benchmarks based on argumentation graphs instantiated using existential rules. As seen in the previous section, existential rules, as a logical language, provide many features (n-ary negative constraints, existential variables in the rule conclusion) that make the instantiated argumentation graph far from simplistic. Furthermore the instantiated graph is reflecting the structure of OBDA inconsistent knowledge bases and it is thus justifying its interest as practical benchmark. Generating such graphs is thus significant for a broader community interested in reasoning in presence of inconsistency on the Semantic Web.

We explain the generation of the *benchmark graphs* in Sect. 3.1. Then, in Sect. 3.2, we run *top argumentation solvers on the benchmark* and discuss the results. The goal of this experimental part is to see how the solvers perform on graphs generated from logical knowledge bases and compare performance with respect to randomly generated graphs.

All experiments presented in this section were performed on a VirtualBox Linux machine running with a clean Ubuntu installation with one allocated processor (100%) of an Intel core i7-6600U 2.60GHz and 8GB of RAM. The result files are available upon request (the files amount to more than 15 GB of data).

## 3.1  Benchmark Generation

*Knowledge Base Generation.* We generated a total of 134 knowledge bases: 108 different knowledge bases for the set of small graphs (denoted $b_1$ to $b_{108}$) and 26 for the set of big graphs accessible online at https://github.com/anonymousIDA/ Knowledge_bases. This was done in order to have graphs of similar sizes as those of the *2015 International Competition on Computational Models of Argumentation* (ICCMA 2015). The ICCMA benchmark contains two sets of graphs: a set composed of small graphs (less than 383 arguments) and a set of big graphs (3783 to 111775 arguments). We define, for a fixed size of generated $\mathcal{F}$ (that varied from 2 to 5), some knowledge bases with binary (respectively ternary when applicable) constraints in order to obtain an incremental coverage of the facts. We then add rules in a similarly incremental manner. Table 1 shows the characteristics of the knowledge bases we selected. For example, if considering 3 facts $a(m), b(m), c(m)$, we chose a representative of binary constraints as $\forall x(a(x) \wedge b(x) \rightarrow \bot)$ or $\forall x(a(x) \wedge b(x) \rightarrow \bot)$ or $\forall x(a(x) \wedge b(x) \rightarrow \bot)$. We then chose $\forall x(a(x) \wedge b(x) \wedge c(x) \rightarrow \bot)$.

**Table 1.** Characteristics of the small knowledge bases.

| Name of the KB | Number of facts | Number of rules | Number of NC | Type of NC | Number of args | Number of attacks |
|---|---|---|---|---|---|---|
| $b_1$ to $b_6$ | 2 to 7 | ∅ | 1 | Binary | 2 to 95 | 2 to 2048 |
| $b_{32}$ | 3 | ∅ | 2 | Binary | 4 | 6 |
| $b_{33}$ to $b_{35}$ | 4 | ∅ | 2 to 3 | Binary | 7 to 9 | 24 to 32 |
| $b_{36}$ to $b_{40}$ | 5 | ∅ | 2 to 3 | Binary | 14 to 19 | 56 to 128 |
| $b_7$ to $b_{12}$ | 2 | 1 to 6 | 1 | Binary | 4 to 30 | 5 to 240 |
| $b_{13}$ to $b_{18}$ | 2 | 2,4 or 6 | 1 | Binary | 6 to 30 | 15 to 450 |
| $b_{19}$ to $b_{28}$ | 2 to 7 | 1 or 3 | 1 | Binary | 11 to 383 | 32 to 32768 |
| $b_{29}$ to $b_{31}$ | 3 | 2 | 1 | Binary | 16 | 27 to 30 |
| $b_{57}$ to $b_{58}$ | 3 | 1 | 2 | Binary | 8 | 13 to 14 |
| $b_{59}$ to $b_{82}$ | 4 | 3 | 2 to 4 | Binary | 22 to 71 | 123 to 896 |
| $b_{41}$ to $b_{56}$ | 3 to 6 | ∅ | 1 to 3 | Ternary | 6 to 55 | 9 to 752 |
| $b_{83}$ to $b_{84}$ | 3 | 1 | 1 | Ternary | 12 | 29 to 39 |
| $b_{85}$ to $b_{87}$ | 3 | 2 | 1 | Ternary | 24 | 93 to 147 |
| $b_{88}$ to $b_{108}$ | 4 | 3 | 1 to 2 | Ternary | 78 to 103 | 990 to 2496 |

*From Knowledge Bases to Argumentation Graphs.* In the argumentation graph generation process, we only kept knowledge bases whose argumentation framework is not automorphic to a previously generated graph. The KB format is *dlgp* (Baget et al. 2015b), allowing translations to and from various Semantic Web languages such as RDF/S, OWL, RuleML or SWRL (Baget et al. 2015a). For graph generation we made use of the *Graal* (Baget et al. 2015c) framework, a Java toolkit for reasoning within the framework of existential rules. Graal was used for storing the existential rule knowledge bases and for computing conflicts. On top of Graal we provided a graph generation program that works in three steps:

1. All possible arguments are generated: $\mathcal{R}$-consistent subsets of $\mathcal{F}$ are used as supports and conclusions are deduced from them (see Definition 1).
2. Non minimal arguments are removed (see Definition 1).
3. Attacks are computed following Definition 2.

The obtained graphs were translated in the *Aspartix (apx)* format (the same format used in ICCMA 2015).

*Example 2.* Let us consider the knowledge base $b_{44} = (\mathcal{F}, \mathcal{R}, \mathcal{N})$, where $\mathcal{F} = \{a(m), b(m), c(m), d(m), e(m)\}, \mathcal{R} = \emptyset$ and $\mathcal{N} = \{\forall x(a(x) \wedge b(x) \wedge c(x) \rightarrow \bot)\}$. The corresponding argumentation graph $\mathcal{AS}_{\mathcal{K}}$ is composed of 26 arguments and 144 attacks and is represented in Fig. 2. We show by this example that some of our generated graphs also possess a sense of "symmetry".

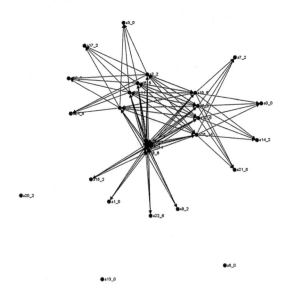

**Fig. 2.** Representation of the argumentation graph corresponding to $b_{44}$.

In the next section, we run the top 6 argumentation solvers on the proposed benchmark and discuss the obtained results.

## 3.2  Benchmark Solvers Results

We recall that the graphs used in the ICCMA 2015 benchmark were separated in three sets: a first set of large graphs (1152 to 9473 arguments) with large grounded extensions and an average density of 1.00%[5], a second set of smaller graphs (141 to 400 arguments) with numerous complete/preferred/stable extensions and an average density of 3.68% and a third set of medium graphs (185 to 996 arguments) with rich structure of strongly connected components and an average density of 7.75%. Our benchmark graphs are denser, having an average density of 31.27% for small graphs and 29.69% for large graphs.

To see if the proposed benchmark graphs behave in a similar manner as the randomly generated graphs of ICCMA 2015, we ran the top six solvers of the competition: *CoQuiAAS, ArgSemSAT (ArgS.SAT), LabSATSolver (LabSATS.), ASGL, ASPARTIX-D* and *ArgTools (ArgT.)*. We used the solvers to complete two computational tasks: SE (given an abstract argumentation framework, determine some extension) and EE (given an abstract argumentation framework, determine all extensions). These two computational tasks were to be solved with respect to the following standard semantics: *Complete Semantics* (CO), *Preferred Semantics* (PR), *Grounded Semantics* (GR) and *Stable Semantics* (ST).

In order to have similar assessment conditions, we used exactly the same ranking method as ICCMA 2015. The solvers were ranked with respect to the number of timeouts on these instances and ties were broken by the actual runtime on the instances. Table 2 shows the average time needed for each solver to complete each task for each semantics in the case of small graphs. There were no errors or time-outs thus the average time reflects the actual ranking (see Tables 3 and 4).

For large instances, many solvers did not support large inputs resulting in several crashes/errors. Ties were broken by the average time of successfully solved instances (Table 5). Please note that for large graphs, for some tasks, some solvers timed out for all instances resulting in equal rankings (*EE-CO: ASGL and ArgTools* for instance).

It is noticeable that *CoQuiAAS* comes first in the two batches of generated graphs. As an explanation, please note that *CoQuiAAS* is based on MiniSAT solver, which is known to work well in the presence of structured information (i.e. symmetries). It might be the case that the generated graphs keep some of their structure even after being translated into a SAT instance which could explain the obtained result.

In order to see how different the solver ranking on the random benchmark used by ICMMA 2015[6] is from the solver ranking on the knowledge base benchmark, we used the normalised Kendall tau distance[7]. The distance outputs 0

---

[5] Graph density for a directed $G = (V, E)$ is equal to $\frac{|E|}{|V|(|V|-1)}$ where $V$ is the set of nodes and $E$ the set of arcs.

[6] http://argumentationcompetition.org/2015/results.html.

[7] This distance is equal to the number of pairwise disagreements between two ranking lists and is normalised by dividing by $\frac{n(n-1)}{2}$, where $n$ is the number of solvers.

**Table 2.** Average time for small instances (in sec).

|       | ArgS.SAT | ASGL | ArgT. | Aspartix-d | CoQuiAAS | LabSATS. |
|-------|----------|------|-------|------------|----------|----------|
| SE-CO | 0,0138 | 0,1719 | 0,0059 | 0,0249 | 0,0031 | 0,3644 |
| SE-PR | 0,0165 | 0,2137 | 0,0059 | 0,4445 | 0,0007 | 0,2906 |
| SE-GR | 0,0339 | 0,2101 | 0,0057 | 0,3217 | 0,0010 | 0,1944 |
| SE-ST | 0,0148 | 0,2194 | 0,0060 | 0,0279 | 0,0018 | 0,2520 |
| EE-CO | 0,0694 | 0,2282 | 0,0096 | 0,0247 | 0,0024 | 0,2908 |
| EE-PR | 0,0517 | 0,1660 | 0,0085 | 0,5763 | 0,0029 | 0,3765 |
| EE-GR | 0,0325 | 0,1861 | 0,0052 | 0,3239 | 0,0016 | 0,2262 |
| EE-ST | 0,0486 | 0,1661 | 0,0065 | 0,0231 | 0,0027 | 0,3151 |

**Table 3.** Corresponding ranking on small graphs.

|       | ArgS.SAT | ASGL | ArgT. | Aspartix-d | CoQuiAAS | LabSATS. |
|-------|----------|------|-------|------------|----------|----------|
| SE-CO | 3 | 5 | 2 | 4 | 1 | 6 |
| SE-PR | 3 | 4 | 2 | 6 | 1 | 5 |
| SE-GR | 3 | 5 | 2 | 6 | 1 | 4 |
| SE-ST | 3 | 5 | 2 | 4 | 1 | 6 |
| EE-CO | 4 | 5 | 2 | 3 | 1 | 6 |
| EE-PR | 3 | 4 | 2 | 6 | 1 | 5 |
| EE-GR | 3 | 4 | 2 | 6 | 1 | 5 |
| EE-ST | 4 | 5 | 2 | 3 | 1 | 6 |

**Table 4.** Number of timeouts on the generated large graphs.

|       | ArgS.SAT | ASGL | ArgT. | Aspartix-d | CoQuiAAS | LabSATS. |
|-------|----------|------|-------|------------|----------|----------|
| SE-CO | 15 | 26 | 16 | 1 | 0 | 11 |
| SE-PR | 18 | 1 | 17 | 26 | 0 | 11 |
| SE-GR | 17 | 0 | 18 | 26 | 0 | 11 |
| SE-ST | 15 | 0 | 18 | 2 | 0 | 11 |
| EE-CO | 22 | 26 | 26 | 9 | 2 | 21 |
| EE-PR | 21 | 26 | 26 | 26 | 15 | 21 |
| EE-GR | 16 | 26 | 17 | 26 | 0 | 11 |
| EE-ST | 15 | 23 | 17 | 1 | 1 | 11 |

if two rankings are identical and 1 if one ranking is the reverse of the other. Table 7 shows the normalised Kendall tau distance between the rankings of the generated graphs and the competition ranking. What comes out is that:

**Table 5.** Ranking on the generated large graphs.

|        | ArgS.SAT | ASGL | ArgT. | Aspartix-d | CoQuiAAS | LabSATS. |
|--------|----------|------|-------|------------|----------|----------|
| SE-CO  | 4        | 6    | 5     | 2          | 1        | 3        |
| SE-PR  | 5        | 2    | 4     | 6          | 1        | 3        |
| SE-GR  | 4        | 2    | 5     | 6          | 1        | 3        |
| SE-ST  | 5        | 2    | 6     | 3          | 1        | 4        |
| EE-CO  | 4        | 5    | 5     | 2          | 1        | 3        |
| EE-PR  | 2        | 4    | 4     | 4          | 1        | 3        |
| EE-GR  | 3        | 5    | 4     | 5          | 1        | 2        |
| EE-ST  | 4        | 6    | 5     | 2          | 1        | 3        |

- Although the ranking of the ICCMA 2015 benchmark and the one for large graphs for the task EE-GR is slightly different (we can not break the tie between ASPARTIX-d and ASGL), they are identical with respect to Kendall tau.
- We have the same normalised Kendall tau distance for the small graphs and the large graphs for the tasks SE-CO, SE-PR and SE-GR.
- The small graphs have a higher normalised Kendall tau distance than the large graphs for the tasks SE-ST, EE-CO, EE-PR and EE-GR.
- The small graphs have a lower normalised Kendall tau distance than the large graphs for the task EE-SET.
- In average, the results are more similar for the large graphs than for the small graphs.

This benchmark is interesting because it shows that for the instantiated graphs we generated, it is strongly advised to use CoQuiAAS as the solver. For relatively small graphs, the choice of the solver can be bypassed as the differences are negligible. However, for larger graphs, we noticed several issues:

**Table 6.** Rankings extracted from the ICCMA 2015 website.

|        | ArgS.SAT | ASGL | ArgT. | Aspartix-d | CoQuiAAS | LabSATS. |
|--------|----------|------|-------|------------|----------|----------|
| SE-CO  | 4        | 2    | 5     | 3          | 1        | 6        |
| SE-PR  | 1        | 4    | 6     | 5          | 3        | 2        |
| SE-GR  | 3        | 5    | 4     | 6          | 1        | 2        |
| SE-ST  | 2        | 5    | 6     | 1          | 4        | 3        |
| EE-CO  | 2        | 5    | 6     | 1          | 3        | 4        |
| EE-PR  | 1        | 4    | 6     | 5          | 2        | 3        |
| EE-GR  | 3        | 5    | 4     | 6          | 1        | 2        |
| EE-ST  | 2        | 4    | 5     | 1          | 3        | 6        |

**Table 7.** Normalised Kendall tau distance between the rankings of the generated graphs and the competition ranking.

|          | Small graphs | Large graphs |
|----------|--------------|--------------|
| SE-CO    | 0.400        | 0.400        |
| SE-PR    | 0.467        | 0.467        |
| SE-GR    | 0.200        | 0.200        |
| SE-ST    | 0.600        | 0.467        |
| EE-CO    | 0.467        | 0.200        |
| EE-PR    | 0.400        | 0.067        |
| EE-GR    | 0.267        | 0.000        |
| EE-ST    | 0.333        | 0.400        |
| Average  | 0.392        | 0.275        |

- It seems that ASGL uses a different algorithm for SE-GR and EE-GR (this is very noticeable by the difference in the number of timeouts).
- ASGL is not suitable for finding complete extensions (Table 6).
- Aspartix-D is not suitable for finding preferred and grounded extensions.
- There are 15 instances that were too big to perform the task EE-PR for all solvers.

## 4   Discussion

This paper starts from the observation that benchmarks of argumentation graphs generated from knowledge bases are currently missing in the literature. We thus propose to consider logic based argumentation frameworks instantiated with existential rules. We provided a tool for generating such graphs out of existential rule knowledge bases. We ran top argumentation solvers on the generated benchmark and analysed their performance with respect to performance on randomly generated graphs.

Note that constructing all the arguments from the knowledge base might result in a big number of arguments. One could reduce the number of arguments by preserving only some of them, i.e. by keeping only the so called *core* (Amgoud et al. 2014). In the present paper, we do not use the notion of a core because we do not want the choice of the core (there are several possibilities) to influence the results of this first study. As it is not convenient to generate too many arguments in practice, investigating benchmark generation using different notions of core is part of future work.

**Acknowledgments.** Srdjan Vesic benefited from the support of the project AMANDE ANR-13-BS02-0004 of the French National Research Agency (ANR).

# References

Amgoud, L., Bodenstaff, L., Caminada, M., McBurney, P., Parsons, S., Prakken, H., Van Veenen, J., Vreeswijk, G.: Final review and report on formal argumentation system. Deliverable D2 (2006)

Amgoud, L., Besnard, P., Vesic, S.: Equivalence in logic-based argumentation. J. Appl. Non Class. Logics **24**(3), 181–208 (2014)

Baget, J.-F., Leclère, M., Mugnier, M.-L., Salvat, E.: On rules with existential variables: walking the decidability line. Artif. Intell. **175**(9–10), 1620–1654 (2011)

Baget, J.-F., Gutierrez, A., Leclère, M., Mugnier, M.-L., Rocher, S., Sipieter, C.: Datalog+, RuleML and OWL 2: formats and translations for existential rules. In: Proceedings of the RuleML 2015 Challenge, the Special Track on Rule-based Recommender Systems for the Web of Data, the Special Industry Track and the RuleML 2015 Doctoral Consortium hosted by the 9th International Web Rule Symposium (RuleML 2015), Berlin, Germany, 2–5 August 2015 (2015a)

Baget, J.-F., Gutierrez, A., Leclère, M., Mugnier, M.-L., Rocher, S., Sipieter, C.: DLGP: an extended Datalog Syntax for Existential Rules and Datalog+/- Version 2.0, June 2015 (2015b)

Baget, J.-F., Leclère, M., Mugnier, M.-L., Rocher, S., Sipieter, C.: Graal: a toolkit for query answering with existential rules. In: Bassiliades, N., Gottlob, G., Sadri, F., Paschke, A., Roman, D. (eds.) RuleML 2015. LNCS, vol. 9202, pp. 328–344. Springer, Cham (2015). doi:10.1007/978-3-319-21542-6_21 (2015c)

Besnard, P., Hunter, A.: Elements of Argumentation. MIT Press, Cambridge (2008)

Bondarenko, A., Toni, F., Kowalski, R.A.: An assumption-based framework for non-monotonic reasoning. In: LPNMR, pp. 171–189 (1993)

Calì, A., Gottlob, G., Lukasiewicz, T.: A general datalog-based framework for tractable query answering over ontologies. In: Proceedings of the Twenty-Eigth ACM SIGMOD-SIGACT-SIGART Symposium on Principles of Database Systems, PODS 2009, 19 June–1 July 2009, Providence, Rhode Island, USA, pp. 77–86, June 2009

Cerutti, F., Dunne, P.E., Giacomin, M., Vallati, M.: Computing preferred extensions in abstract argumentation: a SAT-based approach. In: Black, E., Modgil, S., Oren, N. (eds.) TAFA 2013. LNCS (LNAI), vol. 8306, pp. 176–193. Springer, Heidelberg (2014). doi:10.1007/978-3-642-54373-9_12

Croitoru, M., Vesic, S.: What can argumentation do for inconsistent ontology query answering? In: Liu, W., Subrahmanian, V.S., Wijsen, J. (eds.) SUM 2013. LNCS (LNAI), vol. 8078, pp. 15–29. Springer, Heidelberg (2013). doi:10.1007/978-3-642-40381-1_2

Croitoru, M., Thomopoulos, R., Vesic, S.: Introducing preference-based argumentation to inconsistent ontological knowledge bases. In: Chen, Q., Torroni, P., Villata, S., Hsu, J., Omicini, A. (eds.) PRIMA 2015. LNCS (LNAI), vol. 9387, pp. 594–602. Springer, Cham (2015). doi:10.1007/978-3-319-25524-8_42

Dung, P.M.: On the acceptability of arguments and its fundamental role in non-monotonic reasoning, logic programming and n-Person games. Artif. Intell. **77**(2), 321–358 (1995)

García, A.J., Simari, G.R.: Defeasible logic programming: an argumentative approach. TPLP **4**(1–2), 95–138 (2004)

Gottlob, G., Lukasiewicz, T., Pieris, A.: Datalog+/-: questions and answers. In: Principles of Knowledge Representation and Reasoning: Proceedings of the Fourteenth International Conference, KR 2014, Vienna, Austria, 20–24 July 2014

Leite, J., Son, T.C., Torroni, P., Woltran, S.: Applications of logical approaches to argumentation. Argument Comput. **6**(1), 1–2 (2015)

Magka, D., Krötzsch, M., Horrocks, I.: Computing stable models for nonmonotonic existential rules. In: Proceedings of the 23rd International Joint Conference on Artificial Intelligence, IJCAI 2013, Beijing, China, pp. 1031–1038, 3–9 August 2013

Marnette, B.: Generalized schema-mappings: from termination to tractability. In: Proceedings of the Twenty-Eigth ACM SIGMOD-SIGACT-SIGART Symposium on Principles of Database Systems, PODS 2009, 19 June–1 July 2009, Providence, Rhode Island, USA, pp. 13–22 (2009)

Modgil, S., Prakken, H.: The ASPIC+ framework for structured argumentation: a tutorial. Argument Comput. **5**(1), 31–62 (2014)

Nofal, S., Atkinson, K., Dunne, P.E.: Algorithms for decision problems in argument systems under preferred semantics. Artif. Intell. **207**, 23–51 (2014)

Thomazo, M., Rudolph, S.: Mixing materialization and query rewriting for existential rules. In: ECAI 2014 - 21st European Conference on Artificial Intelligence, Prague, Czech Republic - Including Prestigious Applications of Intelligent Systems (PAIS 2014), pp. 897–902, 18–22 August 2014

Thomazo, M.: Compact rewritings for existential rules. In: Proceedings of the 23rd International Joint Conference on Artificial Intelligence, IJCAI 2013, Beijing, China, pp. 1125–1131, 3–9 August 2013

Zhang, H., Zhang, Y., You, J.-H.: Expressive completeness of existential rule languages for ontology-based query answering. In: Proceedings of the Twenty-Fifth International Joint Conference on Artificial Intelligence, IJCAI 2016, New York, NY, USA, pp. 1330–1337, 9–15 July 2016

# Author Index